A Course in Mathematical Physics 4

A Course in Mathematical Physics

Volume 1

Classical Dynamical Systems

By W. Thirring

Translated by E.M. Harrell

This textbook for physics, mathematics, and applied mathematics students is a unique combination of mathematical rigor and realistic physical approach. Starts with the development of the concept of a manifold and proceeds to a discussion of Hamiltonian systems, canonical transformations, and constants of motion. Among specific problems discussed in detail are nonrelativistic motion of particles and systems, relatilistic motion in electromagnetic and gravitational fields, and the structure of black holes. Numerous helpful examples are included.

1978. ISBN 0-387-**81496**-5

Volume 2

Classical Field Theory

By W. Thirring

Translated by E.M. Harrell

This is a modern and extensive treatment of the classical theories of electromagnetism and gravitation. Among the intriguing areas covered are the mathematical and practical aspects of space time and black hole physics. Other topics include the electromagnetic field of a known charge distribution; the field in the presence of conductors; and gravitation.

1979. ISBN 0-387-**81532**-5

Volume 3

Quantum Mechanics of Atoms and Molecules

By W. Thirring

Translated by E.M. Harrell

Applying the same mathematical rigor as the first two books, this unique and comprehensive volume explores simple systems in quantum mechanics. The author employs the most modern mathematical methods to produce concrete results comparable to experimental fact. Not only does the text introduce basic axioms, from which quantum mechanics are subsequently derived, but it also demonstrates relevant applications in an in-depth and self-contained fashion.

1981. ISBN 0-387-**81620**-8

Volume 4

Quantum Mechanics of Large Systems

By W. Thirring

Translated by E.M. Harrell

A Course in Mathematical Physics concludes with an innovative development of statistical quantum mechanics from a distinctly modern perspective. Much of the material in this volume has never before appeared in book form. Discounting the traditional perturbation-theoretical calculations, Dr. Thirring concentrates on a number of successful treatments of fundamental issues. Topics explored include: properties of entropy, noncommutative ergodic theory, the proof of the existence for thermodynamic functions, and a mathematical analysis of Thomas-Fermi theory.

1983. ISBN 0-387-**87101**-8

Walter Thirring

A Course in Mathematical Physics

4

Quantum Mechanics of Large Systems

Translated by Evans M. Harrell

Springer-Verlag New York Wien

Dr. Walter Thirring
Institute for Theoretical Physics
University of Vienna
Austria

Dr. Evans M. Harrell
The Johns Hopkins University
Baltimore, Maryland
U.S.A.

Translation of Lehrbuch der Mathematischen Physik
Band 4: Quantenmechanik grosser Systeme
Wien–New York: Springer-Verlag 1980

ISBN 3-211-81604-6 Springer-Verlag Wien New York
ISBN 0-387-81604-6 Springer-Verlag New York Wien

Library of Congress Cataloging in Publication Data

Thirring, Walter E., 1927–
Quantum mechanics of large systems.

(A course in mathematical physics; 4)
Translation of: Quantenmechanik grosser Systeme.
Bibliography: p.
Includes index.
1. Statistical thermodynamics. 2. Statistical mechanics. I. Title. II. Series: Thirring, Walter E., 1927– . Lehrbuch der mathematischen Physik. English; 4.
QC20.T4513 vol. 4 [QC311.5] 530.1′5s [530.1′33] 82-19159

With 39 Figures

Typeset by Composition House Ltd., Salisbury, England.
Printed and bound by R. R. Donnelley & Sons, Harrisonburg, VA.
Printed in the United States of America.

9 8 7 6 5 4 3 2 1

ISBN 0-387-81701-8 Springer-Verlag New York Wien
ISBN 3-211-81701-8 Springer-Verlag Wien New York

Preface

In this final volume I have tried to present the subject of statistical mechanics in accordance with the basic principles of the series. The effort again entailed following Gustav Mahler's maxim, "Tradition = Schlamperei" (i.e., filth) and clearing away a large portion of this tradition-laden area. The result is a book with little in common with most other books on the subject.

The ordinary perturbation–theoretic calculations are not very useful in this field. Those methods have never led to propositions of much substance. Even when perturbation series, which for the most part never converge, can be given some asymptotic meaning, it cannot be determined how close the nth order approximation comes to the exact result. Since analytic solutions of nontrivial problems are beyond human capabilities, for better or worse we must settle for sharp bounds on the quantities of interest, and can at most strive to make the degree of accuracy satisfactory.

The last two decades have seen successful and beautiful treatments of many fundamental issues—I have in mind the ordering of the states (2.1), properties of the entropy (2.2), noncommutative ergodic theory (3.1), the proof of the existence of the thermodynamic functions (4.3), and the mathematical analysis of Thomas–Fermi theory (4.1.2), which provides an understanding of the stability of matter. The day is surely not far off when most of the remaining holes in the conceptual structure of quantum statistical mechanics will have been filled in and the questions that are not satisfactorily answered today will be added to the list of achievements.

The successful completion of this course of mathematical physics in a reasonable time required the fortunate conjunction of several circumstances. As with volume III, I had active support from several collaborators, and in particular I am greatly obliged to B. Baumgartner, H. Narnhofer, A. Pflug, and A. Wehrl. Countless other colleagues have helped indirectly by coping with other time-consuming duties for me. The English edition has again greatly benefited from the critical reading of B. Simon. The working conditions at the University of Vienna were invaluable for the completion of this project. Last but not least, the frictionless collaboration of Springer-Verlag in Vienna and my secretary and calligrapher F. Wagner enabled the books to appear quickly and at a reasonable price.

I am aware that the uncompromising way of mathematical physics is not the easiest. Yet I feel that it has been one of the greatest intellectual accomplishments of our era to cast the laws of Nature in a clear mathematical form with rigorously deducible consequences. No amount of labor is too high a price to have paid for this. Let me conclude by also acknowledging and expressing my thanks to the reader who has borne with me to the end of the course.

Walter Thirring

Contents

Symbols Defined in the Text

$\sigma, \sigma^z, \sigma^\pm$	spin components	(1.1.1)
$\mathscr{H}_F$	Fock space	(1.3.1)
Ⓢ, Λ	symmetric and antisymmetric tensor product	(1.3.1)
$a^*(f), a(f)$	creation and annihilation operators	(1.3.2)
$[\ \ ]_+$	anticommutator	(1.3.3; 2)
$\mathscr{A}_B$	C^*-algebra for bosons	(1.3.3; 3)
$\mathscr{A}_F$	C^*-algebra for fermions	(1.3.3; 4)
$a(\mathbf{x})$	annihilation operator	(1.3.3; 7)
Ω	cyclic vector	(1.3.5)
$\mathscr{A}_G, \mathscr{A}_E$	even and gauge-invariant algebras	(1.3.8)
Tr	trace	(1.4.10)
I, II, III	factors of type I, II, III	(1.4.16)
$\rho(n)$	sum of the first n eigenvalues	(2.1.9)
$\precsim$	ordering of density matrices	(2.1.10; 1)
S_α	α-entropy	(2.2.2)
$S(\rho)$	von Neumann entropy	(2.2.4)
$d\Omega_z^N$	Liouville measure on phase space	(2.2.7)
ρ_{qu}	quantum-theoretic phase-space density	(2.2.7)
ρ_{cl}	classical phase-space density	(2.2.7)
$S(\sigma \mid \rho)$	relative entropy	(2.2.22)
$\varepsilon, \sigma, \rho$	energy, entropy and particle densities	(2.3.8)
T	temperature	(2.3.16)
C_V, c_V	total and specific heat capacity at constant volume	(2.3.17; 3)
P	pressure	(2.3.21)
κ	compressibility	(2.3.22; 3)
μ	chemical potential	(2.3.27)
v_u	smeared potential	(2.4.9)
$\varphi(T, \rho)$	Legendre transform of ε	(2.4.14)
$\mathscr{L}$	Legendre transformation	(2.4.15; 2)

z	fugacity	(2.5.9)
tr	trace on the one-particle space	(2.5.10)
v^u	unsmeared potential	(2.5.17)
$F_\sigma(z)$	generalized ζ function	(2.5.20)
m	magnetization per volume	(2.5.37)
$\mathscr{R}$	covariance algebra	(3.1)
U	unitary operators	(3.1)
B_{eff}	effective magnetic field	(3.1.1; 4)
τ_t	time-automorphism	(3.1.2)
τ_t^*	dual time-evolution	(3.1.2)
η, $\eta(a)$, $\eta(\sigma)$	invariant mean	(3.1.14), (3.1.15)
E_0	projection onto eigenvectors of H with eigenvalue 0	(3.1.16; 1)
a_t	transformed operator	(3.2.16; 2)
J	conjugate-linear operator	(3.2.1)
π'	conjugate-linear representation	(3.2.1)
$\tilde{A}$	algebra of analytic elements	(3.2.6; (v))
$\sim$	Fourier transform	(3.2.6; (v))
τ_t^h	perturbed time-automorphism	(3.3.4)
R_h	corresponding operator	(3.3.2)
F_{ab}, G_{ab}	correlation function	(3.3.14)
$\mathbf{X}_k$, Z_k	coordinates and charges of the nuclei	(4.1.3; 1)
$W(\mathbf{x})$	wall potential	(4.1.3; 4)
H_n	Hamiltonian with an effective field	(4.1.6)
C_n	correlation correction	(4.1.6)
$\Xi(H)$	grand canonical partition function	(4.1.8)
$\lVert n \rVert_c$	c-norm	(4.1.10)
h_n	one-particle Hamiltonian	(4.1.17)
v_s	singular part of the potential	(4.1.18)
$a_{p,q}$, $\rho_{p,q}$	annihilation and density operators	(4.1.25)
K, A, R	contributions to the energy	(4.1.33)
$\Phi(\mathbf{x})$	potential	(4.1.36)

Symbols Defined in Earlier Volumes

$\mathscr{B}(\mathscr{H})$	bounded operators	(III: 2.1.24)
$\mathscr{C}_1(\mathscr{H})$	trace-class operators	(III: 2.3.21)
$\mathscr{C}_2(\mathscr{H})$	Hilbert–Schmidt operators	(III: 2.3.21)
$\otimes$	tensor product	(I: 2.4.5)
$\oplus$	direct sum of Hilbert spaces	(III: 2.1.9; 2)
$\mathscr{A}'$	commutant of $\mathscr{A}$	(III: 2.3.4)
$\lVert a \rVert_p$	trace norm	(III: 2.3.21)

1 Systems with Many Particles

1.1 Equilibrium and Irreversibility

Macroscopic bodies act in an irreversible and deterministic manner in contrast with the reversible and indeterministic character of the underlying laws of quantum physics. How can the apparent contradiction be understood?

We have learned to describe systems of finitely many particles with an algebra $\mathscr{A}$ of observables, and information about the systems with a state w on the algebra (cf. (III: 2.2.32)). As our main goal is the study of everyday matter, our framework will be that of nonrelativistic quantum theory. For the purposes of contrast, or of aiding intuition, we shall also have occasion to call upon classical mechanics, where states are measures on phase space, and extremal states are point measures. In either framework time-evolution can be represented as an automorphism $a \to a_t$ for $a \in \mathscr{A}$ in the Heisenberg picture. If desired, time-dependence can alternatively, in the Schrödinger picture, be put upon the state: $w \to w_t$, such that $w_t(a) = w(a_t)$. If the algebra is Abelian (classical mechanics), then the point of an extremal state moves along a classical trajectory in phase-space.

In our earlier experience, systems of N particle are so complex for large N that it becomes impossible to reach precise, quantitative conclusions. It turns out, however, that the theoretical analysis again simplifies in the limit $N \to \infty$. Many properties become independent of the exact number of particles and other detailed characteristics of the physical system, somewhat in analogy to what happens in the central limit theorem of probability theory. This may seem peculiar at first; we have always had $\mathscr{A} = \mathscr{B}(\mathscr{H})$, $\mathscr{H}$ a separable Hilbert space, and time-evolution was given by a unitary group on $\mathscr{H}$. What, then, appears so special about a many-particle system? Just that the information contained in a pure state about a many-particle system is so overwhelming that it would be too ambitious to employ the whole of $\mathscr{B}(\mathscr{H})$ for the observables. Actual measurements could never be made on more than a few observables, so $\mathscr{B}(\mathscr{H})$ has to be cut down to size. For instance, suppose that a device is only equipped to observe one particle at a time, and is unable to detect correlations between particles. Then, rather than taking the entire tensor product of the individual particles as the algebra of observables, it is reasonable to regard $\mathscr{A}$ as a single factor. Accordingly, many states differing on $\mathscr{B}(\mathscr{H})$ reduce to the same state when restricted to $\mathscr{A}$. (The classical situation is similar; the restriction of

$$w(\mathbf{q}_1, \ldots, \mathbf{p}_N)$$

is

$$\int d^3q_2 \ldots d^3q_N \, d^3p_2 \ldots d^3p_N \, w(\mathbf{x}_1, \ldots, \mathbf{p}_N),$$

so whole cylindrical regions of phase-space reduce to a single restricted state.) As a consequence large portions of the space of states on $\mathscr{B}(\mathscr{H})$ are quite similar from the point of view of the reduced algebra $\mathscr{A}$. If, in the Schrödinger picture, the state w_t travels throughout the space of states, then its restriction takes on a certain value with a very high probability, unless prevented by some constants of the motion. This most probable state is called the **equilibrium state** over $\mathscr{A}$.

The irreversible tendency toward equilibrium has always aroused wonder, especially as the basic equations of dynamics are invariant under reversal of the motion (III: 3.3.18). We have even seen in classical mechanics that the trajectory of any point on a compact energy surface returns arbitrarily close to its initial position (I: 2.6.13). In quantum theory the Hamiltonian H of a system confined to a finite volume has purely discrete spectrum. If ε_j and $|j\rangle$ denote the eigenvalues and eigenvectors of H, then the time-dependence of an observable a is given by

$$w_t(a) = \sum_{j,k} c_j^* c_k \exp(it(\varepsilon_j - \varepsilon_k))\langle j|a|k\rangle,$$

where the state w is represented by the vector $\sum_j c_j|j\rangle$. The state $w_t(a)$ is now an almost-periodic function of t; if the sum is finite, and the ε_j are rationally dependent, then it is actually strictly periodic. At any rate, to arbitrarily good accuracy, $w_t(a)$ again becomes nearly $w(a)$ after some sufficiently long delay. The trouble is that the recurrence times are so unimaginably long that they have no physical relevance. Suppose, for instance, that there are N distinct energy differences ω_j. The recurrence time can then be estimated as follows. The factors $\exp(i\omega_j t)$ can be pictured as N clocks with hands moving at N different rates. The question is how long it takes for a certain configuration of clock faces to reappear to within some angular accuracy $\Delta\varphi$. The configuration in the space of angles has measure $(\Delta\varphi/2\pi)^N$, so the recurrence time is on the order of $(\Delta\varphi/2\pi)^{-N}/\omega$, where the reciprocal angular velocity $1/\omega$ is an average of the $1/\omega_j$. Even for just $N = 10$, $1/\omega = 1$ sec., and $(\Delta\varphi/2\pi) = 1/100$, so that w_t returns to w to within 1% accuracy, the recurrence time is 10^{20} sec., which is much longer than the age of the universe.

The approach to equilibrium is connected to a loss of information; to be more precise, information does not get lost, but only less accessible. We have seen that when the wave-packet of a free particle spreads (III: 3.3.3), Δx grows linearly with time, although the state remains pure and thus has maximal information content. The observable with least deviation from the mean is, however, not $x(t)$ but $x(0) \equiv x(t) - pt$.

This behavior can be seen even in classical motion if a minimal spread of the support of the probability distribution function in phase space is hypothesized to account for quantum effects. If, say, the initial probability density $\rho(p, q)$ is concentrated on a part of the energy shell $\{(q; p) | p_1 \leq p \leq p_2\}$ and is not pointlike, and it moves freely on a torus, then it eventually fills the energy shell densely with a "fuzzy" distribution. Faster particles overtake the slower ones, as bicycles racing in a stadium start packed closely together but later draw apart and eventually spread around the whole track (see Figure 1).

The ergodic hypothesis has figured importantly in the history of statistical mechanics; it is the assumption that the trajectory of almost every point winds densely around the energy shell in phase space, so that the time average can be replaced with the average over the energy shell. On the one hand this requires more than is necessary, since it suffices to fill a sufficiently typical part of the energy shell, the average on which equals the average on the whole shell for the reduced algebra of observables. On the other hand, although macroscopic measurements last much longer than the collision time, they last much less than the recurrence time, so one does not wait for the whole energy shell to be sampled. We shall discuss examples in which the equilibrium state is actually attained by the state in a reasonable time after reduction to one particle.

A pictorial description of the situation is as follows. The information about a subsystem (i.e., the opposite of the entropy, to be defined later) as a function on the space of states of the total system consists mainly of a plain with few hills and still fewer mountains. The larger the total system, the further apart the prominences. Even if a path begins on a peak, it soon descends to the plain, and there is only the slightest probability that it will ascend another mountain in any conceivable time. The time of descent to the plain and the recurrence time are of completely different orders of magnitude. It takes only the time corresponding physically to a few collisions to descend to a level near that of the plain, whereas the other mountains lie in the unfathomable distance. This means that equilibrium is reached long before the immense recurrence time required to wind throughout the space of states;

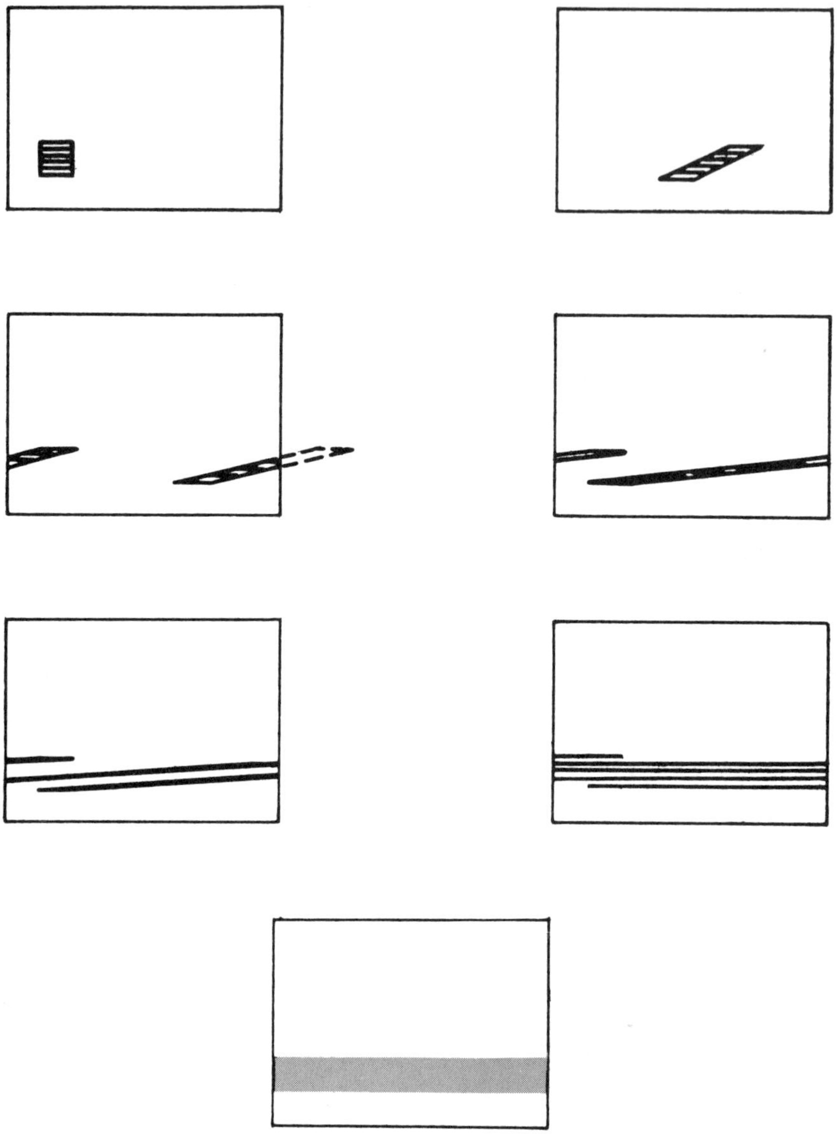

Figure 1 The motion of the density in phase space for a free particle on a torus.

generally, a path soon reaches states that can not be distinguished from equilibrium because of the limits of our measuring abilities. Of course, there is still the question of how one happened, at the beginning, to be at the top of the mountain, but that brings up the one of how the current state of the universe came about and is outside the scope of this book.

Another puzzle is the apparent causal behavior that classical thermodynamics prescribes for macroscopic bodies. According to the arguments that have been advanced, one would rather suspect that the fluctuations of the observables are increased by the loss of information. This is actually true for microscopic variables like the positions and momenta of individual particles. However, if only the so-called macroscopic observables are considered, that is, roughly what was accessible to the more primitive experimental arts of an earlier epoch, then deterministic features arise. Their origin is simply that statistically independent quantities are being averaged: if $a = (1/N)\sum_{j=1}^{N} a_j$, where $w(a_i a_j) = w(a_i)w(a_j)$ for $i \neq j$, then

$$(\Delta a)^2 = \frac{1}{N^2}\left[w\left(\sum_{j,k}(a_j a_k)\right) - \sum_{j,k} w(a_j)w(a_k)\right] = \frac{1}{N^2}\sum_{j=1}^{N}(\Delta a_j)^2.$$

Thus $\Delta a \sim N^{-1/2}$, and for sufficiently large N the deviations from the average are negligible. We shall learn that in the quantum-theoretical formalism such an a approaches a multiple of the identity operator as $N \to \infty$. The limiting coefficient depends on the representation of the algebra.

Let us verify the phenomena described above in two explicitly soluble models. Of necessity they will lack some of the complications arising in reality, but they exhibit the important features. They are embryonic forms of systems of fermions and bosons.

The Chain of Spins (1.1.1)

Let the algebra of observables of the total system be generated by $\boldsymbol{\sigma}_j$, $j = 1, \ldots, N$, where each $\boldsymbol{\sigma}_j$ is a copy of the usual Pauli matrices $\boldsymbol{\sigma}$. Instead of Cartesian components we use $\sigma \equiv \sigma^z$ and $\sigma^{\pm} \equiv (\sigma^x \pm i\sigma^y)/2$, which satisfy the commutation relations

$$[\sigma_j, \sigma_k^{\pm}] = \pm\delta_{jk} 2\sigma_k^{\pm},$$
$$[\sigma_j^+, \sigma_k^-] = \delta_{jk}\sigma_k. \tag{1.1.2}$$

The chain is closed by the identification of $\boldsymbol{\sigma}_{j+N}$ with $\boldsymbol{\sigma}_j$, and the Hamiltonian that determines the time-evolution will be assumed to be of the form

$$H = B\sum_{j=1}^{N}\mu_j\sigma_j + \sum_{n=1}^{N-1}\sum_{j=1}^{N}\sigma_j\sigma_{j+n}\varepsilon(n). \tag{1.1.3}$$

The physical meaning of this is that the spins are coupled with magnetic moments μ_j to an external magnetic field B, and in addition there is an Ising-like spin–spin interaction with the nth neighbor. The strength $\varepsilon(n)$ of this interaction is a function that can be specified later, and the periodicity allows us to assume $\varepsilon(n) = 0$ for $n > N/2$. If the contributions to H are denoted as in

$$H \equiv H_0 + \sum_n H_n, \tag{1.1.4}$$

then the H_k commute with one another and with the σ_j. They are therefore constant in time, and the time-evolution of σ^+ and $\sigma^- = (\sigma^+)^*$ can be calculated easily from the relationship

$$f(\sigma)\sigma^+ = \sigma^+ f(\sigma + 2), \tag{1.1.5}$$

which follows from (1.1.2). We find

$$\begin{aligned}\sigma_k^+(t) &= (\sigma_k^-(t))^* = \sigma_k^+(0) \exp\left\{2it\left[B\mu_k + \sum_n \varepsilon(n)(\sigma_{k+n} + \sigma_{k-n})\right]\right\} \\ &= \sigma_k^+(0) \exp(2itB\mu_k) \prod_n (\cos 2t\varepsilon(n) + i\sigma_{k+n} \sin 2t\varepsilon(n))(\cos 2t\varepsilon(n) \\ &\quad + i\sigma_{k-n} \sin 2t\varepsilon(n)),\end{aligned} \tag{1.1.6}$$

where $a(t) = \exp(iHt)a\exp(-iHt)$.

The time-evolution consists of Larmor precession in the external field and a kind of diffusion along the chain due to the spin–spin interaction. Suppose that the state at $t = 0$ is pure and has the form of a product, where the spins have a 3-component s and σ_k^+ has phase α_k:

$$\langle\sigma_k(0)\rangle = s, \qquad \langle\sigma_k^+(0)\rangle = \tfrac{1}{2}\sqrt{1 - s^2}\exp(i\alpha_k), \qquad \left\langle\prod_j \boldsymbol{\sigma}_j\right\rangle = \prod_j \langle\boldsymbol{\sigma}_j\rangle. \tag{1.1.7}$$

Then

$$\begin{aligned}\langle\sigma_k^+(t)\rangle &= \tfrac{1}{2}\sqrt{1 - s^2}\exp\{i(\alpha_k + 2tB\mu_k)\} f^2(t), \\ f(t) &= \prod_{n=1}^{N/2} (\cos 2t\varepsilon(n) + is \sin 2t\varepsilon(n)).\end{aligned} \tag{1.1.8}$$

If N is finite, then f is almost periodic, and if $N = \infty$, then $f(t)$ will generally tend to zero as $t \to \infty$ (supposing that $\varepsilon(n)$ tends to zero in such a way that the infinite product makes sense). To make this more explicit, let us consider the special case $s = 0$ and $\varepsilon(n) = 2^{-n-1}$. If $N = \infty$, then f satisfies the equation

$$f(t) = \prod_{n=1}^{\infty} \cos 2^{-n}t = \frac{f(2t)}{\cos t}. \tag{1.1.9}$$

Since f is an entire function, this functional equation and the condition $f(0) = 1$ determine f uniquely—differentiate (1.1.9) to get the Taylor series of f. Since the function $(\sin t)/t$ satisfies (1.1.9), it equals f. Hence, as $N \to \infty$, the expectation value of $\sigma^{\pm}$ approaches zero. For finite N it follows from (1.1.9) that

$$f_N(t) = \prod_{n=1}^{N/2} \cos 2^{-n} t = \frac{\sin t}{t}\left[\frac{\sin t 2^{-N/2}}{t 2^{-N/2}}\right]^{-1}. \tag{1.1.10}$$

Therefore, as discussed earlier, the recurrence time $2^{N/2}/\pi$ grows exponentially with N, while the time it takes to reach equilibrium is independent of N.

To summarize, we have ascertained that for $N = \infty$ the initially pure state of the algebra reduced to one spin tends as $t \to \infty$ to $\langle \sigma \rangle = s, \langle \sigma^{\pm} \rangle = 0$, which corresponds to a mixture:

$$\langle \boldsymbol{\sigma} \rangle = \mathrm{Tr}(\rho \boldsymbol{\sigma}), \qquad \rho = \frac{\exp(-\eta\sigma)}{\mathrm{Tr}\exp(-\eta\sigma)}, \qquad \tanh \eta = s. \tag{1.1.11}$$

Even though the expectation values of the $\sigma_k^{\pm}$ go to zero, their fluctuations remain nonzero, since $\sigma_k^+ \sigma_k^- = (1 + \sigma_k)/2$ is constant. The average magnetization

$$\mathbf{M}_N(t) = \frac{1}{N} \sum_k \boldsymbol{\sigma}_k(t) \tag{1.1.12}$$

works differently. In the state (1.1.7) of our example, $\langle M_N^z \rangle = s$, whereas $\langle M_N^{\pm} \rangle$ is $O(N^{-1/2})$, provided either that the initial phases are disordered or that the $\sigma_k^{\pm}$ get out of phase after a while because the μ_k differ. The latter situation can in fact be undone by a sudden reversal of B, in the spin–echo effect. If $N = \infty$, the diffusion caused by suitable $\varepsilon(n)$ is irreversible, and $\lim_{t\to\infty} \langle M_\infty^{\pm}(t) \rangle = 0$. At $t = 0$ the fluctuations are $O(N^{-1/2})$ and remain at this magnitude for all time: If $\sigma_k^+(t)\sigma_{k'}^-(t)$ is calculated by multiplying together two expressions of the form (1.1.6), then it should be recalled that $\sigma^2 = 1$. However, if the function $\varepsilon(n)$ falls off sufficiently rapidly with n, then the σ^2 terms make little difference for large $k - k'$, and the argument given earlier for the deviations of statistically independent quantities remains valid.

Chain of Oscillators (1.1.13)

Now represent the total system by positions and momenta $q_1, \ldots, q_N$, $p_1, \ldots, p_N$, such that $[q_j, p_k] = i\delta_{jk}$, and let the time-evolution be determined by

$$H = \sum_{j=1}^{N} \tfrac{1}{2}(p_j^2 + (q_j - q_{j+1})^2). \tag{1.1.14}$$

This Hamiltonian contains interactions only between nearest neighbors, and the chain can be closed by the condition of periodicity $q_{j+N} = q_j$, $p_{j+N} = p_j$. The masses and force constants have been set to 1, which amounts to measuring the time in units of the natural period of oscillation. The equations of motion are

$$\dot{q}_j = p_j, \qquad \dot{p}_j = q_{j+1} + q_{j-1} - 2q_j. \tag{1.1.15}$$

With a periodic extension of the variables, $\xi_1, \ldots, \xi_{2N}$, such that

$$\xi_{2n} = p_n, \qquad \xi_{2n+1} = q_{n+1} - q_n, \tag{1.1.16}$$

they are put into the form

$$\dot{\xi}_j = \xi_{j+1} - \xi_{j-1}. \tag{1.1.17}$$

The variables ξ_n satisfy

$$\xi_{n+2N} = \xi_n, \qquad \sum_n \xi_{2n+1} = 0.$$

Recall that the Bessel functions satisfy the recursion formula $\dot{J}_n = (J_{n-1} - J_{n+1})/2$; as a consequence we see that the solution of the initial-value problem is

$$\xi_n(t) = \sum_{k=-\infty}^{\infty} \xi_k(0) J_{k-n}(2t). \tag{1.1.18}$$

Remarks (1.1.19)

1. Since $|J_\nu(z)| \sim |z/\nu|^{|\nu|}$ as $|\nu| \to \infty$, the sum over k in (1.1.18) converges for, say, bounded $\{\xi_k(0)\}$.
2. If $N < \infty$, then (1.1.18) still holds provided that $\xi_{k+2N}(0) = \xi_k(0)$.
3. Since the equations of motion are linear, the classical and quantum time-automorphisms are identical.
4. There are still N constants of motion with the variables ξ:

$$I_k = \sum_{j=1}^{2N} \xi_j \xi_{j+n}, \qquad k = 1, \ldots, N.$$

With the auxiliary condition that $\sum_n \xi_{2n+1} = 0$, only $N - 1$ of the constants are independent, and we find that $\sum_n I_{2n+1} = 0$. If $N = \infty$, then I_k remains significant classically, provided that $\{\xi_k\} \in l^2$.

In order to have a useful framework for discussing the questions that will arise as in these two examples, it is convenient for technical reasons to make use of the Weyl algebra (cf. (III, §3.1)). With one particle, the Weyl algebra consists of the operators $W(r + is) = \exp(i(pr + qs))$, $r, s \in \mathbb{R}$, along with their linear combinations and norm-limits. A state on the Weyl algebra is uniquely characterized by the function $E(r, s) \equiv \langle \exp(i(pr + qs)) \rangle$. We shall only concern ourselves with coherent states (III: 3.1.13), which are of the form $W(z')|u\rangle$, where $|u\rangle$ is a Gaussian function, the width of which determines the ratio between Δp and Δq. Since

$$\langle u | W(r + is) | u \rangle = \exp\left[-\frac{1}{4}\left(\omega r^2 + \frac{s^2}{\omega}\right)\right],$$

it follows that

$$(\Delta p)^2 = -\frac{d^2}{dr^2} \ln E_{|r,s=0} = \frac{\omega}{2}, \qquad (\Delta q)^2 = -\frac{d^2}{ds^2} \ln E_{|r,s=0} = \frac{1}{2\omega}.$$

The expectation value in the more general state $W(z')|u\rangle$ can be calculated according to (III: 3.1.2; 1) as

$$\begin{aligned}\langle W(z')u|W(z)|W(z')u\rangle &= \langle u|W(-z')W(z)W(z')|u\rangle \\ &= \langle u|W(z)|u\rangle \exp\left[\frac{i}{2}\operatorname{Im}(z^*z' - z^{*\prime}z)\right] \\ &= \exp\left[-\frac{1}{4}\left(\omega r^2 + \frac{s^2}{\omega}\right) + i(rs' - r's)\right]. \quad (1.1.20)\end{aligned}$$

Thus, the quantities Δp and Δq are the same as with $|u\rangle$, but the expectation values of p and q are now s' and $-r'$.

Let us return to the issue of how the restriction of the many-particle state to a subsystem evolves in time. The operators $\exp[i(r\xi_0(t) + s\xi_1(t))]$, which describe the momentum of a single particle and its position relative to its neighbor, are useful to this end. Since $[\xi_0(t), \xi_1(t)] = i$, they form a Weyl system. A state characterized by

$$\begin{aligned}\left\langle \exp\left[i \sum_{n=-\infty}^{\infty} (\xi_{2n} r_n + \xi_{2n+1} s_n)\right]\right\rangle = \exp\Bigg[-\frac{1}{4}\sum_{n=-\infty}^{\infty}\left(\omega r_n^2 + \frac{s_n^2}{\omega}\right) \\ + i(r_n s_n' - r_n' s_n)\Bigg] \quad (1.1.21)\end{aligned}$$

can be regarded as the generalization of (1.1.20).

Remarks (1.1.22)

1. The exponent on the left is a linear combination of p_k and q_k, as appropriate for a Weyl system for several particles, yet the variables ξ_{2n} and ξ_{2n+1} are not pairs of canonically conjugate variables, since $[\xi_{2n}, \xi_{2n-1}] \neq 0$. Thus (1.1.21) is not simply the tensor product of coherent states of a tensor product of Weyl systems.
2. The significance of (1.1.21) is once again that the variables ξ_{2n} (resp. ξ_{2n+1}) all have deviation ω and expectation values s_n' (resp. $1/\omega$ and $-r_n'$).

With (1.1.21), the desired state on the one-particle system turns out to be

$$\begin{aligned}E(r, s) &\equiv \langle \exp(i(r\xi_0(t) + s\xi_1(t)))\rangle \\ &= \left\langle \exp\left(i \sum_{n=-\infty}^{\infty} [\xi_{2n}(0)(rJ_{2n} + sJ_{2n-1}) + \xi_{2n+1}(0)(rJ_{2n+1} + sJ_{2n})]\right)\right\rangle \\ &= \exp \sum_{n=-\infty}^{\infty} \Bigg\{-\frac{1}{4}\left(\omega(rJ_{2n} + sJ_{2n-1})^2 + (rJ_{2n+1} + sJ_{2n})^2 \frac{1}{\omega}\right) \\ &\quad + is_n'(rJ_{2n} + sJ_{2n-1}) - ir'(rJ_{2n+1} + sJ_{2n})\Bigg\}. \quad (1.1.23)\end{aligned}$$

The sums can be evaluated by recourse to the formulas

$$\sum_{n=-\infty}^{\infty} J_{2n}(2t)J_{2n+j}(2t) = \tfrac{1}{2}(\delta_{0j} + J_j(4t)), \qquad j \in \mathbb{Z},$$

$$\sum_{n=-\infty}^{\infty} J_{2n+1}(2t)J_{2n+1+j}(2t) = \tfrac{1}{2}(\delta_{0j} - J_j(4t)), \tag{1.1.24}$$

which are derived in Problem 2. As $t \to \infty$, only the terms with $j = 0$ remain. Moreover, it can be seen from the integral representations and the Riemann–Lebesgue lemma that the contributions linear in the J_k go to zero as $t \to \infty$. In all, we get

$$\lim_{t\to\infty} E(r, s) = \exp\left[-\frac{1}{4}\left(\omega + \frac{1}{\omega}\right)(r^2 + s^2)\right]. \tag{1.1.25}$$

Remarks (1.1.26)

1. The limiting state corresponds to the mixture $E = \operatorname{Tr} \rho W(z)$, $\rho = \exp[-\eta(p_1^2 + q_1^2)]/\operatorname{Tr} \exp[-\eta(p_1^2 + q_1^2)]$, $\coth \eta = (\omega + 1/\omega)/2$ (Problem 3). As $\omega \to 1$, that is, for minimal mean-square deviation, $\eta \to \infty$, and the state becomes pure. With larger mean-square deviations, $\omega \neq 1$, $(\omega + 1/\omega)/2 > 1$, the limiting state is a mixture.
2. Whereas at $t = 0$ the ratio of Δp to Δq is ω^2, they become equal as $t \to \infty$, i.e., their ratio, 1, becomes the one defined by H. This corresponds to equal amounts of kinetic and potential energy.
3. The reason that the existence of the constants (1.1.19; 4) does not prevent the onset of equilibrium is again the choice of the initial state. Of course, equilibrium can not occur if the system starts off in an eigenstate of a normal mode of oscillation.

These few remarks will serve as our first orientation to irreversible phenomena. We have already studied an example of an irreversible phenomenon in volume II, the emission of light. It is always important to take the limit $N \to \infty$ before $t \to \infty$, as in a finite volume the light returns to the point of emission, and the behavior is almost periodic rather than irreversible. The next section will deal with how the energy is affected by the first limiting process.

Problems (1.1.27)

1. Calculate the entropy $S(t) = -\operatorname{Tr} \rho(t) \ln \rho(t)$ for one spin, where f is given by (1.1.9).
2. Calculate $\sum_{n=-\infty}^{\infty} J_{2n}(x)J_{2n+j}(x)$ and $\sum_{n=-\infty}^{\infty} J_{2n+1}(x)J_{2n+1+j}(x)$.
3. Show that the density matrix ρ has the property stated in (1.1.26; 1).

Solutions (1.1.28)

1. Since Tr $\rho(t) = 1$, the density matrix is of the form $\rho(t) = \frac{1}{2} + \mathbf{c}(t) \cdot \boldsymbol{\sigma}$. Let $c(t) = |\mathbf{c}(t)|$, which $\leq \frac{1}{2}$. The eigenvalues of $\rho(t)$ are $\frac{1}{2} \pm c(t)$, so

$$S(t) = -\left[\frac{1 + c(t)}{2} \ln \frac{1 + c(t)}{2} + \frac{1 - c(t)}{2} \ln \frac{1 - c(t)}{2}\right].$$

Because $\operatorname{Tr} \sigma_i \sigma_j = 2\delta_{ij}$, we find $\mathbf{c}(t) = \frac{1}{2}\langle\boldsymbol{\sigma}\rangle$, and therefore $c(t) = (s^2 + (1 - s^2) f^4(t))^{1/2}$. Observe that f is not monotonic, and hence that S does not increase monotonically from 0 to its equilibrium value,

$$-\left[\frac{1 + s}{2} \ln\left(\frac{1 + s}{2}\right) + \frac{1 - s}{2} \ln\left(\frac{1 - s}{2}\right)\right].$$

2.
$$\exp\left[\frac{z}{2}\left(t - \frac{1}{t}\right)\right] = \sum_{n=-\infty}^{\infty} t^n J_n(z).$$

Putting $z = x + y$ yields

$$\sum_{j=-\infty}^{\infty} t^j J_j(x + y) = \exp\left[\frac{x}{2}\left(t - \frac{1}{t}\right)\right] \exp\left[\frac{y}{2}\left(t - \frac{1}{t}\right)\right]$$
$$= \left(\sum_k t^k J_k(x)\right)\left(\sum_l t^l J_l(y)\right) = \sum_{j=-\infty}^{\infty} t^j \sum_{n=-\infty}^{\infty} J_n(x) J_{j-n}(y),$$

so $J_j(x + y) = \sum_{n=-\infty}^{\infty} J_n(x) J_{j-n}(y)$, which is the addition theorem of Schläfli and Neumann. Putting $y = -x$ and changing j to $-j$ then yields $\sum_n J_n(x) J_{n+j}(x) = \delta_{j0}$, and with $y = x$, there results

$$\sum_n J_n(x) J_{-j-n}(x) = \sum_n (-1)^{n+j} J_n(x) J_{n+j}(x) = J_{-j}(2x) = (-1)^j J_j(2x),$$

from which formulas (1.1.24) follow.

3.
$$\operatorname{Tr} \exp[-\eta(p_1^2 + q_1^2)] = \sum_{n=0}^{\infty} \exp[-\eta(1 + 2n)]$$

and

$$\langle p^2 + q^2 \rangle = \left(-\frac{\partial^2}{\partial r^2} - \frac{\partial^2}{\partial s^2}\right) E(r, s) = -\frac{\frac{\partial}{\partial \eta} \operatorname{Tr} \exp[-\eta(p^2 + q^2)]}{\operatorname{Tr} \exp[-\eta(p^2 + q^2)]}$$

lead to the result.

1.2 The Limit of an Infinite Number of Particles

The first issues to confront for large systems are what happens to macroscopic properties like energy and volume as $N \to \infty$.

The models examined in § 1.1 were only caricatures of reality. We shall now determine the physical properties of large bodies. The first question is how the volume V has to vary as $N \to \infty$, in order to ensure that the potential and kinetic energies will be comparable in magnitude and that the interaction between the particles is correctly accounted for. In particular, when are E and V normal, extensive quantities proportional to N? In order to fix our ideas, we shall pay particular attention to certain special cases, large atoms and macroscopic or cosmic objects. The dominant force is then electrostatic, except that in cosmic matter gravity also has a decisive effect. Heuristic arguments will sometimes be adduced in this section for guidance in finding which quantities have limits as $N \to \infty$ in these systems.

Free Particles (1.2.1)

We begin with a consideration of noninteracting particles confined to a box of side R. The energy consists of the quantum-mechanical zero-point energy plus a thermal component proportional to the temperature T. As we are only interested in the dependence on N for large N, we set $\hbar = k = m = 1$. As explained in (III: 1.2.11) the zero-point energy of a system of fermions is $\sim (\Delta p)^2 \sim (\Delta x)^{-2}$, where Δx is about $RN^{-1/3}$, since the volume available per fermion is only R^3/N. We arrive at

$$E = \frac{N^{5/3}}{2R^2} + \tfrac{3}{2}NT. \tag{1.2.2}$$

If the two contributions are to remain comparable as $N \to \infty$, and if T goes as N^t for some power t, then R must be $\sim N^{1/3 - t/2}$, and EN^{-1-t} will tend to a limiting value. The type of interaction will determine the value of t at which the limit is nontrivial and thus of physical interest. For this to happen the kinetic and potential energies have to remain of the same order of magnitude.

Bosons do not have the solitary temperament, so Δx may be set equal to R. The energy is then on the order of

$$E = \frac{N}{2R^2} + \tfrac{3}{2}NT. \tag{1.2.3}$$

If the two contributions are to have the same dependence on N and we make $T \sim N^t$, then $R \sim N^{-t/2}$ and $E \sim N^{t+1}$. If it is insisted that T remain constant and $R \sim N^{1/3}$, then $E \sim N$, but the zero-point energy drops below the thermal energy. The exact calculation for free bosons in fact reveals that, with a fixed particle density and below a critical temperature, a certain fraction $\lambda(T) > 0$ of the particles are to be found in the ground state with $E_0 \sim N^{1/3}$, and thus N may be replaced with $(1 - \lambda(T))N$. This makes this usual limit also nontrivial.

Large Atoms (1.2.4)

The Hamiltonian of a large atom (with $e^2 = 1$) is

$$H = \sum_{i=1}^{N} \left(\frac{|\mathbf{p}_i|^2}{2} - Z|\mathbf{x}_i|^{-1} \right) + \sum_{i>j} |\mathbf{x}_i - \mathbf{x}_j|^{-1}, \tag{1.2.5}$$

which can, if one wishes, be confined in a box. Recall that in volume III we figured out that if $T = 0$ and $Z = N$, the energy is about $N^{5/3}/2R^2 - N^2e^2/R$, which has a minimum about $-\frac{1}{2}N^{7/3}$ for $R \sim N^{-1/3}$. Therefore, in the limit $N \to \infty$ we should expect to set $t = \frac{4}{3}$. In §4.1 it will not only be proved that these limits converge, but even that the Thomas–Fermi theory becomes exact in that limit. The problem can thus be solved in the limit $N \to \infty$, though the solution is not suitable for a direct numerical comparison of theory and experiment. Since there are corrections of about $N^{-1/3}$, 10% accuracy can not be expected for $N \lesssim 10^3$. On the other hand, relativistic effects become significant when $N \sim 10^2$. The kinetic energy is then $\sim N^{4/3}/R$ and if $Ze^2 > 1$ the energy is no longer bounded below. Hence the picture that emerges of a large atom is only an idealization, but at least one with many instructive aspects.

Systems of bosons depend on N in a different way. They all settle into the ground state, and with $Z \sim N$ the radius goes as N^{-1} and the energy as N^3. The limits of EN^{-3} and $N^3\rho(xN)$ would be expected to exist, where ρ is the one-particle density distribution. For thermal effects to remain significant, T must be chosen $\sim N^2$. This problem is mostly of academic interest, and the convergence of the quantities mentioned above has not yet been proved.

Jellium (1.2.6)

Like an atom, jellium consists of particles repelling one another with a Coulomb force and immersed in the field of an external charge distribution. The difference is that the charge distribution is not concentrated at a point, but rather homogeneously spread with density ξ through a box Λ (Λ will also sometimes denote the volume of Λ). It can be regarded as a model of highly compressed matter, with the homogeneous background charge coming from fast-moving electrons, and the particles with explicit coordinates being the nuclei. It is nevertheless often used to describe electrons in a metal, although it is rather far-fetched to speak of the assemblage of ions as a homogeneous background. The Hamiltonian is

$$H = \sum_{i=1}^{N} \frac{|\mathbf{p}_i|^2}{2} + \sum_{i>j} |\mathbf{x}_i - \mathbf{x}_j|^{-1} - \sum_{i=1}^{N} U(\mathbf{x}_i) + \frac{\xi}{2} \int_\Lambda d^3x U(\mathbf{x}), \tag{1.2.7}$$

where $U(\mathbf{x}) = \xi \int_\Lambda d^3x'/|\mathbf{x} - \mathbf{x}'|$. For the system to be neutral, $\xi \int_\Lambda d^3x = N$. The electrostatic energy of the background has been added in so that the potential energy will remain bounded below by $N(RN^{-1/3})^{-1}$, where R is the linear dimension of Λ. The proof of this relies on the well-known fact of electrostatics that the Coulomb repulsion of two homogeneously charged spheres is less than or equal to that of two point charges at their centers—the inequality occurs when they overlap. Now imagine blowing the charged particles up to homogeneously charged spheres of radius a, and let

$$\left(\frac{4\pi a^3}{3}\right)^{-2} \int_{\substack{|\mathbf{x}-\mathbf{x}_i|\leq a \\ |\mathbf{x}'-\mathbf{x}_j|\leq a}} \frac{d^3x\, d^3x'}{|\mathbf{x}-\mathbf{x}'|} = U_{ij}(a),$$
$$\left(\frac{4\pi a^3}{3}\right)^{-1} \int_{|\mathbf{x}-\mathbf{x}_i|\leq a} d^3x U(\mathbf{x}) = U_i(a). \tag{1.2.8}$$

Then H may be written in the form

$$H = \sum_{i=1}^{N} \frac{|\mathbf{p}_i|^2}{2} + \overbrace{\frac{1}{2}\sum_{i,j=1}^{N} U_{ij}(a) - \sum_{i=1}^{N} U_i(a) + \frac{\xi}{2}\int d^3x U(\mathbf{x})}^{\alpha}$$
$$+ \overbrace{\sum_{i=1}^{N}(U_i(a) - U(\mathbf{x}_i))}^{\beta} - \overbrace{\frac{1}{2}\sum_{i} U_{ii}(a)}^{\gamma} + \overbrace{\sum_{i<j}(|\mathbf{x}_i - \mathbf{x}_j|^{-1} - U_{ij}(a))}^{\delta}. \tag{1.2.9}$$

Contribution α is positive, since it is of the form

$$\int \frac{dx\, dx'}{|\mathbf{x}-\mathbf{x}'|} \rho(\mathbf{x})\rho(\mathbf{x}'),$$

and $1/\sigma$ has a positive Fourier transform. It is easy to show (Problem 1) that $\beta \geq -(2\pi/5)\xi a^2 N$, equality holding provided that all the spheres lie within Λ, and $\gamma = (N/2)(6/5a)$, the self-energy of homogeneously charged spheres. As discussed earlier, $\delta \geq 0$. The lower bound $-N((2\pi/5)\xi a^2 + (3/5a))$ is optimized at $a = (3/4\pi\xi)^{1/3} \equiv r_s$, which is precisely the radius at which the sum of the volumes of the spheres equals that of Λ. This computation leads to the

Lower Bound for the Energy (1.2.10)

$$H \geq \sum_{i=1}^{N} \frac{|\mathbf{p}_i|^2}{2} - \frac{9}{10}\frac{N}{r_s}.$$

Remarks (1.2.11)

1. Nothing has yet been assumed about the shape of Λ or the statistics of the particles. In particular, if Λ is spherical, then by Problem 2,

$$-\sum_{i=1}^{N} U(\mathbf{x}_i) + \frac{\xi}{2}\int_{\Lambda} d^3x U(\mathbf{x}) \leq \frac{N}{2R^3}\sum_{i=1}^{N}|\mathbf{x}_i|^2 - \frac{9}{10}\frac{N^2}{R},$$

where equality holds if $x_i \in \Lambda$ for all i.

2. Despite its great generality, the numerical accuracy of the bound (1.2.10) is surprisingly good. If $\mathbf{x}_i$ are the sites of a simple, face-centered, or body-centered cubic lattice, computer studies have been made of the limit as $N \to \infty$ of the potential energy over Nr_s^{-1}, yielding respectively the values -0.880, -0.895, and -0.896 [3].

Lower bounds for H depending on the particle statistics may be derived from (1.2.10). The energy of free fermions is, as seen earlier, $\sim N^{5/3}/R^2 \sim Nr_s^{-2}$, and with the aid of the more precise proportionality factor,

$$H \geq N(1.1r_s^{-2} - 0.9r_s^{-1}) \geq -\frac{0.81}{4.4}N \quad \text{for all } r_s \in \mathbb{R}^+ \tag{1.2.12}$$

for spin-$\frac{1}{2}$ particles. Even if the volume and consequently r_s are treated as variables, the resultant lower bound is $\sim N$. We shall discover later that with no more than first-order perturbation theory we can obtain an upper bound not much different from (1.2.12): the Pauli exclusion principle makes the electrons stay at a distance r_s apart, and this correlation imitates the energetically favorable configurations of (1.2.11; 2). Since the minimizing radius r_s does not depend on N, in this model $E \sim N$ and $R \sim N^{1/3}$, so the exponent t of (1.2.1) equals zero.

A very different picture emerges of bosons. With the kinetic energy (1.2.3) we find, ignoring precise coefficients, that

$$H \geq \frac{N^{1/3}}{r_s^2} - \frac{N}{r_s}. \tag{1.2.13}$$

The minimizing r_s is $\sim N^{-2/3}$, and so $E \sim N^{5/3}$.

Remarks (1.2.14)

1. It is uncertain whether the lower bound $\sim N^{5/3}$ displays the correct dependence on N. Upper bounds obtained with trial functions include more kinetic energy since the particles have to be correlated in order to attain a sufficiently negative potential energy. Until recently it was only possible to show that $E < -cN^{7/5}$ [1].
2. If the background charge is concentrated at discrete points of a lattice, then trial functions can be thought up that show $E < -cN^{5/3}$, and thus in this case the energy in fact goes as $N^{5/3}$ [2].
3. So far only the electrostatic energy has been accommodated in the background, and minimized according to the density ξ. If the background consists of electrons, then its zero-point energy must also be calculated. In a jellium of deuterium atoms, which are bosons, the energy turns out to be $\sim N$: The background density prevents them from collapsing, and for fixed r_s (1.2.13) is on the order of N.

Real Matter (1.2.15)

Real matter consists of positive and negative point-particles interacting with a Coulomb force, so

$$H = \sum_{i=1}^{N} \frac{|\mathbf{p}_i|^2}{2m_i} + \sum_{i>j} \frac{e_i e_j}{|\mathbf{x}_i - \mathbf{x}_j|} \tag{1.2.16}$$

for particles confined to a box of volume $\Lambda \sim R^3$. We shall often particularize to the situation wherein all negative particles are identical with $m = |e| = 1$ and all positive particles are identical with mass M and charge Z. Provided that Z is not so large that relativistic effects become significant, (1.2.16) gives a reasonably accurate description of ordinary matter. We therefore expect to find that $E \sim -N$ for $R \sim N^{1/3}$.

The proof of this fact, known as the "stability of matter," has to be deferred to §4.3. At this point we shall make do with several

Remarks (1.2.17)

1. Roughly speaking, the difficulty is that the double sum for the kinetic energy contains $\sim N^2$ terms, so many cancellations are needed for the result to be only $\sim N$. If, as in the gravitating system to be described shortly (1.2.19), all the contributions are of like sign, then cancellations certainly do not occur. Similarly, if the total charge $Q \equiv \sum_i e_i$ is $\sim N^{2/3+\varepsilon}$ and the system is restricted to a region of linear dimension $R \sim N^{1/3}$, the energy fails to be extensive. The electrostatic energy Q^2/R is $\leq N$ only if $Q \leq N^{2/3}$.
2. Even requiring that $Q = 0$ will not guarantee that $|E| \sim N$ if all the particles are bosons. To prove this, rewrite (1.2.16) (with $M = Z = 1$) as

$$H = \sum_{i=1}^{N^-} \frac{|\mathbf{p}_i^-|^2}{2} + \sum_{\alpha=1}^{N^+} \frac{|\mathbf{p}_\alpha^+|^2}{2} + \sum_{i>j} |\mathbf{x}_i^- - \mathbf{x}_j^-|^{-1} + \sum_{\alpha>\beta} |\mathbf{x}_\alpha^+ - \mathbf{x}_\beta^+|^{-1}$$
$$- \sum_{i,\alpha} |\mathbf{x}_i^- - \mathbf{x}_\alpha^+|^{-1}, \tag{1.2.18}$$

where $N^+ = N^-$ for a neutral system. Now take the expectation value in a state with $\Psi^+ \otimes \Psi^-$, where $\Psi^\pm$ are the trial functions that led to $E \sim -N^{7/5}$ for Bose-jellium. Although the particles are correlated, the charge density is homogeneous, as for instance

$$\left\langle \Psi^+ \middle| -\sum_{i,\alpha} |\mathbf{x}_i^- - \mathbf{x}_\alpha^+|^{-1} \middle| \Psi^+ \right\rangle = -\xi \sum_i \int_\Lambda \frac{d^3x}{|\mathbf{x}_i^- - \mathbf{x}|}.$$

The last term in (1.2.28) is therefore equivalent to $-\sum_i U(\mathbf{x}_i^-) - \sum_\alpha U(\mathbf{x}_\alpha^+) + 2(\xi/2)\int d^3x U(\mathbf{x})$, and there results the sum of the energies of the positive and negative Bose-jellia. The expectation value is consequently about $-N^{7/5}$, which is an upper bound to the energy by the min–max principle

(III: 3.5.21). This "instability," which corresponds to the ground-state energy being nonextensive and the spatial contraction of many-particle aggregates of charged bosons, does not imply that individual atoms consisting of oppositely charged bosons would be unstable. A single, nonrelativistic atom of He^4 with its electrons subjected to Bose statistics (but with their original mass and charge) would have the same ground-state energy as real He^4, since the two-particle ground-state wave-function is symmetric in the spatial coordinates. The lesson here is that experience with two-electron molecules is not a trustworthy guide to the problem of the stability of matter: Since the Pauli exclusion principle makes no difference, the two electrons might just as well be bosons, but a system of many bosons would be unstable, whereas a many-fermion system is stable.

3. Since He^3 is just as stable as He^4, stability is not a matter of the type of statistics of one of the kinds of charge-carrier. Moreover, the relevant energy is always measured in Rydbergs, using the electronic mass, so matter should remain stable even in the limit of infinite nuclear masses.
4. It could be argued heuristically that the potential energy should go as $-N^{4/3}R^{-1}$, since each charge sees an opposite charge at a distance $RN^{-1/3}$, while charges further away should be screened. If this is added to the kinetic energy $N^{5/3}R^{-2}$ of fermions or NR^{-2} of bosons, the minimum is respectively $\sim -N$ at $R \sim N^{1/3}$ or $\sim -N^{5/3}$ at $R \sim N^{-1/3}$.
5. In relativistic dynamics the kinetic energy is $\sim |\mathbf{p}| \sim 1/\Delta x$, so the system is softer. The heuristic arguments would evaluate the total energy of bosons as $\sim N/R - e^2N^{4/3}/R$, which is unbounded below when N is sufficiently large. Whereas nonrelativistic energies are always semibounded for any fixed N, it may happen that the relativistic energy goes to $-\infty$ for sufficiently large, but still finite, values of N.
6. The instability of a Coulomb system of bosons has nothing to do with the long range of the $1/r$ potential, but comes from its short-range features. If the singularity is chopped off by changing the potential to $V(x) = (1 - \exp(-\mu r))/r$, the system of bosons also becomes stable: Since the Fourier transform of V is

$$\tilde{V}(\mathbf{k}) = \frac{4\pi\mu^2}{|\mathbf{k}|^2(|\mathbf{k}|^2 + \mu^2)} > 0,$$

with $|e_i| = e$, we find that

$$V \equiv \sum_{i>j} e_i e_j V(\mathbf{x}_i - \mathbf{x}_j) = \frac{1}{2}\int \frac{d^3k}{(2\pi)^3}\tilde{V}(\mathbf{k})\left|\sum_j \exp(i\mathbf{k}\cdot\mathbf{x}_j)e_j\right|^2 - \frac{1}{2}\sum_{i=1}^{N} e_i^2 V(\mathbf{0})$$
$$> -\frac{N}{2}e^2V(0) = -\frac{N}{2}e^2\mu,$$

so H is bounded below by $-cN$. It could be argued that nuclei have a form factor, and that if μ is taken as the reciprocal of the nuclear radius, then V would be a more realistic potential than $1/r$. This would lead to a simple proof of stability, but it misses the real point. Since the Rydberg, which is measured in electronvolts (eV), is determined by the mass of the electron, it is the kinetic energy of the electrons rather than the size of the nuclei that matters most for stability. The lower bound from the size of the nuclei alone would be $\sim -N$ MeV.

Cosmic Bodies (1.2.19)

The $1/r$ potentials in an object with gravitationally interacting particles are all attractive, so the situation is drastically different. The ground state of the Hamiltonian

$$H_G = \sum_{i=1}^{N} \frac{|\mathbf{p}_i|^2}{2} - \kappa \sum_{i>j} |\mathbf{x}_i - \mathbf{x}_j|^{-1} \tag{1.2.20}$$

goes as $-N^{7/3}$ for fermions. By the now familiar argument, $E \sim N^{5/3}/R^2 - N^2/R$, which has its minimum value $\sim -N^{-7/3}$ for $R \sim N^{-1/3}$. This can easily be translated into an exact upper bound by the use of trial functions localized in $\mathbb{R}^3$. Lower bounds are harder to come by, since energetically more favorable possibilities have to be ruled out. In this case there is an easier way: Write

$$H_G = \sum_{i=1}^{N} \sum_{j \neq i} \left(\frac{|\mathbf{p}_j|^2}{2(N-1)} - \frac{\kappa}{2} |\mathbf{x}_i - \mathbf{x}_j|^{-1} \right) \equiv \sum_{i=1}^{N} h_i, \tag{1.2.21}$$

so that each h_i is the Hamiltonian of an atom with electrons having no Coulomb repulsion. Particle number i stands for the atomic nucleus, as it has no kinetic energy, and the others are electrons, with mass $N - 1$ and potential $-|\mathbf{x}_i - \mathbf{x}_j|^{-1}/2$. According to (III: 4.5.15) it follows that $h_i \geq -cN^{4/3}$, and indeed the result is a

Bound for the Energy of Gravitating Fermions (1.2.22)

$$H_G > -cN^{7/3}, \qquad c = O(1).$$

Remarks (1.2.23)

1. Fermi statistics were not fully taken into account, since we have only antisymmetrized with respect to $N - 1$ particles when filling the energy levels. Since complete antisymmetrization restricts the set of admissible functions further, (1.2.22) is at any rate a lower bound.
2. The limit as $N \to \infty$ in this case exists with the scaling behavior $t = \frac{4}{3}$ of (1.2.1), as in (1.2.4). This does not mean that the limit with $t = \frac{4}{3}$ fails to exist for ordinary matter, but only that it is trivial. The potential energy goes to zero and the particles remain free.
3. If the particles are bosons, then they can all be put into the ground state, and $E \sim -N^3$. The radius of the ground state then goes as N^{-1}.
4. The Hamiltonian (1.2.20) was for the discussion of electrically neutral particles; if they are instead charged, then κ must be replaced with

$\kappa - e_i e_j$. If we bear normal matter in mind, the gravitational force comes from the protonic mass, and in units where the mass of the proton is 1, $\kappa/e^2 \sim 10^{-36}$. Inequality (1.2.22) then *a fortiori* provides a lower bound, since

$$\frac{1}{2}\sum_i \frac{|\mathbf{p}_i|^2}{2} + \sum_{i>j} \frac{e_i e_j}{|\mathbf{x}_i - \mathbf{x}_j|} + \frac{1}{2}\sum_i \frac{|\mathbf{p}_i|^2}{2} - \sum_{i>j} \frac{\kappa}{|\mathbf{x}_i - \mathbf{x}_j|}$$
$$\geq -2c_e e^4 N - 2c\kappa^2 N^{7/3}.$$

The number of particles determines which N-dependence dominates. Gravity begins to win out when $N \sim (e^2/\kappa)^{3/2} \sim 10^{54}$, which is about the mass of Jupiter, and the energies of larger heavenly bodies are controlled mainly by gravitation. A concrete consequence is that the atoms get squashed and turn into a plasma of nuclei and electrons. This inequality provides a more rigorous foundation for the heuristic considerations of (II: 4.5.1).

We shall see in §4.2 that the system (1.2.20) can be solved in the limit $N \to \infty$, as the Thomas–Fermi theory becomes exact. Thomas–Fermi theory provides an idealization of stars, various corrections again being needed to make it realistic. In particular, if $N \sim 10^{57}$ relativistic effects become important. As with atoms with $Z > 137$, the Hamiltonian is unbounded below, which leads to a catastrophe. Nonetheless, Thomas–Fermi theory reflects the thermodynamic properties of stars rather well.

This section concludes with Table 1 displaying the many possibilities:

Table 1 The N-dependence of the kinetic energy K and the potential energy V when N is large.

			K	V	R_{min}	$E(R_{min})$
Nonrelativistic	electric	Bose	N/R^2	$-N^{4/3}/R$	$N^{-1/3}$	$-N^{5/3}$
		Fermi	$N^{5/3}/R^2$	$-N^{4/3}/R$	$N^{1/3}$	$-N$
	gravitational	Bose	N/R^2	$-N^2/R$	N^{-1}	$-N^3$
		Fermi	$N^{5/3}/R^2$	$-N^2/R$	$N^{-1/3}$	$-N^{2/3}$
Relativistic	electric	Bose	N/R	$-N^{4/3}/R$	0	$-\infty$
		Fermi	$N^{4/3}/R$	$-N^{4/3}/R$	0	$-\infty$
		†			or ∞	or 0
	gravitational	Bose	N/R	$-N^2/R$	0	$-\infty$
		Fermi	$N^{4/3}/R$	$-N^2/R$	0	$-\infty$

† If R_{min} tends to $+\infty$ more rapidly than $N^{1/3}$, then the kinetic energy per particle, $N^{1/3}/R$, becomes arbitrarily small, eventually $\ll m$, and the system is nonrelativistic. Hence R_{min} certainly can not increase faster than $N^{1/3}$. Which energy breaks the stalemate depends on the strength of the charge. If $Z < 137$, the kinetic energy wins out, and if $Z > 137$, the potential energy wins out.

Problems (1.2.24)

1. Calculate the β and γ of (1.2.9).
2. Verify (1.2.11; 1).

Solutions (1.2.25)

1.

$$\gamma: \int_{\substack{|x|\le a\\|x'|\le a}} \frac{d^3x\, d^3x'}{|\mathbf{x}-\mathbf{x}'|} = \int r^2\, dr\, d\Omega r'^2\, dr'\, d\Omega' \sum_{n,m} \left[\frac{r^n}{r'^{n+1}}\Theta(r'-r) + \frac{r'^n}{r^{n+1}}\Theta(r-r')\right]\frac{4\pi}{2n+1}$$

$$\cdot\, Y_n^m(\Omega)Y_n^{m*}(\Omega') = \int_0^a\int_0^a r^2\, dr\, r'^2\, dr'\left(\frac{\Theta(r'-r)}{r'} + \frac{\Theta(r-r')}{r}\right)(4\pi)^2$$

$$= \frac{2a^5}{15}(4\pi)^2.$$

$$\beta: \int_{\substack{|\mathbf{x}|\le a\\ \mathbf{x}'\in\Lambda}} d^3x\, d^3x'\left(\frac{1}{|\mathbf{x}-\mathbf{x}'|} - \frac{1}{|\mathbf{x}'|}\right) = \int_{|\mathbf{x}|,|\mathbf{x}'|\le a} \cdots + \int_{\substack{|\mathbf{x}|\le a\\|\mathbf{x}'|\ge a}} \cdots.$$

The second integral equals 0, as can be seen by expanding $|\mathbf{x}-\mathbf{x}'|^{-1}$ in spherical harmonics. The first integral equals $-(2\pi a^2/5)(4\pi a^3/3)$ if $\{\mathbf{x}': |\mathbf{x}'|\le a\}\subset\Lambda$, and is otherwise greater than or equal to this.

2. $U(\mathbf{x}_i) \le -(3N/2R) + (N/2R)(|\mathbf{x}_i|^2/R^2)$, equality holding for $|x_i| < R$. The self-energy of the background charge is $3N^2/5R$.

1.3 Arbitrary Numbers of Particles in Fock Space

The properties of large systems should not depend on the exact number of particles, so it is convenient to use a representation with a variable number of particles.

We are used to dealing with atomic systems on $\mathscr{H}_n$, the n-particle Hilbert space. As it is impossible to count the particles in a large system, it is convenient to regard the number N of particles as an observable capable of assuming various values. Accordingly, we shall study **Fock space**

$$\mathscr{H}_F = \bigoplus_{n=0}^{\infty} \mathscr{H}_n, \qquad N_{|\mathscr{H}_n} = n, \tag{1.3.1}$$

as the foundation for later analysis. The space $\mathscr{H}_0$ is one-dimensional and spanned by the **vacuum vector** $|0\rangle$. If the particles under consideration are either all bosons or all fermions, then $\mathscr{H}_n$ is either the n-fold symmetric or totally antisymmetric tensor product of $\mathscr{H}_1 = L^2(\mathbb{R}^3, d^3x)$ with itself, which will be denoted $\mathscr{H}_1 \circledS \mathscr{H}_1 \circledS \cdots \circledS \mathscr{H}_1$ or $\mathscr{H}_1 \wedge \mathscr{H}_1 \wedge \cdots \wedge \mathscr{H}_1$. If f_j, $j = 1, 2, \ldots$, is a complete orthonormal set of functions on $\mathscr{H}_1$, then the vectors $|f_{j_1} \circledS f_{j_2} \circledS \cdots \circledS f_{j_n}\rangle$ or respectively $|f_{j_1} \wedge f_{j_2} \wedge \cdots \wedge f_{j_n}\rangle$ are a basis for $\mathscr{H}_n$. In the latter case all the j_k are to be taken different. For bosons the same f's can be collected together and written as $|f_{j_1}^{n_1}, \ldots, f_{j_k}^{n_k}\rangle$, with $\sum_k n_k = N$. The C^* algebra generated on the individual $\mathscr{H}_n$ of the boson Fock space by the symmetrized Weyl operators

$$\sum_\pi \exp\left[i \sum_j (r_{\pi_j} x_j + s_{\pi_j} p_j)\right],$$

where $(\pi_1, \ldots, \pi_n)$ is a permutation of $(1, \ldots, n)$, will be called the **Weyl algebra**, and is represented reducibly on $\mathscr{H}_F$—all bounded functions of N alone belong to the commutant of the representation.

The irreducible **field algebra** on $\mathscr{H}_F$ turns out to be invaluable for the many-body problem:

Definition (1.3.2)

Let $|f_1, f_2, \ldots\rangle \equiv |f_1 \circledS f_2 \ldots\rangle$, and define the **creation and annihilation operators** $a^*(f)$ and $a(f)$ by linear extension of

$$\begin{aligned} a(f_m)|f_{j_1}^{n_1}, \ldots, f_{j_k}^{n_k}\rangle &= \delta_{mj_1}\sqrt{n_1}\,|f_{j_1}^{n_1-1}, f_{j_2}^{n_2}, \ldots, f_{j_k}^{n_k}\rangle \\ &\quad + \delta_{mj_2}\sqrt{n_2}\,|f_{j_1}^{n_1}, f_{j_2}^{n_2-1}, \ldots, f_{j_k}^{n_k}\rangle + \cdots \\ &\quad + \delta_{mj_k}\sqrt{n_k}\,|f_{j_1}^{n_1}, f_{j_2}^{n_2}, \ldots, f_{j_k}^{n_k-1}\rangle \qquad \text{(for bosons)}, \\ a(f_m)|f_{j_1} \wedge \cdots \wedge f_{j_n}\rangle &= \delta_{mj_1}|f_{j_2} \wedge \cdots \wedge f_{j_n}\rangle - \delta_{mj_2}|f_{j_1} \wedge f_{j_3} \wedge \cdots \wedge f_{j_n}\rangle \\ &\quad + \cdots + (-1)^{n+1}|f_{j_1}, \ldots, f_{j_{n-1}}\rangle \qquad \text{(for fermions)}, \\ a^*(f_m)|f_{j_1}^{n_1}, \ldots, f_{j_k}^{n_k}\rangle &= \delta_{mj_1}\sqrt{n_1 + 1}\,|f_{j_1}^{n_1+1}, f_{j_2}^{n_2}, \ldots, f_{j_k}^{n_k}\rangle \\ &\quad + \delta_{mj_2}\sqrt{n_2 + 1}\,|f_{j_1}^{n_1}, f_{j_2}^{n_2+1}, \ldots, f_{j_k}^{n_k}\rangle + \cdots \\ &\quad + \delta_{mj_k}\sqrt{n_k + 1}\,|f_{j_1}^{n_1}, \ldots, f_{j_k}^{n_k+1}\rangle \\ &\quad + \left(1 - \sum_{l=1}^k \delta_{mj_l}\right)|f_m f_{j_1}^{n_1}, \ldots, f_{j_k}^{n_k}\rangle \qquad \text{(for bosons)}, \end{aligned}$$

$$a^*(f_m)|f_{j_1} \wedge \cdots \wedge f_{j_n}\rangle = |f_m \wedge f_{j_1} \wedge \cdots \wedge f_{j_n}\rangle \qquad \text{(for fermions)},$$

and $a(\alpha f + \beta g) = \alpha a(f) + \beta a(g)$ for f and $g \in \mathscr{H}_1$.

Remarks (1.3.3)

1. The prototypes of the a's for bosons are the a and a^* of a harmonic oscillator (III: 3.3.5; 2), and for fermions they are the matrices $\sigma^\pm$ of (1.1.2). The formal analogy is not just superficial; the operators $a(f)$ show up when one quantizes coupled oscillators and then passes to a continuous limit, in the procedure known as **field quantization**, or **second quantization**.

2. Formally, the a's satisfy the commutation or anticommutation relations:
$$[a(f), a^*(g)] = (f|g) \quad \text{(the scalar product on } \mathscr{H}_1\text{)},$$
$$[a(f), a(g)] = 0 \quad \text{for bosons},$$
$$a(f)a^*(g) + a^*(g)a(f) \equiv [a(f), a^*(g)]_+ = (f|g),$$
$$[a(f), a(g)]_+ = 0 \quad \text{for fermions}.$$
Conversely, (1.3.2) can be derived from the commutation relations and $a(f)|0\rangle = 0$. The commutation relations are invariant under unitary transformations of the f_j, so (1.3.2) is independent of the choice of the basis. In the spirit of the GNS Construction, vector states may be identified with operators:
$$|f_{j_1}^{n_1}, \ldots, f_{j_k}^{n_k}\rangle = (n_1! \ldots n_k!)^{-1/2} a^*(f_{j_1})^{n_1} \ldots a^*(f_{j_k})^{n_k}|0\rangle,$$
or
$$|f_{j_1} \wedge \cdots \wedge f_{j_k}\rangle = a^*(f_{j_1}) \ldots a^*(f_{j_k})|0\rangle.$$
3. As in (III: 3.1.10; 2) the commutation relations reveal that the operators $a(f)$ are unbounded. To get a C^* algebra, it is necessary to use the bounded operators $\exp[i(\alpha a(f) + \alpha^* a^*(f))]$; the algebra they generate is called $\mathscr{A}_B$.
4. The anticommutation relations for fermion fields are the same as those of $\sigma^\pm$, for which reason their $a(f)$ are bounded: $\|a(f)\Psi\|^2 + \|a^*(f)\Psi\|^2 = \langle\Psi|(a^*(f)a(f) + a(f)a^*(f))\Psi\rangle = (f|f)\|\Psi\|^2$, so $\|a(f)\| \leq \|f\|$. Because $\langle 0|a(f)a^*(f)|0\rangle = \|f\|^2$, this means $\|a(f)\| = \|a^*(f)\| = \|f\|$. The operators $a(f)$ generate a C^* algebra $\mathscr{A}_F$, which is the norm-closure of the polynomials in a and a^*.
5. It follows from Remark 4 that the mapping $f \to a^*(f)$ is an isometric homomorphism of the Banach-space structure of $\mathscr{H}_1$ to that of $\mathscr{A}_F$. (The mapping $f \to a(f)$ is continuous but antilinear, that is, $a(\lambda f + \mu g) = \lambda^* a(f) + \mu^* a(g)$.) For every unitary transformation $U \in \mathscr{B}(\mathscr{H}_1)$ there is a linear transformation $a(f) \to a(Uf)$, which can be extended to an automorphism u:
$$u(a(f_1) \ldots a(f_k)a^*(g_1) \ldots a^*(g_j)) = \\ = a(Uf_1) \ldots a(Uf_k)a^*(Ug_1) \ldots a^*(Ug_j). \qquad (1.3.4)$$
In particular, for every strongly continuous unitary group $U(t)$ there is a norm-continuous group of automorphisms u_t on $\mathscr{A}_F$ (i.e., the mapping $t \to u_t(a)$ from $\mathbb{R}$ to $\mathscr{B}(\mathscr{H}_F)$ is continuous in norm for all a). Therein lies a difference from the Weyl algebra, for which, although the free time-evolution $\exp[i(rp + sx)] \to \exp[i(rp + s(x + pt))]$ is strongly continuous in t, it is not continuous in norm. The time-evolution on $\mathscr{A}_B$ is also not continuous in norm, so the property of continuity can not be expressed without reference to a representation. In this regard the field algebra of

fermions is much the nicer, owing ultimately to its being modeled on the matrices

$$\begin{pmatrix} 0 & 1 \\ 0 & 0 \end{pmatrix}, \quad \begin{pmatrix} 0 & 0 \\ 1 & 0 \end{pmatrix}.$$

Fermion fields will consequently be preferred when investigating more problematic cases.

6. The algebras $\mathscr{A}_F$ and $\mathscr{A}_B$ may be thought of as constructed from local algebras $\mathscr{A}_\Lambda$, containing only those $a(f)$ and $a^*(f)$ for which supp $f \subset \Lambda$. Clearly, $\mathscr{A}_\Lambda \subset \mathscr{A}_{\Lambda'}$ when $\Lambda \subset \Lambda'$. Since $\mathscr{H}_1$ is the norm-closure of $\bigcup_{\Lambda \subset \mathbb{R}^3} L^2(\Lambda, d^3x)$, $\mathscr{A}_F$ equals the norm-closure of $\bigcup_{\Lambda \subset \mathbb{R}^3} \mathscr{A}_\Lambda$.
7. It is common for annihilation operators to be introduced at single points, for which formally $[a(\mathbf{x}), a^*(\mathbf{x}')] = \delta^3(\mathbf{x} - \mathbf{x}')$, $a(f^*) = \int d^3x a(\mathbf{x}) f(\mathbf{x})$, $a^*(f) = \int d^3x' a^*(\mathbf{x}') f(\mathbf{x}')$. Although $a(\mathbf{x})$ is densely defined as an operator, it is not closeable, so $a^*(\mathbf{x})$ exists only in the sense of a quadratic form and not as an operator (Problem 8). The object $a^*(\mathbf{x})$ is called an operator-valued distribution.
8. Since a annihilates a particle and a^* creates one, the spaces $\mathscr{H}_n$ are not invariant subspaces of Fock space. It can in fact be shown that $\mathscr{A}_F$ and $\mathscr{A}_B$ are irreducibly represented on $\mathscr{H}_F$ (Problem 1). The algebra $\mathscr{A}_F$ is said to be **quasilocal**.

Remark (1.3.3; 5) implies that such things as translations and free time-evolution correspond to norm-continuous one-parameter groups of automorphisms on $\mathscr{A}_F$. The question arises as to whether they can be presented as strongly continuous, one-parameter unitary groups on $\mathscr{H}_F$. If the representation called for is just like the GNS representation of (III: 2.3.9) with the vacuum $|0\rangle$ as a cyclic, and also invariant, vector, then the answer is yes (however, see Problems 6 and 7):

The Unitary Representability of the Automorphism (1.3.5)

Let u_g be a group of automorphisms of a C^* algebra $\mathscr{A}$, w be an invariant state (i.e., $w(u_g(a)) = w(a)$ for all g), and π_w be the representation constructed with w. Then the group of automorphisms has a unique unitary representation U_g on the Hilbert space $\mathscr{H}_F$, such that

$$\pi_w(u_g(a)) = U_g \pi_w(a) U_g^{-1}, \qquad U_g\Omega = \Omega, \tag{1.3.6}$$

where Ω is the cyclic vector.

Proof

If we let $U_g \pi_w(a)\Omega = \pi_w(u_g(a))\Omega$, then the U_g thereby defined satisfies the stated requirements. It is unique, since if there existed another $\tilde{U}_g$ with the same properties, then it would follow that $(\tilde{U}_g U_g^{-1} - \mathbf{1})\Omega = 0$, $\tilde{U}_g U_g^{-1} \in \pi(\mathscr{A})'$. Now, because Ω is cyclic for $\pi(\mathscr{A})$, it separates $\pi(a)'$, and therefore

$\tilde{U}_g U_g^{-1} = \mathbf{1}$ (cf. Problem 5). (**Separating** means that for $a' \in \pi(\mathscr{A})'$, $a'|\Omega\rangle = 0$ implies $a' = 0$.) □

Remarks (1.3.7)

1. If the group is topological and the realization as a group of automorphisms is weakly continuous, then U_g is strongly continuous,

$$\|(U_g - \mathbf{1})\pi_w(a)\Omega\|^2 = 2w(a^*a) - w(a^*u_g(a)) - w(u_g(a^*)a) \to 0$$

as g approaches the identity.
2. Our representation of $\mathscr{A}_F$ (1.3.2) is a π_w such that $w(a) = \langle 0|a|0\rangle$ for $a \in \mathscr{A}_F$. Therefore Ω is the vacuum vector $|0\rangle$,and is invariant under the transformations brought up in (1.3.3; 5). It follows that the Euclidean group and free time-evolution can be represented by strongly continuous unitary groups of operators on Fock space. They consequently have self-adjoint generators (Problem 2), which are, however, not bounded. Even the operators U_g do not belong to $\mathscr{A}_F$. To prove this fact we shall make use of

Definition (1.3.8)

The C^* algebra obtained by closing the even polynomials in a and a^* in norm is denoted $\mathscr{A}_G$. The norm-closure of the polynomials having the same number of a's as a^*'s in each summand is $\mathscr{A}_E$.

Remarks (1.3.9)

1. $\mathscr{A}_F \supset \mathscr{A}_G \supset \mathscr{A}_E$. In the Fock representation, $\mathscr{A}_E = \{N\}' \cap \mathscr{A}_F$.
2. Because $[ab, c] = a[b, c]_+ - [a, c]_+ b = a[b, c] + [a, c]b$, if $d \in \mathscr{A}_{\Lambda G}$ and $c \in \mathscr{A}_{\bar{\Lambda}}$, $\bar{\Lambda} \cap \Lambda = \emptyset$, then $[d, c] = 0$.

Asymptotic Commutativity (1.3.10)

Let $V(t) \in \mathscr{B}(L^2\mathbb{R}^3))$ be a one-parameter, unitary group of operators with absolutely continuous spectrum, such that $V(t) \rightharpoonup 0$ as $t \to \infty$, and let $u_t(a(f)) \equiv a(V(t)f)$. Then $\lim_{t\to\infty} \|[a, u_t(b)]\| = 0$ for all $a \in \mathscr{A}_G$ and $b \in \mathscr{A}_F$; this state of affairs is described by saying that $\mathscr{A}_G$ is **asymptotically Abelian** *with respect to u_t.*

Proof

First note that $\|[a(f), u_t(a^*(g))]_+\| = \|[a^*(f), u_t(a(g))]_+ = |(V(t)g|f)| \to 0$ as $t \to \infty$. If d is an even polynomial and c is any polynomial in $a(f)$ and $a^*(g)$, then with Remark (1.3.9; 2) it follows that the commutator vanishes asymptotically. Because the algebraic operations are continuous in norm, this extends to $\mathscr{A}_G$ and $\mathscr{A}_F$. □

Corollaries (1.3.11)

1. Since the generators of the spatial translation group and the free time-evolution have purely continuous spectrum, for them $V(t) \rightharpoonup 0$, and the appropriate commutators involving them go to zero.
2. The corresponding one-parameter groups of unitary operators on Fock space, $U_t \in \mathscr{B}(\mathscr{H}_F)$, can not belong to $\mathscr{A}_F$. Since every U_t commutes with N, it must belong to $\mathscr{A}_E$, and hence $\|[U_t, u_{t'}(a)]\| < \varepsilon$ for all $\varepsilon \in \mathbb{R}^+$, $a \in \mathscr{A}_F$, and sufficiently large t'. Note that $\|U_t U_{t'} a U_{t'}^{-1} - U_{t'} a U_{t'}^{-1} U_t\| = \|U_t a U_t^{-1} - a\|$ which obviously can not be arbitrarily small for all t. It is even true that $\mathscr{A}_F \cap \bigcup_t U_t = U_0$.
3. Since $\mathscr{A}_F$ is irreducible, $\mathscr{A}_F'' = \mathscr{B}(\mathscr{H}_F)$ (III: 2.3.4), so U_t is certainly attainable as the strong limit of elements of $\mathscr{A}_F$, or even $\mathscr{A}_E$.

Remarks (1.3.12)

1. Since commuting observables are jointly diagonable, and hence can be measured simultaneously, if V is a group of translations, this implies that measurements separated by a large spatial distance do not interfere with each other. The local character of the algebra is important for this, and it does not apply to the Weyl operators, as $\exp[i(rp + sx)]$ and $\lim_{a\to\infty} \exp[i(r'p + s'(x + a))]$ do not commute. Even the bicommutant $\mathscr{A}_F''$ in the Fock representation is not asymptotically Abelian—for instance, the generators of the Euclidean group belong to the strong closure of $\mathscr{A}_F$ and are constant with respect to the free time-evolution but do not commute. Therefore A_F'' is not asymptotically Abelian with respect to free time-evolution.
2. The point of (1.3.10) for the time-evolution is that as time passes the disturbance due to a measurement diffuses so widely that local observables are not affected at much later times. This does not apply to the observables x and p, as p and $x + pt$ fail to commute even at large t. Observe that we have as yet proved commutativity only for free time-evolution; the question of whether it also holds for more realistic time-evolutions remains open.
3. This phenomenon does not occur for compact groups like the rotations; for them U is a sum of finite-dimensional representations, for which it is impossible that $U \rightharpoonup 0$.

Global Observables (1.3.13)

The particle-number operator N was defined in (1.3.1). It is unbounded and thus $\notin \mathscr{B}(\mathscr{H}_F)$, which $\supset \mathscr{A}_F$. Its domain of self-adjointness is

$$D_N = \left\{\psi_0 \oplus \psi_1 \oplus \cdots \oplus \psi_n \oplus \cdots \in \mathscr{H}_F : \sum_{n=1}^{\infty} n^2 \|\psi_n\|^2 < \infty\right\}.$$

Moreover, unitary gauge transformations $U(\alpha) = \exp(iN\alpha) \in \mathscr{B}(\mathscr{H}_F)$ also do not belong to $\mathscr{A}_F$, but can be attained as strong limits of elements of $\mathscr{A}_E$. In the Fock representation,

$$U(\alpha) = s\text{-}\lim_{M \to \infty} \exp\left(i\alpha \sum_{j=1}^{M} a^*(f_j)a(f_j)\right),$$

where $\{f_j\}$ is an orthonormal basis. Although $U(\alpha)$ does not depend on the basis, it can only be defined in certain representations.

Remark (1.3.14)

Since N is conserved in all of the systems treated here, it is not physically possible to measure the relative phase of states of different N. This means that N creates a superselection rule in the sense of (III: 2.3.6; 7), and the algebra of observables should, properly speaking, be $\{N\}' = \mathscr{A}_E''$. The representation of this algebra on $\mathscr{H}_F$ is reducible, as its commutant is $\{N\}'' \neq \{\lambda \cdot \mathbf{1}\}$.

Observables at a Point (1.3.15)

One frequently considers the particle density and current at a point,

$$\rho(\mathbf{x}) = a^*(\mathbf{x})a(\mathbf{x}) = \sum_{j,k} a^*(f_j)a(f_k)f_j^*(\mathbf{x})f_k(\mathbf{x}),$$

$$\mathbf{j}(\mathbf{x}) = -\frac{1}{2mi}(a^*(\mathbf{x})\nabla a(\mathbf{x}) - (\nabla a^*(\mathbf{x}))a(\mathbf{x}))$$

$$= \sum_{j,k} a^*(f_j)a(f_k)\left(\frac{1}{2mi}(f_j^*(\mathbf{x})\nabla f_k(\mathbf{x}) - (\nabla f_j^*(\mathbf{x}))f_k(\mathbf{x}))\right).$$

The f_k in these formulas must be chosen as an orthonormal basis of C^1 functions, in which case these observables are densely defined as quadratic forms. They are not, however, closeable: Their restrictions to $\mathscr{H}_1$ are the quadratic forms of

$$\psi^*(\mathbf{x})\psi(\mathbf{x}) \quad \text{and} \quad \frac{1}{2mi}(\psi^*(\mathbf{x})\nabla\psi(\mathbf{x}) - (\nabla\psi^*(\mathbf{x}))\psi(\mathbf{x})),$$

the former of which is recognizable as the prototype of this phenomenon as encountered in (III: 2.5.18; 3). Matrix elements with, say, $\rho(\mathbf{x})$ may be understood as distributional limits of matrix elements of the bounded operators $a^*(f)a(f)$ as $f \to \delta^3(\mathbf{x})$. Similarly, the continuity equation $\dot{\rho} + \nabla \cdot \mathbf{j} = 0$ holds at least for matrix elements *if*, evolving freely in time, $i\dot{f} = -\Delta f/2m$.

Problems (1.3.16)

1. Show that the representations of $\mathscr{A}_F$ and $\mathscr{A}_B$ on $\mathscr{H}_F$ are irreducible.
2. Construct the generators of free time-evolution and of translation.
3. Find dense domains of definition for the quadratic forms $\rho(\mathbf{x})$ and $\mathbf{j}(\mathbf{x})$.
4. Define the number of particles in the volume V, $N_V = \int_V d^3x\rho(\mathbf{x})$, as an unbounded, self-adjoint operator.
5. For $\mathscr{A} \subset \mathscr{B}(\mathscr{H})$ and $\Omega \in \mathscr{H}$, show that Ω is cyclic for $\mathscr{A}$ iff Ω separates $\mathscr{A}'$.
6. The mapping $a \to b: b(f) = a(f) + L(f)$ is an automorphism α_L of the Bose algebra whenever L is a linear, but not necessarily continuous, functional. Show that α_L is unitarily implementable on $\mathscr{H}_F$, i.e., there exists a $U_L \in \mathscr{B}(\mathscr{H}_F)$ such that $\mathbf{1} = U_L^* U_L = U_L U_L^*$ and $U_L a(f) U_L^{-1} = b(f)$, iff L is continuous, which means that it can be written as $L(f) = (\rho | f)$ for some $\rho \in \mathscr{H}_1$.
7. Let $b(f) = a(\Phi f) + a^*(\overline{\Psi} f)$, $\Phi, \Psi \in \mathscr{B}(\mathscr{H}_1)$, Φ invertible. Show

   (i) that $a \to b$ is an automorphism of the Bose (resp. Fermi) field algebra if

   $$\Phi\Phi^* \mp \Psi\Psi^* = 1 = \Phi^*\Phi \mp (\Psi^*\Psi)^{\ell},$$

   $$\Phi\Psi^{\ell} \mp \Psi\Phi^{\ell} = 0 = (\Psi^*\Phi)^{\ell} \mp \Psi^*\Phi,$$

   where $\overline{\Psi} = \Psi^{*\ell}$; and

   (ii) that it can be represented as a unitary operator on $\mathscr{H}_F$ iff $\Phi^{-1}\Psi \in \mathscr{C}_2(\mathscr{H}_1)$.
8. Show that although the $a(\mathbf{x})$ of (1.3.3; 7) is densely defined, it is not closeable, and the domain of definition of its adjoint $a^*(\mathbf{x})$ contains only the zero vector.

Solutions (1.3.17)

1. Let b be an operator such that $[b, a(f)] = [b, a^*(f)] = 0$ for all $f \in \mathscr{H}_F$. From the commutation relations of (1.3.3; 2) and $a(f)|0\rangle = 0$, it follows that $\langle 0 | a(f_1) \ldots a(f_m) b a^*(g_1) \ldots a^*(g_n)|0\rangle = \langle 0|b|0\rangle \cdot \langle 0|a(f_1) \ldots a^*(g_n)|0\rangle$, which implies that $\langle x | bx\rangle = \langle 0|b|0\rangle \|x\|^2$ on a dense set, and therefore $b = \langle 0|b|0\rangle \cdot \mathbf{1}$.
2. With Theorem (1.3.5) and the fact that the $\mathscr{H}_n$ are invariant, by reasoning as in (1.3.13) we find that the two generators are

   $$\text{s-}\lim_{M\to\infty} \sum_{i,j}^{M} \int \nabla f_j^*(\mathbf{x}) \cdot \nabla f_i^*(\mathbf{x}) a^*(f_j) a(f_i)\, d^3x$$

   and

   $$\text{s-}\lim_{M\to\infty} i \sum_{k,j}^{M} \int \nabla f_j^*(\mathbf{x}) f_k(\mathbf{x}) a(f_k)\, d^3x,$$

   where the strong limit is defined as in (III: 2.5.8; 3). Formally, these can be written as $\int d^3x \nabla a^*(\mathbf{x}) \cdot \nabla a(\mathbf{x})$ and $i \int d^3x a^*(\mathbf{x}) \overleftrightarrow{\nabla} a(\mathbf{x})$.
3. For $\rho(\mathbf{x})$, linear combinations of $\prod_j a^*(f_j)|0\rangle$ with continuous f_j. For $\mathbf{j}(\mathbf{x})$, the f_k have to be continuously differentiable.

4. $N_V = \sum_{j,k} a^*(f_j)a(f_k)\int_V d^3x f_j^*(\mathbf{x})f_k(\mathbf{x})$, $0 \leq N_V \leq N$, is a Hermitian operator on D_N (1.3.13), and hence the domain of its Friedrichs extension contains D_N.

5. "If": Let P be the projection onto the orthogonal complement of $\{a|\Omega\rangle\}$ for $a \in \mathscr{A}$. Then $P \in \mathscr{A}'$ and $P|\Omega\rangle = 0$, so $P = 0$.
"Only if": Let $a' \in \mathscr{A}'$, $a'|\Omega\rangle = 0$. Then $a'a|\Omega\rangle = 0$ for all $a \in \mathscr{A}$, which implies that $a' = 0$ on a dense set, so $a' = 0$.

6. The mapping $a \to b$ is unitarily implementable on $\mathscr{H}_F$ iff there exists a vector $|0_b\rangle \in \mathscr{H}_F$ such that $b(f)|0_b\rangle = 0$ for all $f \in \mathscr{H}_1$. It is clear that the existence of U implies that of $|0_b\rangle = U|0\rangle$. On the other hand, the mapping

$$\prod_{i=1}^{n} a_i^*|0\rangle \to \prod_{i=1}^{n} b_i^*|0_b\rangle,$$

where $a_i = a(f_i)$, $b_i = b(f_i)$, and $\{f_i\}$ is an orthonormal basis, defines a unitary operator U, since this set of vectors is total. (Every vector is cyclic for an irreducible representation.) If L is not continuous, then ker L is dense in $\mathscr{H}_1$, and therefore $a(f)|0_b\rangle = 0$ for a dense set of f's. This implies that $|0_b\rangle = |0\rangle$ and thus that $L \equiv 0$, which is continuous. Therefore $|0_b\rangle \notin \mathscr{H}_F$. If, however, $L(f) = (g|f)$, $g \in \mathscr{H}_1$, it is possible to choose $f_1 = g/\|g\|$. Because $a\exp[-a^*\|g\|] = \exp[-a^*\|g\|](a - \|g\|)$, the vector $|0_b\rangle = \exp[-a_1^*\|g\|]|0\rangle$ formally satisfies $b_k|0_b\rangle = (a_k + \delta_{k1}\|g\|)|0_b\rangle = 0$. It is also normalizable provided that

$$\infty > \langle 0|\exp[-\|g\|a_1]\exp[-\|g\|a_1^*]|0\rangle = \sum_{n=0}^{\infty} \frac{1}{(n!)^2}\|g\|^{2n} n! = \exp\|g\|^2,$$

so $\langle 0_b|0_b\rangle < \infty$ if $\|g\|^2 < \infty$.

7. (i) In matrix notation, for $b = \Phi a + \Psi a^*$, (i) must hold: $1 = [b, b^*]_\mp = \Phi\Phi^* \mp \Psi\Psi^*$, and $0 = [b, b]_\mp = \Phi\Psi^t \mp \Psi\Phi^t$. Written as block matrices, this becomes

$$\begin{pmatrix} \Phi & \Psi \\ \Psi^{*t} & \Phi^{*t} \end{pmatrix}\begin{pmatrix} \Phi^* & \mp\Psi^t \\ \mp\Psi^* & \Phi^t \end{pmatrix} = \mathbf{1}.$$

For invertibility it is necessary that

$$\begin{pmatrix} \Phi^* & \mp\Psi^t \\ \mp\Psi^* & \Phi^t \end{pmatrix}\begin{pmatrix} \Phi & \Psi \\ \Psi^{*t} & \Phi^{*t} \end{pmatrix} = \mathbf{1},$$

which produces the second line of the conditions.

(ii) The Fock vacuum $|0_b\rangle$ satisfies $0 = (\Phi^{-1}b)_k|0_b\rangle = (a_k + M_{kl}a_l^*)|0_b\rangle$, where $M = \Phi^{-1}\Psi$. Because $[a, a^*Ma^*] = 2Ma^*$, it can be written formally as $|0_b\rangle = c\exp[-a^*Ma^*/2]|0\rangle$. (Observe that by (i), $M = M^t$ (resp. $M = -M^t$).) To determine the normalization constant c, we shall calculate

$$\langle 0|\exp[-\tfrac{1}{2}aNa]\exp[-\tfrac{1}{2}a^*Ma^*]|0\rangle$$

when $M = \pm M^t$, $N = \pm N^t$, $[M, N^*] = 0$ and M and N are for the moment real. They can then simultaneously be put into the normal forms

$$\begin{pmatrix} n_1 & & & \\ & n_2 & & \\ & & n_3 & \\ & & & \ddots \end{pmatrix}, \quad \begin{pmatrix} m_1 & & & \\ & m_2 & & \\ & & m_3 & \\ & & & \ddots \end{pmatrix}$$

and respectively

$$\begin{pmatrix} & n_1 & & & \\ -n_1 & & & & \\ & & & n_2 & \\ & & -n_2 & & \\ & & & & \ddots \end{pmatrix}, \quad \begin{pmatrix} & m_1 & & & \\ -m_1 & & & & \\ & & & m_2 & \\ & & -m_2 & & \\ & & & & \ddots \end{pmatrix}$$

with real, orthogonal transformations. The transformations preserve the commutation relations of the field operators, so we may use this basis to calculate

$$\langle 0| \exp\left[-\frac{n_1}{2} a_1^2\right] \exp\left[-\frac{m_1}{2} a_1^{*2}\right]|0\rangle = \sum_{n=1}^{\infty} \frac{(n_1 m_1)^n}{4^n (n!)^2}(2n)! = (1 - n_1 m_1)^{-1/2}$$

and, respectively, for fermions,

$$\langle 0| \exp[-n_1 a_2 a_1] \exp[-m_1 a_1^* a_2^*]|0\rangle = 1 + n_1 m_1.$$

Therefore,

$$\langle 0| \exp[-\tfrac{1}{2} aNa] \exp[-\tfrac{1}{2} a^* M a^*]|0\rangle = \prod_i (1 - n_i m_i)^{-1/2} = \left(\operatorname{Det}\begin{pmatrix} 1 & M \\ N & 1 \end{pmatrix}\right)^{-1/2}$$

and, respectively,

$$\prod_i (1 + n_i m_i) = \left(\operatorname{Det}\begin{pmatrix} 1 & M \\ N & 1 \end{pmatrix}\right)^{1/2}.$$

This can be continued analytically to complex matrix elements, and, in particular, in our case,

$$|c|^2 \left(\operatorname{Det}\begin{pmatrix} 1 & M \\ M^* & 1 \end{pmatrix}\right)^{\mp 1/2} = 1.$$

The determinant is finite for $M \in \mathscr{C}_2$. Observe that in the case of bosons, $\Phi^*\Phi \geq 1$, and so $\Phi = V(\Phi^*\Phi)^{1/2}$ is always invertible. The result for fermions is valid for M acting on either even or odd dimensional spaces.

8. The dense domain of definition of $a(\mathbf{x})$ consists of vectors with continuous, bounded f's. For example, for fermions,

$$\begin{aligned} a(\mathbf{x})|f_{j_1} \wedge \cdots \wedge f_{j_n}\rangle = f_{j_1}(x)|f_{j_2} \wedge \cdots \wedge f_{j_n}\rangle - f_{j_2}(\mathbf{x})|f_{j_1} \wedge f_{j_3} \wedge \cdots \wedge f_{j_n}\rangle + \cdots \\ + (-1)^{n+1} f_{j_n}(\mathbf{x})|f_{j_1} \wedge \cdots \wedge f_{j_n}\rangle. \end{aligned}$$

The operator $a(\mathbf{x})$ is not closeable. Suppose that $f_\lambda(\mathbf{x}') = \exp[-|\mathbf{x} - \mathbf{x}'|^2 \lambda]$; then $|f_\lambda\rangle \to 0$ as $\lambda \to \infty$, but $a(\mathbf{x})|f_\lambda\rangle = |0\rangle \nrightarrow 0$. Formally, $a^*(\mathbf{x})$ creates a particle with wave-function $f(\mathbf{x}') = \delta^3(\mathbf{x} - \mathbf{x}')$. Since this is not normalizable, $a^*(\mathbf{x})$ makes every vector $|f_{j_k} \wedge \cdots \wedge f_{j_n}\rangle$ infinitely long.

1.4 Representations with $N = \infty$

Systems of N *particles are represented on a Hilbert space that is the tensor product of* N *Hilbert spaces for single particles. The infinite tensor product opens the door to the new mathematical features of field theory.*

The scalar product on an N-fold tensor product of spaces $\mathscr{H}_1$ was defined multiplicatively by

$$\langle x|x\rangle = \prod_{i=1}^{N} (x_i|x_i), \qquad |x\rangle = |x_1) \otimes |x_2) \ldots |x_n), \qquad x_i \in \mathscr{H}_1. \quad (1.4.1)$$

If $N = \infty$, the vectors $|x\rangle$ that can be used in this formula are initially only those for which the infinite product converges. The product might well converge to 0 even though $(x_i|x_i) > 0$ for all i. In order to form the quotient space with respect to the zero vectors, it will first be necessary to form the equivalence class not only of vectors with some factor zero but also containing the vectors for which the product

$$\prod_{i=1}^{\infty}(x_i|x_i)$$

converges to zero. On the quotient space, (1.4.1) defines a separating norm, so the space can be completed to a Hilbert space $\mathscr{H}$, with the linear structure defined in the usual way.

This does not yet, however, suffice to define the scalar product of different vectors $|x\rangle$ and $|y\rangle$. Though only vectors such that $(x_i|x_i) = (y_i|y_i) = 1$ for all i need to be considered, there are still two possibilities, namely

$$\text{(I)} \qquad \prod_{i=1}^{\infty} |(x_i|y_i)| \to c > 0,$$

and

$$\text{(II)} \qquad \prod_{i=1}^{\infty} |(x_i|y_i)| \to 0,$$

where $\to$ means unconditional convergence. In case (II), $\prod_{i=1}^{\infty}(x_i|y_i) \to 0$ as well, and the vectors may be considered orthogonal. Possibility (I), on the other hand, does not guarantee that $\prod_i (x_i|y_i)$ converges. If $(x_j|y_j) = \exp(i\varphi_j)|(x_j|y_j)|$, then their product is said to converge if not only $\prod_i |(x_i|y_i)|$ but also $\sum_i |\varphi_i|$ converges. One now encounters the convention that vectors may be deemed orthogonal whenever $\sum_i |\varphi_i| \to \infty$ (case (I_b)). Let us thus agree on a

Definition of the Scalar Product (1.4.2)

$$\langle x|y\rangle = c \quad \text{provided that } \prod_i (x_i|y_i) \to c \neq 0, \quad \text{(case (Ia));}$$

$$\langle x|y\rangle = 0 \quad \text{provided that } \prod_i (x_i|y_i) \to 0 \quad \text{(case (II), or in the divergent sense } (\mathrm{I}_b)).$$

Remarks (1.4.3)

1. It is easy to see that the scalar product this defines on $\mathscr{H}$ obeys all the rules of the game.
2. The space $\mathscr{H}_1$ has been assumed separable, yet even if $\mathscr{H}_1 = \mathbb{C}^2$, the larger space $\mathscr{H}$ is nonseparable. Let $|\mathbf{n}) \in \mathbb{C}^2$ be defined such that $(\mathbf{n}|\mathbf{n}) = 1$, $(\mathbf{n}|\boldsymbol{\sigma}|\mathbf{n}) = \mathbf{n} \in \mathbb{R}^3$, $|\mathbf{n}|^2 = 1$, and $|\mathbf{n}\rangle = |\mathbf{n}) \otimes |\mathbf{n}) \otimes \ldots$. Then $\langle \mathbf{n}|\mathbf{n}'\rangle = 1$ if $\mathbf{n} = \mathbf{n}'$ and is otherwise 0, showing that there is an uncountable orthonormal system of vectors.
3. Possibilities (Ia) and (I) create equivalence relations between vectors, because the convergence of $\prod_i (x_i|y_i)$ and $\prod_i (y_i|z_i)$ implies that of $\prod_i (x_i|z_i)$, and, likewise, that of $\prod_i |(x_i|y_i)|$ and $\prod_i |(y_i|z_i)|$ implies that of $\prod_i |(x_i|z_i)|$ (Problem 2). It is accordingly necessary to distinguish between strong (Ia) and weak (I) equivalence classes:

$$\text{(Ia)}: \prod_i{}' (x_i|y_i) \to c \neq 0, \qquad \text{(I)}: \prod_i{}' |(x_i|y_i)| \to c > 0.$$

The symbol $\prod'$ means that any finite number of factors 0 are to be left out. The equivalence classes span linear subspaces, so $\mathscr{H}$ can be decomposed into (uncountably) many weak equivalent classes, for which vectors of different classes are orthogonal. Each weak equivalence class can be further decomposed into mutually orthogonal strong equivalence classes. Since the latter differ only by phase factors within a given weak equivalence class, they contain the same physical information.

Representations of $\mathscr{A}$ on Infinite Tensor Products (1.4.4)

For the reasons stated in §1.1 and §1.3 we shall be interested in the algebra generated by the operators $\mathscr{B}(\mathscr{H}_i)$. More precisely, let $\mathscr{A}$ be the algebra generated by $\mathscr{B}(\mathscr{H}_1) \otimes \mathbf{1} \otimes \mathbf{1} \ldots, \mathbf{1} \otimes \mathscr{B}(\mathscr{H}_2) \otimes \mathbf{1} \ldots$, etc., and let $\mathscr{A}''$ be its strong (= weak) closure. The first thing to notice is that an element a of $\mathscr{A}$ sends no vector of $\mathscr{H}$ out of its strong equivalence class; since other than a finite number of entries there is always an infinite $\mathbf{1} \otimes \mathbf{1} \otimes \mathbf{1} \ldots$, nothing alters the convergence of $\prod_{i=1}^{\infty} (x_i|y_i)$. The representation of $\mathscr{A}$ on $\mathscr{H}$ is consequently reducible to a high degree; every strong equivalence class is an invariant subspace. The formation of the weak closure changes nothing, since $\langle x|a_n y\rangle = 0$ for $|x\rangle$ and $|y\rangle$ in different equivalence classes, and if $a_n \rightharpoonup a$, then clearly $\langle x|ay\rangle = 0$. Thus every strong equivalence class provides a representation of $\mathscr{A}$ and of $\mathscr{A}''$, and it is a peculiarity of the infinite tensor product that these representations are inequivalent so long as they arise from different weak equivalence classes.

Example (1.4.5)

Return to the simple case of (1.4.3; 2), and define $\boldsymbol{\sigma}_j \cdot \mathbf{n} = \sigma_j$, and $\sigma_j^{\pm}$ in analogy with (1.1.2) such that $\sigma_j^- |\mathbf{n}) = |-\mathbf{n})$, $\sigma_j^+ |-\mathbf{n}) = |\mathbf{n})$, $\sigma_j^+ |\mathbf{n}) = \sigma_j^- |-\mathbf{n}) = 0$. Let $\mathscr{A}$ be the algebra generated by σ_j and $\sigma_j^{\pm}$, $j = 1, 2, \ldots$, let

$\pi_{\mathbf{n}}$ be its representation on the strong equivalence class of $|\mathbf{n}\rangle$, and define $\mathscr{A}_{\mathbf{n}} \equiv \pi_{\mathbf{n}}(\mathscr{A})$. The representation is constructed like the Fock representation, the operators $\pi_{\mathbf{n}}(\sigma_j^{\pm})$ corresponding to creation and annihilation operators and $|\mathbf{n}\rangle$ to the vacuum: $\pi_{\mathbf{n}}(\sigma_j^{+})|\mathbf{n}\rangle = 0$ for all j. The vectors $\pi_{\mathbf{n}}(\sigma_{j_1}^{-} \ldots \sigma_{j_k}^{-})|\mathbf{n}\rangle$ are total for the (strong) equivalence class, and the representation $\mathscr{A}_n$ is irreducible (likewise for $\mathscr{A}_n''$ *a fortiori*).

Remarks (1.4.6)

1. These representations of the σ's are always equivalent on finite tensor products; the Hilbert space constructed with the GNS procedure contains every vector $|\mathbf{n}'\rangle$, in contrast to the infinite case, where the σ's never send vectors out of equivalence classes, which, however, contain no vectors $|\mathbf{n}'\rangle$ with $\mathbf{n}' \neq \mathbf{n}$.
2. The mean magnetization

$$\mathbf{s} = \lim_{N\to\infty} \sum_{j=1}^{N} \frac{1}{N} \pi_{\mathbf{n}}(\boldsymbol{\sigma}_j)$$

exists as a strong limit, so $\mathbf{s} \in \mathscr{A}_n''$. As $N \to \infty$ the commutator of this observable with any element of the algebra goes to zero in the norm topology, so $\mathbf{s}$ is in the center of $\mathscr{A}_n''$. In any irreducible representation, $\mathbf{s}$ must be a multiple of the identity, and is thus the same as $\mathbf{n}$, its expectation value in the state $|\mathbf{n}\rangle$. If $\mathbf{n} \neq \mathbf{n}'$, then $\pi_{\mathbf{n}}$ and $\pi_{\mathbf{n}'}$ are inequivalent: If there existed a unitary transformation U mapping the equivalence classes of $\mathbf{n}$ and $\mathbf{n}'$ onto each other and such that $U\pi_{\mathbf{n}}(\boldsymbol{\sigma}_j)U^{-1} = \pi_{\mathbf{n}'}(\boldsymbol{\sigma}_j)$, then this could be extended to a transformation of the strong closures $\mathscr{A}_n''$ and $\mathscr{A}_{n'}''$, and when applied to $\mathbf{s}$ it would imply that $U\mathbf{n}U^{-1} = \mathbf{n}'$. This is impossible, since two different multiples of the identity can not be unitarily related.
3. On the space $\mathscr{H}$ there exists a unitary transformation sending $|\mathbf{n}\rangle$ to $|\mathbf{n}'\rangle$. Let $n_j' = M_{jk} n_k$, $MM^t = 1$; then the transformation $|\mathbf{n}) \to |M\mathbf{n})$ (on every factor of $|\mathbf{n}\rangle$) is clearly the unitary transformation that brings this about. Upon restriction to an equivalence class, its action is

$$U\pi_{\mathbf{n}}(\sigma_j)U^{-1} = \pi_{\mathbf{n}}(\sigma_k)M_{kj},$$

in contrast to the previous U, and so it creates an isomorphism between $\pi_{\mathbf{n}}(\mathscr{A})$ and $\pi_{\mathbf{n}'}(\mathscr{A})$.
4. Within a given representation the rotation

$$\pi_{\mathbf{n}}(\sigma_j) \to \pi_{\mathbf{n}}(\sigma_k)M_{kj}$$

represents an automorphism of the C^* algebra generated by the σ's, and as such it preserves norms. Yet it can not be extended continuously to the weak closure. If there were such an extension, then $n_j \cdot \mathbf{1} \to n_k M_{kj} \cdot \mathbf{1}$, but $\lambda \cdot \mathbf{1}$ is invariant under every automorphism. Consequently, in the representation space of $\pi_{\mathbf{n}}$ there exists no unitary transformation $U^{-1}\pi_{\mathbf{n}}(\sigma_j)U = M_{jk}\pi_{\mathbf{n}}(\sigma_k)$, as it would extend to $\pi_{\mathbf{n}}(\mathscr{A})''$. Formally, it would turn $|\mathbf{n}\rangle$ into $|\mathbf{n}'\rangle$, but there is no vector $|\mathbf{n}'\rangle$ in the representation space of $\pi_{\mathbf{n}}$ (cf. Problems (1.3.16; 6) and (1.3.16; 7)).

5. Let $M(t)$ be a one-parameter group of rotations on $\mathbb{R}^3$—for definiteness about the 3-axis—and let $U(t)$ be its representation on $\mathscr{H}$ as discussed in Remark 3. On a formal level, $\sum_{j=1} \sigma_j^3$ could be regarded as the generator of the group. The unitary operators $U(t)$ map the equivalence class of $|\mathbf{n}\rangle$ into itself only if $\mathbf{n}$ points in the 3-direction, and in that case the restriction of $U(t)$ to this equivalence class belongs to $\mathscr{A}_{\mathbf{n}}''$. Although it is not possible to define $\sum_{j=1}^{\infty} \sigma_j^3$ densely, $\sum_{j=1}^{\infty} (\sigma_j^3 - \mathbf{1})$ is essentially self-adjoint in the representation $\pi_{\mathbf{n}}$ on the dense set specified in (1.4.5) and is the generator of the rotations about the 3-axis. In other representations there is no workable definition of this operator, as all its matrix elements are infinite. It is natural to ask at this point what the generator of $U(t)$ looks like. It turns out, though, that $U(t)$ has no generator: By Stone's theorem (III: 2.4.24) the existence of a generator is equivalent to strong continuity of $U(t)$, but $U(t)$ is not even weakly continuous, for if $\mathbf{n}$ does not point in the 3-direction, then $\langle \mathbf{n}|U(t)|\mathbf{n}\rangle = 1$ if $t = 0$ and is otherwise 0. It is true that the mapping $t \to U(t)$ is weakly measurable, but the generalization of Stone's theorem for weakly measurable groups works only on separable Hilbert spaces.
6. "Local" rotations of m spins are generated by $\sum_{j=1}^{m} \sigma_j^3$ and always exist.

The representations of the σ's on the individual strong equivalence classes studied until now have all been irreducible, and correspond to GNS constructions using a pure state (cf. (III: 2.3.10; 5)). We shall also see in (2.1.6; 5) that mixed states likewise correspond to vectors in a larger Hilbert space on which the algebra is represented reducibly. That space is the tensor product of the irreducible representation space with another Hilbert space. The key fact to bear in mind when constructing such representations of the σ's is that the infinite tensor product is no longer associative; for instance $\mathbb{C}^4 \otimes \mathbb{C}^4 \otimes \mathbb{C}^4 \otimes \cdots = (\mathbb{C}^2 \otimes \mathbb{C}^2) \otimes (\mathbb{C}^2 \otimes \mathbb{C}^2) \otimes (\mathbb{C}^2 \otimes \mathbb{C}^2) \otimes \cdots \neq \mathbb{C}^2 \otimes \mathbb{C}^2 \otimes \mathbb{C}^2 \otimes \ldots$: The vector

$$\frac{1}{\sqrt{2}}\left[\begin{pmatrix}1\\0\end{pmatrix} \otimes \begin{pmatrix}0\\1\end{pmatrix} + \begin{pmatrix}0\\1\end{pmatrix} \otimes \begin{pmatrix}1\\0\end{pmatrix}\right] \otimes \frac{1}{\sqrt{2}}\left[\begin{pmatrix}1\\0\end{pmatrix} \otimes \begin{pmatrix}0\\1\end{pmatrix} + \begin{pmatrix}0\\1\end{pmatrix} \otimes \begin{pmatrix}1\\0\end{pmatrix}\right] \otimes \ldots$$

on the left has no counterpart on the right. For this reason we shall not simply take the tensor product of the space examined in Example (1.4.5) with another Hilbert space, but shall instead proceed as follows.

Thermal Representations (1.4.7)

If there is only one spin, i.e., $\mathscr{A}$ is generated by $\mathbf{1}$ and $\boldsymbol{\sigma}$, then the GNS representation using the state given in (1.1.11) becomes a reducible representation on

$\mathbb{C}^4$: $\pi(\mathscr{A}) = \mathscr{B}(\mathbb{C}^2) \otimes \mathbf{1}$, $\pi(\boldsymbol{\sigma}) = \boldsymbol{\sigma} \otimes \mathbf{1}$, $\pi(\mathscr{A})' = \mathbf{1} \otimes \mathscr{B}(\mathbb{C}^2)$, $Z = \pi(\mathscr{A}) \cap \pi(\mathscr{A})' = \{\alpha \cdot \mathbf{1}\}$,

$$\Omega = \begin{pmatrix}1\\0\end{pmatrix} \otimes \begin{pmatrix}1\\0\end{pmatrix} \sqrt{\frac{1+s}{2}} + \begin{pmatrix}0\\1\end{pmatrix} \otimes \begin{pmatrix}0\\1\end{pmatrix} \sqrt{\frac{1-s}{2}}, \qquad 0 < s < 1,$$

$$\langle \boldsymbol{\sigma} \rangle = \langle \Omega | \boldsymbol{\sigma} \Omega \rangle = (0, 0, s).$$

Despite being reducible ($\mathscr{A}' \neq \{\alpha \cdot \mathbf{1}\}$), this representation is a factor (its center is $Z = \{\alpha \cdot \mathbf{1}\}$). Accordingly, when passing to infinitely many spins we consider the representation on $\mathbb{C}^4 \otimes \mathbb{C}^4 \otimes \mathbb{C}^4 \otimes \cdots$ constructed with $\Omega \otimes \Omega \otimes \Omega \otimes \ldots$. We find, analogously, that

$$\pi(\mathscr{A}) = (\mathscr{B}(\mathbb{C}^2) \otimes \mathbf{1}) \otimes (\mathscr{B}(\mathbb{C}^2) \otimes \mathbf{1}) \otimes \ldots,$$
$$\pi(\mathscr{A})' = (\mathbf{1} \otimes \mathscr{B}(\mathbb{C}^2)) \otimes (\mathbf{1} \otimes \mathscr{B}(\mathbb{C}^2)) \otimes \cdots + \text{weak limits}$$

$$\pi(\mathscr{A})'' \text{ is the weak closure of } \mathscr{A}, \text{ and } Z = \{\alpha \cdot \mathbf{1}\},$$

which is a reducible factor representation.

Remarks (1.4.8)

1. This representation is not equivalent to any of those found in (1.4.5); as mentioned above, the vector $\Omega \otimes \Omega \otimes \Omega \otimes \ldots$ has no counterpart in the earlier representations π_n, since the corresponding functional in π_n would then be strongly continuous. The state defined by $\Omega \otimes \Omega \otimes \Omega \otimes \ldots$ on $\mathscr{A}$.

$$\langle (\boldsymbol{\sigma}_{j_1} \cdot \mathbf{n}_1)(\boldsymbol{\sigma}_{j_2} \cdot \mathbf{n}_2) \ldots (\boldsymbol{\sigma}_{j_k} \cdot \mathbf{n}_k) \rangle = s^k n_1^z n_2^z \ldots n_k^z$$

is a (norm) continuous linear functional, and therefore extensible to the whole C^* algebra generated by $\mathscr{A}$, but it still need not be strongly continuous in a representation: For instance, in the representation using π_n,

$$P_N = \prod_{i=N}^{2N} \frac{1 + \boldsymbol{\sigma}_i \cdot \mathbf{n}}{2}$$

converges strongly to $\mathbf{1}$, but $\langle P_N \rangle = ((1 + sn^z)/2)^N \to 0 \neq 1$. Recall that a refinement of the topology on the range space or a coarsening of the topology on the domain space may destroy the continuity of a mapping.

2. The fact that with only one spin, $\langle \boldsymbol{\sigma} \rangle = \operatorname{Tr} \boldsymbol{\sigma} \exp(-\eta\sigma_3)/\operatorname{Tr} \exp(-\eta\sigma_3)$, might mislead one into thinking that for infinitely many spins, in the notation of (1.1.1),

$$\langle \cdot \rangle = \operatorname{Tr} \cdot \rho, \qquad \rho = \frac{\exp(-\eta \sum_j \sigma_j)}{\operatorname{Tr} \exp(-\eta \sum_k \sigma_k)}.$$

What goes wrong is that

$$\frac{\exp(-\eta \sum_{j=1}^N \sigma_j)}{\operatorname{Tr} \exp(-\eta \sum_{j=1}^N \sigma_j)} \Rightarrow 0 \quad \text{as } N \to \infty.$$

3. In the thermal representation (1.4.7) it is of course possible to write $\langle\cdot\rangle = \mathrm{Tr} \cdot P_\Omega$, where P_Ω is the projection onto the cyclic vector, but $P_\Omega \notin \mathscr{A}''$.

Decomposition of the Representations (1.4.9)

Because of the analogy between $\sigma^\pm$ and the operators a and a^* for fermions, the phenomena we have discussed are also characteristic of systems of infinitely many fermions. It is not so important that the σ's commute whereas the a's anticommute; the distinction can be gotten around with the right transformation. For a system of bosons the individual factors of the tensor product are already infinite-dimensional, which causes additional complications. In either case there are a great number of inequivalent representations; the uniqueness theorem (III: 3.1.5) for finite systems does not hold any more. Thus it would be desirable to find a point of view that organizes them somehow. The concept of a factor was introduced in (III: 2.3.4), as an algebra with a trivial center, $Z = \{\alpha \cdot \mathbf{1}\}$. On a finite-dimensional space it amounts to a direct sum of equivalent irreducible representations. The first step in any decomposition is to collect the equivalent irreducible representations together in factors and then write the whole representation as a sum of various factors. In the finite-dimensional case this appears as shown in Figure 2.

It will be observed that the projections onto the space $\mathscr{H}_{ik}$ of the irreducible representations belong to $\pi(\mathscr{A})'$ and the projections onto the spaces $\mathscr{H}_i$ of the factors belong to the center. Both $\pi(\mathscr{A})$ and $\pi(\mathscr{A})'$ map

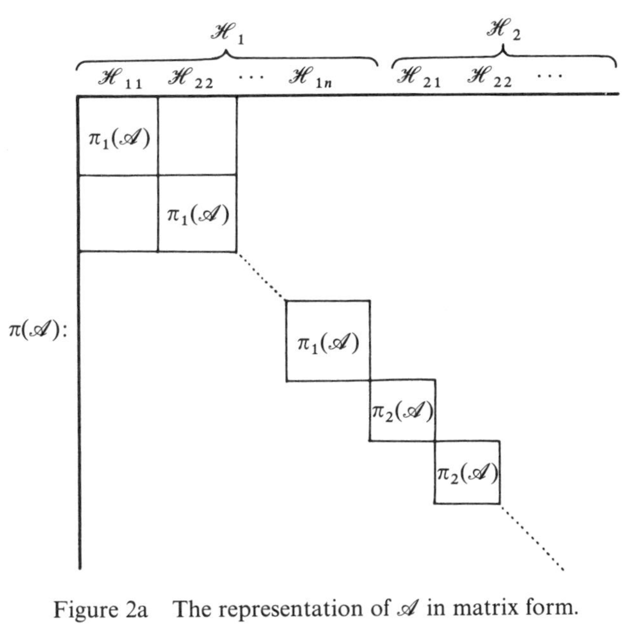

Figure 2a The representation of $\mathscr{A}$ in matrix form.

$\pi(\mathscr{A})'$:

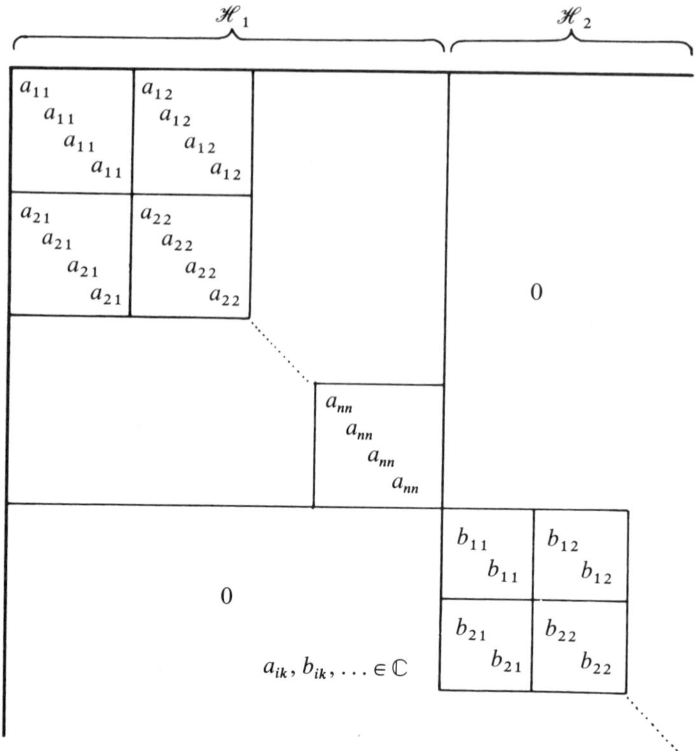

$Z = \pi(\mathscr{A}) \cap \pi(\mathscr{A})'$:

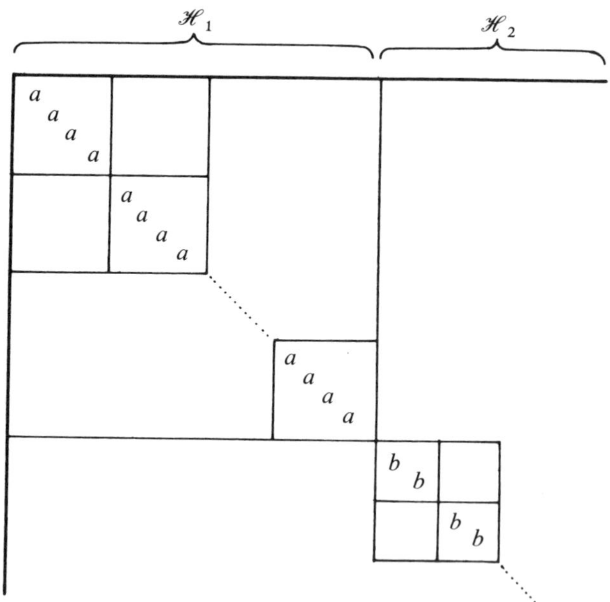

Figures 2b, c The representation of $\mathscr{A}'$ and the center Z in matrix form.

$\mathscr{H}_i$ into itself. The elements of the center become multiples of the identity when projected onto $\mathscr{H}_i$; they can assume different values only on different $\mathscr{H}_i$. The decomposition into factors is thus uniquely fixed by Z and consequently by $\pi(\mathscr{A})$. The further decomposition into irreducible representations is not likewise fixed; some arbitrariness is connected with the spaces $\mathscr{H}_{ik}$. If, for example, $\mathscr{H}_1 = \mathscr{H}_{11} \otimes \mathbb{C}^n = \mathscr{H}_{11} \otimes e_1 \oplus \mathscr{H}_{11} \otimes e_2 \oplus \cdots \oplus \mathscr{H}_{11} \otimes e_n$, then the choice of the basis $\{e_i\}$ for $\mathbb{C}^n$ remains free, since the space is the same for every choice of orthogonal basis. Different bases correspond to the different maximally Abelian subalgebras of $\pi(\mathscr{A})'$ that they diagonalize.

The passage to an infinite dimension requires the generalization of sums to integrals. The spectral theorem (III: 2.3.11) states that a Hermitian operator $a \in \mathscr{B}(\mathscr{H})$ may be represented as a multiplication operator on some space $L^2(d\mu, \operatorname{Sp}(a))$. If there is degeneracy, then a spectral value $\alpha \in \operatorname{Sp}(a)$ is associated not with a single complex number but with a many-dimensional Hilbert space $\mathscr{H}_\alpha$. If $v(\alpha)$ denotes the component of $v \in \mathscr{H}$ in $\mathscr{H}_\alpha$, then the scalar product on $\mathscr{H}$ can be written as

$$\langle v | w \rangle = \int d\mu(\alpha) \langle v(\alpha) | w(a) \rangle.$$

The action of a on v is $(av)(\alpha) = \alpha v(\alpha)$. The center $Z = \pi(\mathscr{A}) \cap \pi(\mathscr{A})'$ is a commutative algebra, and its elements may be simultaneously diagonalized, and so any $z \in Z$ may be written as $(zv)(\alpha) = f(\alpha)v(\alpha)$, where f assigns a complex number to α. Any element a of $\mathscr{A}$ can then be represented by $[\pi(a)v](\alpha) = \pi_\alpha(a)v(\alpha)$, $\pi_\alpha(a) \in \mathscr{B}(\mathscr{H}_\alpha)$, and $b \in \pi(\mathscr{A})' \Rightarrow (bv)(\alpha) = b(\alpha)v(\alpha)$, $b(\alpha) \in \mathscr{B}(\mathscr{H}_\alpha)$, $[b(\alpha), \pi_\alpha(a)] = 0$ for all $a \in \mathscr{A}$. In a finite number of dimensions every $\mathscr{H}_\alpha$ can be written $\mathscr{H}_\alpha = \mathscr{H}_\alpha^{(1)} \otimes \mathscr{H}_\alpha^{(2)}$, $\pi_\alpha(\mathscr{A}) = \mathscr{B}(\mathscr{H}_\alpha^{(1)}) \otimes \mathbf{1}_{\mathscr{H}_\alpha^{(2)}}$, and $b(\alpha)$ is of the form $\mathbf{1}_{\mathscr{H}_\alpha^{(1)}} \otimes b$, $b \in \mathscr{B}(\mathscr{H}_\alpha^{(2)})$. This is as far as the finite-dimensional analogy goes; it will not be possible to write every factor π_α in the form $\mathscr{B}(\mathscr{H}) \otimes \mathbf{1}$.

Classification of Factors (1.4.10)

We pause now to take stock of the factors, which will function as basic building blocks. The possibility that comes to mind first for a preliminary, rough classification is to define a trace. In (III: 2.3.19) the trace was defined as a mapping from $\mathscr{A}_+$, the positive operators, to $\overline{\mathbb{R}}^+$, and it was extended to a linear mapping from the trace class $\mathscr{C}_1(\mathscr{H})$ to $\mathbb{C}$. The trace is discontinuous in all topologies weaker than the trace topology given by $\|\cdot\|_1$. It may even occur that the only element of an algebra $\mathscr{A}$ in the trace class is the zero operator, as for example with the factor $\mathscr{B}(\mathscr{H}) \otimes \mathbf{1}$, where $\mathbf{1}$ is the identity on an infinite-dimensional space. In this case there is plainly the possibility of defining a trace by $\Phi(a \otimes \mathbf{1}) = \operatorname{Tr}_1 a$, which has all the necessary properties. This observation suggests an abstract

Definition of the Trace (1.4.11)

Let $\mathscr{A}_+$ be the positive cone of a strongly closed algebra $\mathscr{A}$, i.e., a von Neumann algebra. A **trace** is a mapping $\Phi: \mathscr{A}_+ \to \overline{\mathbb{R}}^+$ with the following properties.

(i) $\Phi(\lambda_1 a_1 + \lambda_2 a_2) = \lambda_1 \Phi(a_1) + \lambda_2 \Phi(a_2)$ for $a_i \in \mathscr{A}_+$ and $\lambda_i \in \mathbb{R}^+$;

(ii) $\Phi(a) = \Phi(uau^{-1})$ for all $a \in \mathscr{A}_+$ and all unitary $u \in \mathscr{A}$.

The trace Φ is said to be

faithful, if $\Phi(a) = 0$ and $a \in \mathscr{A}_+ \Leftrightarrow a = 0$;
finite, if $\Phi(a) < \infty$ for all $a \in \mathscr{A}_+$;
semifinite, if for all $a \in \mathscr{A}_+$ there exists a nonzero $b < a$ such that $\Phi(b) < \infty$;
normal, if for every increasing filter (see (III: 2.2.21)) $F \subset \mathscr{A}_+$ with supremum s, $\Phi(s) = \sup_{a \in F} \Phi(a)$.

Examples (1.4.12)

1. $\Phi(a) = 0$ for all $a \in \mathscr{A}_+$. The trace is unfaithful, finite, and normal.
2. $\Phi(0) = 0$, $\Phi(a) = \infty$ for all $a \neq 0$. The trace is faithful, not semifinite, and normal (purely infinite).
3. Let $\mathscr{A}$ be the $n \times n$ matrices and $\Phi(a) = \operatorname{Tr} a$. The trace is faithful, finite, and normal.
4. $\mathscr{A} = \mathscr{B}(\mathscr{H})$, $\mathscr{H}$ infinite-dimensional, and $\Phi(a) = \operatorname{Tr}(a)$. The trace is faithful, semifinite, and normal.
5. $\mathscr{A} = \mathscr{B}(\mathscr{H}_1) \oplus \mathscr{B}(\mathscr{H}_2)$, $\Phi(a \oplus b) = \alpha \operatorname{Tr} a + \beta \operatorname{Tr} b$, α and $\beta \in \mathbb{R}^+$. The trace is faithful only if α and β are nonzero and finite only if the $\mathscr{H}_i$ are finite-dimensional. In all cases it is semifinite and normal. (Note that although Φ is invariant under unitary transformations belonging to $\mathscr{A}$ for $\alpha \neq \beta$, it is not invariant under all unitary transformations in $\mathscr{B}(\mathscr{H}_1 \oplus \mathscr{H}_2)$.)
6. Let $\mathscr{A}$ be the algebra of multiplication operators $L^\infty(\mathbb{R}, d\mu)$ on $L^2(\mathbb{R}, d\mu)$, and $\Phi(a) = \int d\mu(x) a(x) \rho(x)$ for some non-negative, measurable ρ. If $\rho > 0$ a.e., then Φ is faithful; if $\rho \in L^1(\mathbb{R}, d\mu)$, then Φ is finite; and if $\rho < \infty$ a.e., then Φ is semifinite. In all cases the trace is normal.
7. Let $\mathscr{A}$ be the algebra of multiplication operators l^∞ on l^2, and $\Phi(a) = \lim_{i \to \infty} a_i$ when the limit exists, and otherwise let the trace be defined by linear extension with the Hahn–Banach theorem. The trace is finite and neither faithful nor normal: If $F = \{(a_i)$, where $a_i = 1$ for finitely many i and otherwise $= 0\}$, then $s = (a_i = 1)$, and $\Phi(s) = 1$, but $\Phi(a) = 0$ for all $a \in F$.

Remarks (1.4.13)

1. Property (ii) may be replaced with (ii)$'$: $\Phi(aa^*) = \Phi(a^*a)$ for all $a \in \mathscr{A}$ (Problem 3).
2. It can be shown in general that $\{a \in \mathscr{A}_+ : \Phi(a) < \infty\}$ consists of the positive elements of a two-sided self-adjoint ideal $\mathscr{M}_\Phi$, onto which Φ can be extended as a linear form (also denoted Φ). It is discontinuous in every topology that is strictly coarser than the one defined by the norm $\|a\|_\Phi = \Phi((a^*a)^{1/2})$. All continuous linear functionals on $\mathscr{M}_\Phi$ with this topology are of the form $a \to \Phi(ab)$, $a \in \mathscr{M}_\Phi$, $b \in \mathscr{A}$ (Problem 4), and nonzero for $b \neq 0$.
3. Property (ii) implies for $a \in \mathscr{M}_\Phi$ and any unitary $u \in \mathscr{A}$ that $\Phi(ua) = \Phi(au)$. Moreover, since every element of $\mathscr{A}$ is a linear combination of unitary operators, $\Phi(ab) = \Phi(ba)$, $a \in \mathscr{M}_\Phi$, $b \in \mathscr{A}$.
4. The requirement of normality originates in the theory of integration, where monotonic convergence can be permuted with integration. The trace can consequently be regarded as a generalization of the integral to noncommutative integrands.
5. If Φ is normal, then $\mathscr{A}$ may be written as $\mathscr{A} = \mathscr{A}_1 \oplus \mathscr{A}_2 \oplus \mathscr{A}_3$, where $\Phi_{|\mathscr{A}_3}$ is faithful and semifinite, $\Phi_{|\mathscr{A}_1} = 0$, and $\Phi_{|\mathscr{A}_2}$ is purely infinite (Problem 5). As we shall be interested solely in normal traces and shall ignore the trivial cases of Examples 1 and 2, we may confine our attention to faithful, semifinite traces.

The ordering of operators induces an ordering of traces, whereby $\Phi \leq \Psi$ shall mean $\Phi(a) \leq \Psi(a)$ for all $a \in \mathscr{A}_+$. For the ordering of the trace there is a theorem on

The Form of a Dominating Trace (1.4.14)

Let Φ and Ψ be normal, semifinite traces on a von Neumann algebra $\mathscr{A}$. Then $\Phi \leq \Psi$ iff there exists $b \in \mathscr{A} \cap \mathscr{A}'$, $0 < b \leq 1$, such that $\Phi(a) = \Psi(ab)$ for all a.

Proof

Let $\mathscr{M}_\Psi$ be the ideal on which $\Psi < \infty$, given the norm $\|a\| = \Psi((aa^*)^{1/2})$. The mapping $a \to \Phi(a)$ is then a continuous linear form on $\mathscr{M}_\Psi$, and by Remark (1.4.13; 2) it is $\Psi(ab)$ for some $b \in \mathscr{A}$. To prove that $b \in \mathscr{A}'$, observe that for all $a \in \mathscr{M}_\Phi$ and $c \in \mathscr{A}$, $0 = \Phi(ac - ca) = \Psi(acb - cab) = \Psi(a[c, b])$, so, according to (1.4.13; 2), $[c, b] = 0$. □

Corollary (1.4.15)

Any two faithful, normal, semifinite traces on the same factor are proportional. More specifically, if Φ_1 and Φ_2 are two such traces, then $\Phi_1 < \Phi_1 + \Phi_2$ and

$\Phi_2 < \Phi_1 + \Phi_2$. Since the center of the factor consists of multiples of the identity, $\Phi_i = \lambda_i(\Phi_1 + \Phi_2)$, $0 < \lambda_i < 1$, so $\Phi_1 = \lambda_1 \lambda_2^{-1} \Phi_2$.

Because the trace is essentially unique on any factor, it may be asked whether the trace of a projection is an integer c, which would allow a reasonable definition of the dimension of the subspaces onto which they project.

The Types of Factors (1.4.16)

Factors of Type I

The range of the trace of the projections of factors of type I is $c \in \mathbb{Z}^+$, and they are of the form $\mathscr{B}(\mathscr{H}) \otimes \mathbf{1}$, with $\mathscr{H}$ separable, i.e., a sum of identical copies of an irreducible algebra of operators. The trace is given by $\Phi(a \otimes \mathbf{1}) = c \operatorname{Tr} a$, and if the dimension of $\mathscr{H}$ is n, then it is finite for $n < \infty$ and not finite but only semifinite for $n = \infty$. This creates a distinction between subtypes I_n and I_∞.

Factors of Type II

On Factors of Type II there is a semifinite, normal, faithful trace the range of which when applied to the projections is either $[0, 1]$ or $\mathbb{R}^+$. Depending on whether the trace is finite or only semifinite, one distinguishes between subtypes II_1 and II_∞. An example of type II_1 is the algebra of infinitely many spins (1.1.2) represented with the GNS construction using the state Φ: $\Phi(\mathbf{1}) = 1$, $\Phi(\prod \boldsymbol{\sigma}_j) = 0$ ((1.4.8) with $s = 0$). This state has the properties of a trace; commutativity (1.4.11(ii)) holds trivially, and this representation is a factor. Since the factor is obviously not isomorphic to anything of the form $\mathscr{B}(\mathscr{H}_n) \otimes \mathbf{1}$, $n < \infty$, and the trace is finite, it must be of type II_1. It is reducible but not of type I, since it can not be written as a direct sum of identical irreducible algebras. Type II_∞ factors are of the form type $\mathrm{I}_\infty \otimes$ type II_1, where the trace is defined multiplicatively on the tensor product.

Factors of Type III

They have no normal, faithful, semifinite trace. The infinite spin algebra (1.1.2) again provides an example, this time with the GNS representation using the state (1.1.11) with $s \neq 0$, in other words (1.4.8).

Remarks (1.4.17)

1. The type with the properties familiar from finite matrices is I, while types II and III are less intuitive. All three types occur in the GNS representation of the spin algebra with a state of the form (1.1.11), I_∞ with $s = 1$, II_1 with

$s = 0$, and III with $0 < s < 1$. To the malicious delight of many mathematicians the initial impression that type III is the rule for infinite systems has panned out with the passage of time. Types I and II turn out to be peripheral possibilities.

2. It was ascertained in (III: 2.3.6; 5) that factor representations with maximally Abelian subalgebras are irreducible. As a result, representations of types II and III have no maximally Abelian subalgebras.
3. If a factor includes an irreducible subrepresentation, then a semifinite, normal trace can be defined on it, mapping the projections to a discrete set of values, and it must therefore be of type I. It was remarked in (III: 2.3.10; 5) that the GNS construction yields an irreducible representation iff the state it builds on is pure. This means that no vector in the Hilbert space of a representation of type II or III corresponds to a pure state on the algebra.
4. Any operator a of an algebra of type III is of course bounded, so Tr ρa is well defined for any $\rho \in \mathscr{C}_1(\mathscr{H})$, only ρ can not come from the algebra, which contains no element of a trace class (other than 0).

Let us end the section by recapitulating the physical significance of the new mathematical phenomena that make an appearance in infinite systems.

1. **Inequivalent Representations**
   Since vectors that differ globally are always orthogonal, globally different situations lead to inequivalent representations. Within a given representation different elements of the algebra produce vectors that differ only locally.

2. **Non-normal States**
   Expectation values with a vector of a different, inequivalent representation constitute a state on the algebra, but one that fails to be strongly continuous with respect to the original representation, and hence it is not normal. They are representations of different global circumstances, and thus assign different values to global observables like densities, which are only defined with strong limits.

3. **Factors**
   Whereas $\pi(\mathscr{A})$ describes microscopic observables, $\pi(\mathscr{A})''$ covers macroscopic observables as well. Factors associate certain numerical values to the global observables lying in the center $\pi(\mathscr{A})'' \cap \pi(\mathscr{A})'$—factors are the macroscopically pure states. In factors, Khinchin's ergodic theorem applies to them, stating that these global quantities exhibit no fluctuation. Even if vectors of a factor are pure with respect to this subalgebra, they may produce mixed states. The ground state is associated with type I, finite temperature with type III, and infinite temperature with type II.

4. **Unitary Representation of the Time-Evolution**
   If the algebra changes globally as time passes, then a representation may change at any moment into an inequivalent representation, and it is not

possible to represent the time-evolution with a group of unitary transformations within the representation. Yet if the representation is based on a time-invariant state, then the other vectors of the representation differ only locally, and thus do not change in time, from the global point of view. This establishes the possibility of a unitary time-evolution.

Problems (1.4.18)

1. Show that with vectors $|x^{(1)}\rangle, \ldots, |x^{(n)}\rangle$ and $\lambda_1, \ldots, \lambda_n \in \mathbb{C}$, Definition (1.4.2) implies that $\sum_{i,k} \lambda_i^* \lambda_k \langle x^{(i)} | x^{(k)} \rangle \geq 0$. (Hint: it suffices to show this for the case where the $|x^{(i)}\rangle$ are strongly equivalent. Prove that $\sum_{i,k} \lambda_i^* \lambda_k \prod_{j=1}^N (x_j^{(i)} | x_j^{(k)}) \geq 0$ for any N and take the limit $N \to \infty$.)

2. (i) Show that $|x\rangle$ and $|y\rangle$ are equivalent iff $\sum_i |1 - (x_i | y_i)| < \infty$ and weakly equivalent iff $\sum_i |1 - |(x_i | y_i)|| < \infty$.
   (ii) Conclude from (i) that the $\underset{\text{strong}}{\sim}$ of $|x\rangle \underset{\text{strong}}{\sim} |y\rangle$ has all the properties of an equivalence relation, namely reflexivity, symmetry, and transitivity. (Hint: use the inequality $|1 - (x|z)| \leq 4[|1 - (x|y)| + |1 - (y|z)|]$, which holds for unit vectors. This 4 is a generous constant.)
   (iii) Show that $|x\rangle \sim |y\rangle$ iff there exists a sequence $\{\varphi_j\}$ such that $|x\rangle \underset{\text{strong}}{\sim} |y'\rangle$, $|y'\rangle \equiv \exp(i\varphi_1)|y_1) \otimes \exp(i\varphi_2)(|y_2) \otimes \ldots$.
   (iv) Show that $\underset{\text{weak}}{\sim}$ is also an equivalence relation.

3. Show that condition (ii) of the definition of the trace (1.4.11), i.e., $\Phi(a) = \Phi(UaU^{-1})$, may be replaced with: $\Phi(a^*a) = \Phi(aa^*)$ for all a in a von Neumann algebra $\mathscr{A}$.

4. Show that for a faithful, normal, semifinite trace Φ, all continuous linear forms on $a \in \mathscr{M}_\Phi$ may be written as $a \to \Phi(ab)$ for some $b \in \mathscr{A}$. (Hint: use the inequality $|\Phi(ab)| \leq \Phi(|ab|) \leq \|b\| \Phi(|a|)$.)

5. Show that with any normal trace Φ, $\mathscr{A}$ can be written $\mathscr{A} = \mathscr{A}_1 \oplus \mathscr{A}_2 \oplus \mathscr{A}_3$, where $\Phi_{|\mathscr{A}_1} \equiv 0$, $\Phi_{|\mathscr{A}_2}$ is faithful and semifinite, and $\Phi_{|\mathscr{A}_3}$ is purely infinite. (Use the following corollaries of von Neumann's density theorem (III: 2.3.24; 4):

   (I) Let $\mathscr{M} \subset \mathscr{A}$ be a strongly closed, two-sided ideal. Then $\mathscr{M}$ contains a projection operator P such that $P \in \mathscr{A} \cap \mathscr{A}'$ and $P \geq Q$ for all projection operators $Q \in \mathscr{M}$.
   (II) Let $\mathscr{N}$ be a two-sided ideal and suppose a is in the positive part of the weak closure of $\mathscr{N}$. Then there exists an increasing filter $\subset \mathscr{N}^+$ having a for its supremum.)

Solutions (1.4.19)

1. The $n \times n$ matrix

$$\begin{pmatrix} (x_j^{(1)} | x_j^{(1)}) & \cdots & (x_j^{(1)} | x_j^{(n)}) \\ \vdots & & \\ (x_j^{(n)} | x_j^{(1)}) & \cdots & (x_j^{(n)} | x_j^{(n)}) \end{pmatrix}$$

is Hermitian and non-negative, and is thus a sum of projections, i.e., matrices of the form

$$\begin{pmatrix} h_1^* h_1 & \dots & h_1^* h_n \\ \vdots & & \\ h_n^* h_1 & \dots & h_n^* h_n \end{pmatrix}.$$

This implies

$$(x_j^{(i)} | x_j^{(k)}) = \sum_{l_j=1}^{n} h_i^{l_j *} h_k^{l_j},$$

and

$$\sum_{i,k} \lambda_i^* \lambda_k \prod_{j=1}^{N} (x_j^{(i)} | x_j^{(k)}) = \sum_{i,k} \lambda_i^* \lambda_k \sum_{l_1, \dots, l_N} h_i^{l_1 *} h_k^{l_1} \dots h_i^{l_N *} h_k^{l_N} \geq 0,$$

since

$$\sum_{i,k} \lambda_i^* \lambda_k h_i^{l_1 *} \dots h_k^{l_N} = \Big| \sum_k \lambda_k h_k^{l_1} \dots h_k^{l_N} \Big|^2 \geq 0.$$

2. (i) follows from the theory of infinite products [12].
   (ii) To prove the inequality, choose a basis for the subspace spanned by $|x)$, $|y)$, and $|z)$ such that they correspond to the vectors $(\alpha, \beta, 0)$, $(1, 0, 0)$ and $(\gamma, \delta, \varepsilon)$, where $|\alpha|^2 + |\beta|^2 = |\gamma|^2 + |\delta|^2 + |\varepsilon|^2 = 1$. Then $(x|y) = \alpha^*$, $(y|z) = \gamma$, $(x|z) = \alpha^*\gamma + \beta^*\delta$. $|1 - \alpha^*\gamma - \beta^*\delta| \leq |1 - \alpha^*\gamma| + |\beta||\delta| \leq 2|1 - \alpha| + 2|1 - \gamma| + (1 - |\alpha|^2)^{1/2}(1 - |\gamma|^2)^{1/2} \leq 2(|1 - \alpha|^{1/2} + |1 - \gamma|^{1/2})^2 \leq 4[|1 - \alpha| + |1 - \gamma|]$. The reflexivity and symmetry of the equivalence relation are trivial, and transitivity follows from (i) together with the inequality.
   (iii) $\Rightarrow$: Choose $\varphi_j = -\arg(x_j | y_j)$.
   $\Leftarrow$: This is trivial.
   (iv) follows from (ii) and (iii).
3. (ii) $\Rightarrow$ (ii′): With a polar decomposition, $a = V|a|$, where $a^*a = |a|^2 = V^*V|a|^2$, $aa^* = V|a|^2V^*$. Let $\mathscr{M}_\Phi$ be the trace-class ideal: $a \in \mathscr{M}_\Phi \Rightarrow a^*a \in \mathscr{M}_\Phi$ and $aa^* \in \mathscr{M}_\Phi$ $\Rightarrow Va^*a \in \mathscr{M}_\Phi$, since $V = w\text{-}\lim_{\varepsilon \downarrow 0} a(|a|^2 + \varepsilon)^{-1/2} \in \mathscr{A}$, which, with Remark (1.4.13; 3) implies $\Phi(V^*Va^*a) = \Phi(Va^*aV^*)$.
   (ii′) $\Rightarrow$ (ii): Let $a \geq 0$. $\Phi(UaU^{-1}) = \Phi(Ua^{1/2}a^{1/2}U^*) = \Phi(a^{1/2}U^*Ua^{1/2}) = \Phi(a)$, and every operator is a linear combination of positive operators.

4. To prove the inequality, let a and b be non-negative. $\Phi(ab) = \Phi(a^{1/2}ba^{1/2}) \leq \|b\|\Phi(a)$, since for any a and b, $a^{1/2}ba^{1/2} \leq a^{1/2}\|b\|a^{1/2}$. Thus $|\Phi(ab)|^2 \leq \Phi(|a^*||b|)\Phi(|a||b^*|)$ and is consequently $\leq \||b|\|\Phi(|a^*|)\||b^*|\|\Phi(|a|) = \|b\|^2\Phi(|a|)^2$, in which the Cauchy–Schwarz inequality $|\Phi(ab)|^2 \leq \Phi(aa^*)\Phi(bb^*)$ (see (III: 2.2.20; 1)) was used in the form $|\Phi(ab)|^2 = |\Phi(U|a|V|b|)|^2$ (with the polar decompositions $a = U|a|$ and $b = V|b|$). This $= |\Phi(|b|^{1/2}U|a|^{1/2}|a|^{1/2}V|b|^{1/2})|^2 \leq \Phi(|b|^{1/2}U|a|^{1/2} \times |a|^{1/2}U^*|b|^{1/2}) \cdot \Phi(|b|^{1/2}V^*|a|^{1/2}|a|^{1/2}V|b|^{1/2}) = \Phi(|b|U|a|U^*)\Phi(V|b|V^*|a|) = \Phi(|b||a^*|)\Phi(|b^*||a|)$. Now let $ab = W|ab|$; then $\Phi(|ab|) = \Phi(W^*ab) \leq \|bW^*\| \times \Phi(|a|) \leq \|b\|\Phi(|a|)$. The first part of the inequality follows from $|\Phi(ab)| = |\Phi(ab \cdot \mathbf{1})| \leq \|\mathbf{1}\|\Phi(|ab|) = \Phi(|ab|)$.

   It is a corollary of the inequality that the norm of the mapping $a \to \Phi(ab)$ is $\|b\|$. This allows $\mathscr{A}$ to be identified with a closed subspace of $\mathscr{M}_\Phi^*$. To see that $\mathscr{A} = \mathscr{M}_\Phi^*$, first suppose $a \in \mathscr{M}_\Phi^+$. Then the mapping $\mathscr{A} \to \mathbb{C}: b \to \Phi(ab)$ is normal, entailing ultraweakly continuous (see (2.1.4)), which implies that for any $a \in \mathscr{M}_\Phi$, $b \to \Phi(ab)$ is ultraweakly continuous. Because of the inequality again, the norm of this mapping is $\Phi(|a|)$, which implies that $\mathscr{M}_\Phi$ can be imbedded isometrically and isomorphically in the predual $\mathscr{A}_*$, i.e., the space of ultraweakly continuous linear functions. Thus $\mathscr{M}_\Phi \subset \mathscr{A}_*$. We shall see in (2.1.3) and (2.1.4) that $\mathscr{C}_1 = \mathscr{B}(\mathscr{H})_*$ and $\mathscr{C}_1^* = \mathscr{B}(\mathscr{H})$. Since $\mathscr{A}$ is ultraweakly closed, $\mathscr{A}_\perp = \mathscr{C}_1/\mathscr{C}_1^\perp$ with $\mathscr{C}_1^\perp = \{\rho \in \mathscr{C}_1 : \operatorname{Tr} \rho a = 0 \text{ for all } a \in \mathscr{A}\}$, so $\mathscr{A} = (\mathscr{C}_1/\mathscr{C}_1^\perp)^*$. Therefore $\mathscr{M}_\Phi^* \subset (\mathscr{A}_*)^* = \mathscr{A}$, which implies $\mathscr{M}_\Phi^* = \mathscr{A}$. Remark: $\mathscr{M}_\Phi$ is dense in $\mathscr{A}_*$ but not in general closed.

5. For more about types I and II, see Chapter I, §3 of [4]. The set $\{a \in \mathscr{A}^+ : \Phi(a) = 0\}$ is the positive part of a two-sided ideal $\mathscr{N}$. Let $\mathscr{M}$ be the trace class, let $\bar{\mathscr{N}}$ and $\bar{\mathscr{M}}$ be the strong closures of $\mathscr{N}$ and $\mathscr{M}$, and P_1 and P_2 be respectively the largest projections they contain (see Corollary I). The Hilbert space $\mathscr{H}$ can be decomposed as $\mathscr{H}_1 \oplus \mathscr{H}_2 \oplus \mathscr{H}_3$, where $\mathscr{H}_1 \equiv P_1\mathscr{H}$, $\mathscr{H}_2 \equiv (P_2 - P_1)\mathscr{H}$, $\mathscr{H}_3 \equiv (1 - P_2)\mathscr{H}$ in which case $\mathscr{A} = \mathscr{A}_1 \oplus \mathscr{A}_2 \oplus \mathscr{A}_3$, where $\mathscr{A}_i \equiv \mathscr{A}_{|\mathscr{H}_i}$, since P_1 and P_2 belong to $\mathscr{A} \cap \mathscr{A}'$.
   It is obvious that $\Phi_{|\mathscr{A}_1} = 0$. To see that $\Phi_{|\mathscr{A}_2}$ is semifinite, apply Corollary II: Let $a \in \bar{\mathscr{M}}^+ \setminus \mathscr{M}^+$; then there exists an operator $b \in \mathscr{M}^+$, $b \le a$, such that $\Phi(b) > 0$. The remaining claims are trivial.

Thermostatics 2

2.1 The Ordering of the States

The heuristic concepts of purer and more chaotic states can be made mathematically precise with reference to a lattice structure of the classes of equivalent density matrices.

States are by definition (III: 2.2.18) normed, positive linear functionals on an algebra $\mathscr{A}$ of observables. If the dimension of the underlying space is finite, $\mathscr{A} = \mathscr{B}(\mathbb{C}^n)$, then all linear functionals are of the form $\mathscr{A} \ni a \to \operatorname{Tr} \rho a \equiv (\rho | a)$, $\rho \in \mathscr{B}(\mathbb{C}^n)$, and $\mathscr{B}(\mathbb{C}^n)$ is its own dual space. The inequality of (1.4.18; 4),

$$|(\rho | a)| \leq \|a\| \, \|\rho\|_1, \qquad \|\rho\|_1 = \operatorname{Tr}(\rho^* \rho)^{1/2} \tag{2.1.1}$$

then holds, and is optimal in the sense that

$$\sup_{\|\rho\|_1 = 1} |(\rho | a)| = \|a\|, \qquad \sup_{\|a\| = 1} |(\rho | a)| = \|\rho\|_1. \tag{2.1.2}$$

If the dimension of $\mathscr{H}$ is infinite, the inequality applies initially to the operators of finite rank (cf. (III: 2.3.21)), denoted $\mathscr{E}$ or $\mathscr{E}_1$, depending on whether the norm $\| \;\|$ or $\| \;\|_1$ is used. In these topologies continuous, linear functionals are of the form

$$\mathscr{E} \ni a \to \operatorname{Tr} \rho a \quad \text{with } \|\rho\|_1 < \infty$$

or

$$\mathscr{E}_1 \ni a \to \operatorname{Tr} \rho a \quad \text{with } \|\rho\| < \infty.$$

The linearity and continuity of the functionals thus defined are obvious, and it can be seen as follows that all functionals with these properties are of that form. By what was said earlier, a linear functional on $\mathscr{E}$ determines the restriction of an operator ρ to any finite-dimensional subspace. To guarantee that $|(\rho|a)| \leq c\|a\|$ for all $a \in \mathscr{E}$ or $|(\rho|a)| \leq c\|a\|_1$ for all $a \in \mathscr{E}_1$, by (2.1.2) it is necessary to ensure that $\|\rho\|_1$ or respectively $\|\rho\|$ is bounded. If the spaces $\mathscr{E}$ and $\mathscr{E}_1$ are now completed, becoming the Banach spaces $\mathscr{C}$ and $\mathscr{C}_1$, of (III: 2.3.21), then their dual spaces are unaffected—the dual spaces of a space and of a dense subspace are the same. The state of affairs is analogous to that of l^0, l^1, and l^∞, the spaces of sequences (x_i) satisfying respectively $\lim_{i\to\infty} x_i = 0$, $\sum_i |x_i| < \infty$, and $\sup_i |x_i| < \infty$:

Duality for the Subspaces of $\mathscr{B}(\mathscr{H})$ (2.1.3)

$\mathscr{C}^* = \mathscr{C}_1$, $\mathscr{C}_1^* = \mathscr{B}(\mathscr{H})$, where $\mathscr{C}$ and $\mathscr{B}(\mathscr{H})$ are given the norm $\| \;\|$, and $\mathscr{C}_1$ the norm $\| \;\|_1$. These norms on $\mathscr{C}_1$ and $\mathscr{B}(\mathscr{H})$ produce the strong topology on the dual spaces, as can be seen from a comparison of (2.1.2) with (III: 2.1.21).

The Banach space $\mathscr{C}$ is thus not reflexive, so $\mathscr{B}(\mathscr{H})^*$ is strictly larger than $\mathscr{C}_1$. If a Banach space $\mathscr{E}$ is nonreflexive, then the same is true of $\mathscr{E}^*$, $\mathscr{E}^{**}$, etc.: Let $a \in \mathscr{E}^{**}$ but $a \notin \mathscr{E}$. The functional $w: e + \lambda a \to \lambda$ defined on $\{E + \lambda a\}$ can be extended continuously to $\mathscr{E}^{**}$ by the Hahn–Banach theorem. Therefore, $w \in \mathscr{E}^{***}$, but $w_{|\mathscr{E}} = 0$. Hence $\mathscr{C}_1$ and $\mathscr{B}(\mathscr{H})$ are also not reflexive; $\mathscr{B}(\mathscr{H})^*$ is strictly larger than $\mathscr{C}_1$. All trace-class operators provide linear functionals on the bounded operators by $a \to \operatorname{Tr} \rho a$, and these linear functionals are even continuous if $\mathscr{B}(\mathscr{H})$ is equipped with a weaker topology than the one from $\| \;\|$: If the neighborhood basis is defined by

$$U_{\rho,\varepsilon}(a) = \{a' \in \mathscr{B}(\mathscr{H}): |\operatorname{Tr} \rho(a - a')| < \varepsilon\}, \tag{2.1.4}$$

and ρ ranges only over $\mathscr{E}$, then this is the weak topology. If ρ is allowed to range over $\mathscr{C}_1$, then it is known as the ultraweak topology, and is genuinely finer than the weak topology but coarser than the $\| \;\|$-topology. The linear functionals $a \to \operatorname{Tr} \rho a$ for $\rho \in \mathscr{C}_1$ are, however, obviously continuous if $\mathscr{B}(\mathscr{H})$ has the ultraweak topology. These functionals have in addition the property of normality (III: 2.2.21): the order of taking weakly continuous linear functionals and suprema over bounded sets can be interchanged, since by Vigier's theorem (III: 2.3.24; 11) the supremum is the limit of a strongly, and therefore also weakly, convergent sequence. Since the weak and ultraweak topologies are equivalent on bounded sets, normality carries over to ultraweakly continuous, linear functionals. A somewhat deeper theorem ([4], I, §4, Theorem I) states that these include all normal linear functionals on $\mathscr{B}(\mathscr{H})$. We summarize by stating the

Characterization of Normal States (2.1.5)

The following properties are equivalent for a state w on $\mathscr{B}(\mathscr{H})$:

(i) w is normal (III: 2.2.21);
(ii) w is given by a **density matrix** ρ such that $w(a) = \mathrm{Tr}\, \rho a, \rho \geq 0, \mathrm{Tr}\, \rho = 1$;
(iii) w is ultraweakly continuous.

Remarks (2.1.6)

1. The density matrices form a norm-closed, convex subset of the unit sphere of $\mathscr{C}_1$, the trace-class operators with the trace norm $\| \; \|_1$.
2. If the system is classical, then instead of $\mathscr{B}(\mathscr{H})$ there is an Abelian von Neumann algebra, and we are familiar with the normal traces in the guise of probability measures. Specifically, on the L^∞ functions on phase space they are of the form $\rho(p, q)\, d\Omega$, $d\Omega$ being Liouville measure (I: 3.1.2; 3), $\rho \in L^1$, $\rho \geq 0$, $\int d\Omega \rho = 1$. Yet it may be that $|\rho| = \sup_{p,q} |\rho(p, q)| \not< 1$: Suppose that χ_A is the characteristic function of a set A such that $\Omega(A) \equiv \int d\Omega \chi_A < 1$; then an example is furnished by $\rho = \chi_A/\Omega(A)$.
3. All states constructed with a vector of $\mathscr{H}$ are pure, normal, and even weakly continuous—the density matrix for them is a one-dimensional projection. Conversely, any one-dimensional projection yields a pure state on $\mathscr{B}(\mathscr{H})$.
4. The spectrum of a density matrix is discrete, as it is in the trace class (and hence compact). The sum of the eigenvalues ρ_i is 1.
5. The density matrix can be thought of as a combination of the vectors that diagonalize it, or as a pure state on a larger Hilbert space $\mathscr{H}_g \equiv \mathscr{H} \otimes \mathscr{H}$, in which $\mathscr{B}(\mathscr{H})$ is imbedded as $\mathscr{B}(\mathscr{H}) \otimes \mathbf{1}$. The vector of $\mathscr{H}_g$ corresponding to $\rho = \sum_j |j\rangle\langle j| \rho_j$ is $\sum_j |j\rangle \otimes |j\rangle \sqrt{\rho_j}$ (cf. (1.4.7)). If $\mathscr{H}$ is separable, then the weak topology on $\mathscr{H}_g$ induces the ultraweak topology of $\mathscr{B}(\mathscr{H})$ on $\mathscr{B}(\mathscr{H}) \otimes \mathbf{1}$.
6. The normal states are weak-* dense in the positive unit sphere of $\mathscr{B}(\mathscr{H})^*$ (see (III: 2.1.19)), but are a proper subset rather than the whole of it. Hence they are not also weak-* compact.

Traces offer many advantages for doing calculations, owing to the commutativity property (1.4.13; 3). Inequalities for ordinary numbers often extend to traces, even when noncommutativity prevents them from extending directly to operators. Some of these inequalities will be used frequently later, and so are listed below. It will always be assumed that whatever the trace is taken of belongs to the trace class, though many of them have the generalization that if the lesser side of an inequality becomes infinite, then so does the greater side. For greater flexibility general forms are presented, while the name attached refers to the original version. The symbol Tr will always mean the trace on $\mathscr{B}(\mathscr{H})$. These inequalities apply trivially to factors of type I, and many also apply to type II.

Basic Inequalities (2.1.7)

1. Peierls's Inequality. Let k be a convex function from $\mathbb{R}$ to $\mathbb{R}^+$ and $\{|i\rangle\}$ be a not necessarily complete, orthonormal set. Then
$$\operatorname{Tr} k(a) = \sup_{\{|i\rangle\}} \sum_i k(\langle i|a|i\rangle).$$
2. Convexity. Let k be a convex function from $\mathbb{R}$ to $\mathbb{R}$ and $0 \le \alpha \le 1$. Then
$$\operatorname{Tr} k(\alpha a + (1-\alpha)b) \le \alpha \operatorname{Tr} k(a) + (1-\alpha) \operatorname{Tr} k(b).$$
3. The Peierls–Bogoliubov Inequality. Let k be a strictly monotonically increasing, convex, differentiable function $\mathbb{R} \to \mathbb{R}$ (and thus the inverse function k^{-1} exists), and suppose k/k' is convex. Then
$$k^{-1}(\operatorname{Tr} k(\alpha a + (1-\alpha)b)) \le \alpha k^{-1}(\operatorname{Tr} k(a)) + (1-\alpha)k^{-1}(\operatorname{Tr} k(b)).$$
4. Monotony. If m is a monotonically increasing function $\mathbb{R} \to \mathbb{R}$,
$$a \ge b \Rightarrow \operatorname{Tr} m(a) \ge \operatorname{Tr} m(b).$$
5. Klein's Inequality. Let f, g, and h be functions $\mathbb{R} \to \mathbb{R}$ such that for all $\alpha \in \operatorname{Sp} a$, $\beta \in \operatorname{Sp} b$, and $c_k \in \mathbb{R}$,
$$\sum_k c_k f_k(\alpha) g_k(\beta) h_k(\alpha) \ge 0.$$
Then
$$\operatorname{Tr} \sum_k c_k f_k(a) g_k(b) h_k(a) \ge 0.$$
6. Hölder's Inequality. Suppose that k_1 and k_2 are convex, strictly monotonic functions $\mathbb{R} \to \mathbb{R}$, the mapping $(\alpha, \beta) \to k_1^{-1}(\alpha) k_2^{-1}(\beta)$ is concave, and $\mathscr{H}$ has dimension $N < \infty$. Then
$$\left|\frac{1}{N} \operatorname{Tr} \mathrm{ab}\right| \le k_1^{-1}\left(\operatorname{Tr} \frac{1}{N} k_1(|a|)\right) k_2^{-1}\left(\operatorname{Tr} \frac{1}{N} k_2(|b|)\right).$$
7. The Cauchy–Schwarz Inequality. $|\operatorname{Tr}(ab)^2| \le \operatorname{Tr} a^*abb^*$.
8. Lieb's Theorem. Let a and b be non-negative, $a, b, c \in \mathscr{B}(\mathscr{H})$, and $0 \le \alpha \le 1$. Then the functions $a \to \operatorname{Tr} \exp(c + \ln a)$ and $(a, b) \to \operatorname{Tr} a^\alpha c b^{1-\alpha} c^*$ are concave.

Proof

1. By the spectral theorem and Jensen's inequality, for any unit vector $|i\rangle$, $\langle i|k(a)|i\rangle \ge k(\langle i|a|i\rangle)$, and therefore $\sum_i \langle i|k(a)|i\rangle \ge \sum_i k(\langle i|a|i\rangle)$.

Equality holds if the $|i\rangle$ are eigenvectors of a. It suffices to take the supremum over finite sets $\{|i\rangle\}$.

2. Let $|i\rangle$ be the eigenvectors of $\alpha a + (1-\alpha)b$. By Peierls's inequality,

$$\begin{aligned}\operatorname{Tr} k(\alpha a + (1-\alpha)b) &= \sum_i k(\alpha\langle i|a|i\rangle + (1-\alpha)\langle i|b|i\rangle)\\ &\le \alpha\sum_i k(\langle i|a|i\rangle) + (1-\alpha)\sum_i k(\langle i|b|i\rangle)\\ &\le \alpha \operatorname{Tr} k(a) + (1-\alpha)\operatorname{Tr} k(b).\end{aligned}$$

Note that the inequality $k(\alpha a + (1-\alpha)b) \le \alpha k(a) + (1-\alpha)k(b)$ can be false in the sense of operator ordering.

3. If k/k' is convex, then for sequences of numbers $\{\beta_i\}$ and $\{\gamma_i\}$,

$$k^{-1}\left(\sum_i k(\beta_i\alpha + \gamma_i(1-\alpha))\right) \le \alpha k^{-1}\left(\sum_i k(\beta_i)\right) + (1-\alpha)k^{-1}\left(\sum_i k(\gamma_i)\right)$$

by Problem 2. Hence, as with Inequality 2,

$$\begin{aligned}k^{-1}(\operatorname{Tr} k(\alpha a + (1-\alpha)b)) &= k^{-1}\left(\sum_i k(\alpha\langle i|a|i\rangle + (1-\alpha)\langle i|b|i\rangle)\right)\\ &\le \alpha k^{-1}\left(\sum_i k(\langle i|a|i\rangle)\right)\\ &\quad + (1-\alpha)k^{-1}\left(\sum_i k(\langle i|b|i\rangle)\right)\\ &\le \alpha k^{-1}(\operatorname{Tr} k(a)) + (1-\alpha)k^{-1}(\operatorname{Tr} k(b)),\end{aligned}$$

using Inequality 1 again.

4. If $a \ge b$, then the min–max principle implies for their ordered eigenvalues that $a_i \ge b_i$, so $\sum_i m(a_i) \ge \sum_i m(b_i)$. Once again, the inequality $m(a) \ge m(b)$ may fail for operators.

5. Let a_i and b_i be the eigenvalues of a and b, and c_{ij} be the scalar product of the eigenvectors of a with those of b. Then

$$\operatorname{Tr}\sum_k c_k f_k(a)g_k(b)h_k(a) = \sum_{i,j}|c_{ij}|^2\sum_k c_k f_k(a_i)g_k(b_j)h_k(a_i) \ge 0.$$

6. Let a_i and b_i be the ordered eigenvalues of $|a|$ and $|b|$, and let $|i\rangle$ denote the eigenvectors of a. By the min–max principle (III: 3.5.21),

$$\begin{aligned}\operatorname{Tr} ab &= \sum_{i,j}\langle i|a|j\rangle\langle j|b|i\rangle \le \sum_i (a_i - a_{i+1})\sum_{k=1}^{i}\langle k\|b\|k\rangle\\ &\le \sum_i (a_i - a_{i+1})\sum_{k=1}^{i} b_k = \sum_i a_i b_i.\end{aligned}$$

The inequality

$$\frac{1}{N}\sum_{i=1}^{N} k_1^{-1}(\alpha_i)k_2^{-1}(\beta_i) \le k_1^{-1}\left(\frac{1}{N}\sum_{i=1}^{N}\alpha_i\right)k_2^{-1}\left(\frac{1}{N}\sum_{i=1}^{N}\beta_i\right),$$

$$\text{for } \alpha_i \equiv k_1(a_i) \text{ and } \beta_i \equiv k_2(b_i)$$

is just the assumption of concavity.

7. By the Cauchy–Schwarz inequality (III: 2.2.20; 1) for states,

$$|\mathrm{Tr}\, abab|^2 \le \mathrm{Tr}\, abb^*a^*\, \mathrm{Tr}\, b^*a^*ab = (\mathrm{Tr}\, a^*abb^*)^2.$$

The order of the operations is important; it is not true in general that $\mathrm{Tr}(ab)^2 \le \mathrm{Tr}\, a^*ab^*b$.

8. The proof of this rather deep proposition in the noncommutative case is too laborous to be repeated here—see [5]. □

Corollaries (2.1.8)

1. For any orthonormal system $\{|i\rangle\}$, $\beta F(H) \equiv -\ln \mathrm{Tr}\exp(-\beta H) \le -\ln \sum_i \exp(-\beta\langle i|H|i\rangle)$.
2. The function $H \to \mathrm{Tr}\exp(-\beta H)$ is convex.
3. In fact, even $H \to \ln \mathrm{Tr}\exp(-\beta H)$ is convex, so $F(H)$ is concave. By recourse to $(\partial/\partial\alpha)\,\mathrm{Tr}\, f(H + \alpha V)_{|\alpha=0} = \mathrm{Tr}\, Vf'(H)$, and the fact that F is majorized by any tangent, one finds that

$$F(H_0) + \langle V\rangle_H \le F(H_0 + V) \le F(H_0) + \langle V\rangle_{H_0},$$

where $\langle a\rangle_H = \mathrm{Tr}\, a\exp(-\beta H)/\mathrm{Tr}\exp(-\beta H)$.

4. $H_1 \ge H_2 \Rightarrow F(H_1) \ge F(H_2)$.
5. If k is convex, then $\mathrm{Tr}(k(a) - k(b) - (a - b)k'(b)) \ge 0$, so $\mathrm{Tr}(a\ln a - a\ln b - (a - b)) \ge 0$, too. If $f_1(\alpha) = \int_0^\alpha d\alpha' g(\alpha')$ and $f_2(\beta) = \int_0^\beta d\beta' g^{-1}(\beta')$, then by Young's inequality, $\alpha\beta \le f_1(\alpha) + f_2(\beta)$, and therefore $\mathrm{Tr}\, ab \le \mathrm{Tr}\, f_1(a) + \mathrm{Tr}\, f_2(b)$. In particular, if p and q are ≥ 1 and related by $1/p + 1/q = 1$, and a and b are nonnegative, then $\mathrm{Tr}\, ab \le (1/p)\,\mathrm{Tr}\, a^p + (1/q)\,\mathrm{Tr}\, b^q$.
6. With $k_1(\alpha) = \alpha^p$, $k_2(\beta) = \beta^q$, Corollary 5 can be improved to $\mathrm{Tr}\, ab \le (\mathrm{Tr}\,|a|^p)^{1/p}(\mathrm{Tr}|b|^q)^{1/q}$; since this no longer involves N, it also holds when $N = \infty$. By iteration,

$$\left\|\prod_{i=1}^{n} a_i\right\|_p \le \prod_{i=1}^{n} \|a_i\|_{p_i},$$

$$\|a\|_p = (\mathrm{Tr}\,|a|^p)^{1/p}, \quad \text{where } \sum_i \frac{1}{p_i} = \frac{1}{p}, \quad p, p_i \ge 1.$$

As $p \to \infty$, $\|a\|_p \to \|a\|$, so $|\mathrm{Tr}\, ab| \le \|a\|\mathrm{Tr}|b|$; the trace class is a two-sided ideal of $\mathscr{B}(\mathscr{H})$ (cf. (III: 2.3.20; 3)).

7. If a and b are Hermitian, then $\mathrm{Tr}(ab)^2 \leq \mathrm{Tr}\, a^2b^2$, $a = a^*$, $b^{-1} = b^*$: $|\mathrm{Tr}(ab)^2| \leq \mathrm{Tr}\, a^2$. By iterating this, $|\mathrm{Tr}(ab)^{2^p}| \leq \mathrm{Tr}(abb^*a^*)^{2^{p-1}} = \mathrm{Tr}(|a|^2|b^2|)^{2^{p-1}} \leq \cdots \leq \mathrm{Tr}|a|^{2^p}|b|^{2^p}$. Because of the Trotter product formula $\exp(a + b) = s\text{-}\lim_{n\to\infty}(\exp(a/n)\exp(b/n))^n$ (see (III; 2.4.9)), $|\mathrm{Tr}\exp(\alpha a + \beta b)| \leq \mathrm{Tr}|\exp(\alpha a)||\exp(\beta b)|$, for $\alpha, \beta \in \mathbb{C}$, and initially for Hermitian operators of finite rank. It then extends to $\exp(\alpha a + \beta b) \in \mathscr{B}_1(\mathscr{H})$, $\exp(\alpha a) \in \mathscr{B}_1(\mathscr{H})$, $\exp(\beta b) \in \mathscr{B}(\mathscr{H})$ and thereby yields a generalization of Corollary 3 known as the Golden–Thompson–Symanzik inequality [6], $\exp(-\beta\langle V\rangle_{H_0}) \leq \mathrm{Tr}\exp[-\beta(H_0 + V - F(H_0))] \leq \langle\exp(-\beta V)\rangle_{H_0}$.
8. The function $(a, b) \to \lim_{\alpha\downarrow 0}\mathrm{Tr}(1/\alpha)(a - a^{1-\alpha}b^\alpha) = \mathrm{Tr}\, a(\ln a - \ln b)$ is convex.

Our next task is to give the density matrices an ordering that indicates which of two ρ's corresponds to the more chaotic state. The ordering must of course be independent of the basis, and so it can depend only on the eigenvalues ρ_i. If the eigenvalues are thought of as ordered by their magnitudes, then pure states are associated with sequences $(1, 0, 0, \ldots)$, i.e., with the greatest possible first eigenvalue. Because $\sum_{i=1}^\infty \rho_i = 1$, two density matrices might not be strictly ordered by the natural ordering of Hermitian operators. However, by the min–max principle (III: 3.5.21),

$$\rho(n) \equiv \sum_{i=1}^n \rho_i = \sup_{\mathscr{H}_n} \mathrm{Tr}_{\mathscr{H}_n}\rho,$$

which permits the following

Definition of the Ordering of the Density Matrices (2.1.9)

A density matrix $\tilde{\rho}$ is said to be **more mixed**, or **more chaotic**, than ρ if $\tilde{\rho}(n) \leq \rho(n)$ for all n. In symbols, $\tilde{\rho} \succeq \rho$ (or $\rho \preceq \tilde{\rho}$).

Remarks (2.1.10)

1. This clearly defines a preordering of the density matrices, i.e., $\rho \preceq \rho$; and if $\rho \preceq \tilde{\rho}$ and $\tilde{\rho} \preceq \tilde{\tilde{\rho}}$, then $\rho \preceq \tilde{\tilde{\rho}}$. If two density matrices are equivalent, that is, $\rho \preceq \tilde{\rho}$ and $\tilde{\rho} \preceq \rho$, then $\rho_i = \tilde{\rho}_i$, and so they are related by $\tilde{\rho} = V\rho V^*$. If the space is finite-dimensional, then V can be chosen unitary, and otherwise it is only an isometric mapping $(\mathrm{Ker}\,\rho)^\perp \to (\mathrm{Ker}\,\tilde{\rho})^\perp$; if $\mathrm{Dim}\,\mathrm{Ker}\,\rho \neq \mathrm{Dim}\,\mathrm{Ker}\,\tilde{\rho}$, then it has no unitary extension.
2. If the equivalent density matrices are classed together, then (2.1.9) gives the classes a lattice structure, characterized by the sequences of numbers $\{\rho(n)\}$. The sequence $\{\min(\rho(n), \tilde{\rho}(n))\}$ yields the equivalence class of the purest states more mixed than either ρ or $\tilde{\rho}$. The concave hull of $\max(\rho(n), \tilde{\rho}(n))$ with respect to n characterizes the most mixed states

purer than either ρ or $\tilde{\rho}$. The sequences thus defined are positive, increasing, and concave in n, and tend to 1 as $n \to \infty$ (or equal 1 when $n = \text{Dim}\,\mathscr{H}$). Their successive differences are therefore decreasing sequences of positive numbers summing to 1, which correspond to an equivalence class of density matrices. The lattice contains a class of purest elements, namely the extremal states. If the dimension of $\mathscr{H}$ is finite, then there is also a most mixed state with $\rho = \mathbf{1}/\text{Dim}\,\mathscr{H}$, but if it is infinite, there is none.

3. The ordering and convexity are compatible on the space of states in the sense that if $\rho \precsim \mu$ and $\rho \precsim \nu$ then $\rho \precsim \alpha\mu + (1-\alpha)\nu$ for $0 \le \alpha \le 1$:

$$\sup_{\mathscr{H}_n} \text{Tr}_{\mathscr{H}_n}(\alpha\mu + (1-\alpha)\nu) \le \alpha \sup_{\mathscr{H}_n} \text{Tr}_{\mathscr{H}_n} \mu + (1-\alpha) \sup_{\mathscr{H}_n} \text{Tr}_{\mathscr{H}_n} \nu \le \rho(n).$$

4. Since the operators $\rho(n)$ are suprema of the weakly continuous functions $\text{Tr}_{\mathscr{H}_n}\rho$, they are weakly lower semicontinuous. Moreover, it will be shown later (2.4.19; 1) that sequences of density matrices converging weakly to a density matrix are convergent even in the trace norm. Hence the maps $\rho \to \rho(n)$ are actually weakly continuous, and the limit belongs to the same mixing class.
5. The ordering of the density matrices is not total—for instance

$$\begin{pmatrix} \frac{1}{2} & & \\ & \frac{1}{2} & \\ & & 0 \end{pmatrix} \quad \text{and} \quad \begin{pmatrix} \frac{3}{4} & & \\ & \frac{1}{8} & \\ & & \frac{1}{8} \end{pmatrix}$$

are not related by it.

Examples (2.1.11)

1. In the Schrödinger picture the time-evolution of a system is given by $\rho \to \rho_t \equiv U(t)\rho U^{-1}(t)$, which shows that density matrices remain in their equivalence classes.
2. The time-average $(1/T)\int_0^T dt\rho_t$ is more mixed than the original density matrices. This operation involves combinations and weak limits, which can only make density matrices more chaotic.
3. If the time-evolution of a density matrix is a linear transformation of the eigenvalues, $\rho_i(t) = M_{ik}(t)\rho_k(0)$, then for $\text{Tr}\,\rho = 1$ and $\rho \ge 0$ it must be true that $\sum_i M_{ik} = 1$ for all k, and $M_{ik} \ge 0$ for all i and k. If, for finite dimension N, it is also required that the chaotic state $\rho_i = 1/N$ be stationary for all i, then, moreover, $\sum_k M_{ik} = 1$ for all i. The matrix M is then said to be **doubly stochastic**. Such matrices clearly form a convex set, and are consequently convex combinations of the extremal elements by the Krein–Milman theorem. The extremal elements have entries $M_{ik} = 0$ or 1, and so $1 = \sum_i M_{ik} = \sum_k M_{ik}$ implies that each row and each column has exactly one 1; this makes them permutation matrices, mapping any ρ to an equivalent ρ. Therefore, $\rho(t) \succsim \rho(0)$, as $\rho(t)$ is a convex combina-

tion of ρ's equivalent to $\rho(0)$. This kind of time-evolution thus increases the mixing. Its differential version $\dot{\rho}(t) = M\rho(0)$ is a **master equation** $\dot{\rho}_i = \sum_k W_{ik}(\rho_k - \rho_i)$, where W satisfies $\sum_i W_{ik} = \sum_i W_{ki}$.

4. If an observable has one-dimensional projections P_i, then the state is immediately converted to $\tilde{\rho} \equiv \sum_i P_i \rho P_i$ when the observable is measured. Once it is perceived that the kth eigenvalue has been measured, ρ becomes P_k. The first stage of the measurement increases the mixing of the state, $\tilde{\rho} \succeq \rho$. This follows from the min–max principle: If $P_i|i\rangle = |i\rangle$, then

$$\tilde{\rho}(n) = \sum_{i=1}^{n} \langle i|\rho|i\rangle \leq \rho(n) = \sup_{\mathscr{H}_n} \mathrm{Tr}_{\mathscr{H}_n} \rho.$$

The second stage makes the state pure. This can be interpreted in that the interaction with the measuring apparatus extracts information, which unmixes the state upon transmission to the human mind.

5. The "coarse-grained" density matrix $\tilde{\rho} \equiv \sum_i P_i \lambda_i$, $\lambda_i = \mathrm{Tr}\, \rho P_i$, is more mixed than $\sum_i P_i \rho P_i$ by Problem 1, and *a fortiori* $\tilde{\rho} \succeq \rho$.
6. Suppose that the function k is convex from $\mathbb{R}^+$ to $\mathbb{R}^+$ and $k(0) = 0$; then clearly the smaller eigenvalues are suppressed to a greater degree in $k(\rho)$. In fact, $\rho \succeq k(\rho)/\mathrm{Tr}\, k(\rho)$ by Problem 3, and the resulting states are purer. In particular, if $k(x) = x^{\beta'/\beta}$, $\beta' > \beta$, then $\exp(-\beta H)/\mathrm{Tr}\exp(-\beta H) \succeq \exp(-\beta' H)/\mathrm{Tr}\exp(-\beta' H)$. The physical significance is that the mixing of the canonical density matrices is greater at higher temperatures.

We have seen that convex combinations of $U\rho U^{-1}$ and weak limits increase the mixing of ρ. This exhausts the possibilities:

Theorem (2.1.12)

$\tilde{\rho} \succeq \rho$ iff $\tilde{\rho}$ is in the weakly closed convex hull of $\{U\rho U^{-1}\}$.

Remark (2.1.13)

The weak closure of $\{a \in \mathscr{B}^+(\mathscr{H}),\ \|a\| = 1\}$ is $\{a \in \mathscr{B}^+(\mathscr{H}),\ \|a\| \leq 1\}$, and density matrices may converge weakly to zero. This means that the set of density matrices is not closed, which causes technical difficulties in the proof, which is put off to Problem 4 for that reason.

Corollary (2.1.14)

If $\tilde{\rho} \succeq \rho$, then for any convex function k, $\mathrm{Tr}\, k(\tilde{\rho}) \leq \mathrm{Tr}\, k(\rho)$.

Proof

If $\tilde{\rho} = \sum_i c_i U_i \rho U_i^{-1}, 0 \leq c_i \leq 1, \sum_i c_i = 1$, and the sum is finite, then by the convexity inequality (2.1.7; 2), $\mathrm{Tr}\ k(\tilde{\rho}) \leq \sum_i c_i \mathrm{Tr}\, k(U_i \rho U_i^{-1}) = \mathrm{Tr}\, k(\rho)$. Moreover, $\rho \to \mathrm{Tr}\, k(\tilde{\rho})$ is weakly lower semicontinuous, so the limiting case of an infinite sum is likewise bounded by $\mathrm{Tr}\, k(\rho)$. □

Corollary (2.1.14) gives rise to the possibility of defining mappings of the density matrices to the real numbers, monotonic with respect to the ordering $\succeq$, and so enables the degree of disorder to be measured. For instance, if $k(\rho) = \rho^2$, then $\mathrm{Tr}\, k(\rho)$ can equal 1 only for pure states, and is otherwise smaller. The next section will discuss some other properties distinguished by the function $-k(\rho) = -\rho \ln \rho$ used to define the entropy. For now, note that the converse of (2.1.14) is also true:

Theorem (2.1.15)

$\tilde{\rho} \succeq \rho$ iff for every convex function k, $\mathrm{Tr}\, k(\tilde{\rho}) \leq \mathrm{Tr}\, k(\rho)$.

Proof

Because of (2.1.14), we need only show that if $\tilde{\rho} \not\succeq \rho$, then there exists a function k such that $\mathrm{Tr}\, k(\tilde{\rho}) \geq \mathrm{Tr}\, k(\rho)$. Let m be the first integer such that $\tilde{\rho}_1 + \tilde{\rho}_2 + \cdots + \tilde{\rho}_m > \rho_1 + \rho_2 + \cdots + \rho_m$, and let $k(x) = (x - \rho_m)$, when $x \geq \rho_m$, and otherwise 0. Then $k(\rho_1) = \rho_1 - \rho_m, \ldots, k(\rho_m) = \rho_m - \rho_m = 0 = k(\rho_{m+1}) = k(\rho_{m+2}) = \cdots$. By assumption, $\tilde{\rho}_1 + \tilde{\rho}_2 + \cdots + \tilde{\rho}_{m-1} \leq \rho_1 + \rho_2 + \cdots + \rho_{m-1}$, so $\tilde{\rho}_m > \rho_m$, which implies $k(\tilde{\rho}_i) = \tilde{\rho}_i - \rho_m > 0$ for all $i \leq m$. Therefore, $\mathrm{Tr}\, k(\rho) = \rho_1 + \rho_2 + \cdots + \rho_m - m\rho_m < \tilde{\rho}_1 + \tilde{\rho}_2 + \cdots + \tilde{\rho}_m - m\rho_m \leq \mathrm{Tr}\, k(\tilde{\rho})$. □

Since expectation values in mixed states are averages of different spectral values, they do not reach the extremes of the spectrum so easily. This observation creates a new way to define the ordering relationship.

Theorem (2.1.16)

(i) $\tilde{\rho} \succeq \rho \Leftrightarrow \sup_{\substack{U \\ U^* = U^{-1}}} \mathrm{Tr}\, U\tilde{\rho}U^{-1}a \leq \sup_{\substack{U \\ U^* = U^{-1}}} \mathrm{Tr}\, U\rho U^{-1}a$ for all $a \in \mathscr{B}^+(\mathscr{H})$,

(ii) $\tilde{\rho} \succeq \rho \Leftrightarrow \inf_{\substack{U \\ U^* = U^{-1}}} \mathrm{Tr}\, U\tilde{\rho}U^{-1}a \geq \inf_{\substack{U \\ U^* = U^{-1}}} \mathrm{Tr}\, U\rho U^{-1}a$ for all $a \in \mathscr{B}^+(\mathscr{H})$.

Proof

See Problem 5. □

Corollary (2.1.17)

Let $(\Delta_\rho a)^2 \equiv \operatorname{Tr} \rho a^2 - (\operatorname{Tr} \rho a)^2 = \inf_\lambda \operatorname{Tr} \rho(a-\lambda)^2$. *Then* $\tilde\rho \succeq \rho$ *implies that*

$$\inf_U \Delta_{U\tilde\rho U^{-1}} a \geq \inf_U \Delta_{U\rho U^{-1}} a \quad \textit{for all } a.$$

This means that if one is interested in the least deviation Δa of a within the equivalence classes of ρ and $\tilde\rho$, then it is smaller for the state that is less mixed.

The various aspects of the relationship can be summarized as follows:

Conditions for Density Matrices to be Compared (2.1.18)

The ordering relationship $\tilde\rho \succeq \rho$ is equivalent to each of the following:

(i) $\tilde\rho(n) \leq \rho(n)$ for all n;
(ii) $\tilde\rho = w\text{-}\lim_\alpha \sum_i c_{i\alpha} U_{i\alpha} \rho U_{i\alpha}^{-1}$, $c_{i\alpha} > 0$, $\sum_i c_{i\alpha} = 1$, $U_{i\alpha}^{-1} = U_{i\alpha}^*$;
(iii) $\operatorname{Tr} k(\tilde\rho) \geq \operatorname{Tr} k(\rho)$ for every concave function k;
(iv) $\begin{matrix}\sup \\ \inf \\ U\end{matrix} \operatorname{Tr} U\tilde\rho U^{-1} a \begin{matrix}\leq \sup \\ \leq \inf \\ U\end{matrix} \operatorname{Tr} U\rho U^{-1} a$, $a \in \mathscr{B}^+(\mathscr{H})$, $U^{-1} = U^*$.

Problems (2.1.19)

1. Let P_i be pairwise orthogonal projections of dimensions $n_i < \infty$ and $\sum_i P_i = \mathbf{1}$. Show that $\sum_i (1/n_i) P_i \operatorname{Tr} P_i \rho \succeq \sum_i P_i \rho P_i$.

2. Let $k(x) > 0$, $k' > 0$, $k'' > 0$, k/k' convex. Show that the mapping $(\beta_1, \ldots, \beta_n) \to k^{-1}(\sum_{i=1}^n k(\beta_i))$ of $\mathbb{R}^n$ to $\mathbb{R}$ is convex. (Hint: note that: (i) A mapping $f(\beta_1, \ldots, \beta_n)$ is convex if $\chi''(0) \geq 0$, where χ is the function $\chi(t) = f(\beta_1 + u_1 t, \ldots, \beta_n + u_n t)$ and $(u_1, \ldots, u_n)$ and $(\beta_1, \ldots, \beta_n)$ are arbitrary. (ii) If the function $K(\delta)/\delta$ increases monotonically, then $K(\sum_i \delta_i) \geq \sum_i K(\delta_i)$, $\delta_i > 0$.)

3. Let k be a convex, monotonically increasing function, $k(x) \geq 0$ for $x \geq 0$, and $k(0) = 0$. Show that $\rho \succeq k(\rho)/\operatorname{Tr} k(\rho)$.

4. Show that $\bar\rho \succeq \rho \Leftrightarrow \bar\rho \in \overline{\operatorname{Conv}\{U\rho U^{-1}\}}^{\text{weak}}$.

   (i) Let $\mathscr{K}(\rho) = \{a \geq 0$: a is compact, and $\alpha_1 + \cdots + \alpha_n \leq \rho(n)$ for all n, where α_i are the eigenvalues in increasing order$\}$. Show that $\mathscr{K}(\rho)$ is convex and weakly compact.
   (ii) Let $\mathscr{E}(\rho) = \{a \in \mathscr{K}(\rho)$: $\alpha_1 = \rho_1, \ldots, \alpha_n = \rho_n, \alpha_{n+1} = \cdots = 0$ or $\alpha_i = \rho_i$ for all $i\}$. Show that $\mathscr{E}(\rho)$ contains the extremal points of $\mathscr{K}(\rho)$.
   (iii) Show that $\mathscr{E}(\rho) \subset \overline{\{U\rho U^{-1}\}}^{\text{weak}}$.
   (iv) Finish the proof by applying the Krein–Milman theorem: Every compact, convex set equals the closure of the convex hull of its extremal points.

5. Prove Theorem (2.1.16).

Solutions (2.1.20)

1. Let $d\mu(U)$ be the invariant measure on the compact group $U(n)$, normalized to 1. For all $a \in \mathscr{B}(\mathbb{C}^n)$, $\mathbf{1}_{|\mathbb{C}^n}(1/n) \operatorname{Tr} a = \int d\mu U a U^{-1}$, since the right side is invariant under all U and hence proportional to $\mathbf{1}_{|\mathbb{C}^n}$, and $\operatorname{Tr} \int d\mu U a U^{-1} = \operatorname{Tr} a$. Similarly,

$$\frac{1}{n} P \operatorname{Tr} P\rho + (\mathbf{1} - P)\rho(\mathbf{1} - P) = \int d\mu_P U_P \rho U_P^{-1}$$

$$= \int d\mu_P U_P(P\rho P + (\mathbf{1} - P)\rho(\mathbf{1} - P))U_P^{-1},$$

if the operators U_P vary over the unitary transformations of $\mathscr{H}$ equaling $\mathbf{1}$ on $(\mathbf{1} - P)\mathscr{H}$. Therefore, $(1/n)P \operatorname{Tr} P\rho + (\mathbf{1} - P)\rho(\mathbf{1} - P) \succeq P\rho P + (\mathbf{1} - P)\rho(\mathbf{1} - P)$, which proves the claim by iteration.

2. (i) is trivial, and (ii) follows from

$$\sum_i \delta_i \geq \delta_k \Rightarrow \delta_k K\Big(\sum_i \delta_i\Big) \geq \Big(\sum_i \delta_i\Big)K(\delta_k), \Big(\sum_k \delta_k\Big)K\Big(\sum_i \delta_i\Big) \geq \Big(\sum_i \delta_i\Big)\sum_k K(\delta_k).$$

Now let $\chi(t) \equiv k^{-1}(\sum k(\beta_i + u_i t))$. The function $\chi(t)$ is convex iff $\chi''(t) \geq 0$. $[k'(\chi)]^3\chi'' = [k'(\chi)]^2[\sum_i u_i^2 k''(\beta_i)] - k''(\chi)[\sum_i u_i k'(\beta_i)]^2$ (where $\chi \equiv \chi(0), \chi'' \equiv \chi''(0)$), so it remains to show that $[k'(\chi)]^2 \sum_i u_i^2 k''(\beta_i) \geq k''(\chi)[\sum_i u_i k'(\beta_i)]^2$. By the Cauchy–Schwarz inequality, $[\sum_i u_i k'(\beta_i)]^2 = [\sum_i u_i \sqrt{k''(\beta_i)}\sqrt{k'(\beta_i)^2/k''(\beta_i)}]^2 \leq [\sum_i u_i^2 k''(\beta_i)] \times [\sum_i k'(\beta_i)^2/k''(\beta_i)]$, and the desired inequality is certainly satisfied if $\psi(\chi) \equiv k'(\chi)^2/k''(\chi) \geq \sum_i k'(\beta_i)^2/k''(\beta_i) = \sum_i \psi(\beta_i)$. By (ii), this is the case if $K(\delta)/\delta$ increases monotonically, where K is defined by $\delta_i = k(\beta_i)$, $K(\delta_i) = \psi(\beta_i)$. Finally, $K(\delta)/\delta$ increases monotonically $\Leftrightarrow k'^2/kk''$ increases monotonically $\Leftrightarrow k/k'$ is convex.

3. If $0 \leq x \leq y$, then $x = (x/y)y + (1 - (x/y))0$, and hence $k(x) \leq (x/y)k(y)$, $yk(x) \leq xk(y)$. Consequently

$$\sum_{i=1}^{m} k(\rho_i)\Big(\sum_{j=1}^{m} \rho_j + \sum_{j=m+1}^{\infty} \rho_j\Big) \geq \sum_{j=1}^{m} \rho_j\Big(\sum_{i=1}^{m} k(\rho_i) + \sum_{i=m+1}^{\infty} k(\rho_i)\Big),$$

i.e.,

$$[k(\rho_1) + \cdots + k(\rho_m)]\Big(\sum_{i=1}^{\infty} k(\rho_i)\Big)^{-1} \geq [\rho_1 + \cdots + \rho_m]\Big(\sum_{i=1}^{\infty} \rho_i\Big)^{-1}.$$

Remark: If k is concave, then $\rho \precsim k(\rho)/\operatorname{Tr} k(\rho)$.

4. (i) By (2.1.10; 3) the set $\mathscr{K}(\rho)$ is convex. Moreover, $\alpha_1 + \cdots + \alpha_n = \sup_{\mathscr{H}_n} \operatorname{Tr}_{\mathscr{H}_n} a$ is weakly lower semicontinuous in a, so $\mathscr{K}(\rho)$ is weakly closed and, since $\|a\| = \alpha_1 \leq \rho_1 = \|\rho\| \leq \operatorname{Tr} \rho = 1$, also weakly compact.
   (ii) By considering all the possibilities, one realizes that it is possible to write any $a \in \mathscr{K}(\rho)$ as $\alpha\rho_1 + (1 - \alpha)\rho_2$, $0 < \alpha < 1$, with $\rho_i \in \mathscr{K}(\rho)$, unless $a \in \mathscr{E}(\rho)$.
   (iii) Let $a = \sum_{i=1}^{n} \rho_i |1, i\rangle\langle 1, i|$, $\rho = \sum \rho_i |2, i\rangle\langle 2, i|$, where $\{|1, i\rangle\}$ and $\{|2, i\rangle\}$ are two orthonormal systems. Let $U|2, i\rangle = |1, i\rangle$, $U_l|1, n + i\rangle = |1, n + l - i\rangle$ for $1 \leq i \leq l - 1$, $U_l|1, i\rangle = |1, i\rangle$ otherwise. $a = s\text{-}\lim_{l\to\infty} U_l U \rho U^{-1} U_l^{-1}$.

(iv) By the Krein–Milman theorem, $\mathscr{K}(\rho) = \overline{\text{Conv}\,\mathscr{E}(\sigma)}^{\text{weak}} = \overline{\text{Conv}\{U\rho U^{-1}\}}^{\text{weak}}$ (by (iii)), and $\tilde{\rho} \in \mathscr{K}(\rho)$, if $\tilde{\rho} \succeq \rho$.

5. By a replacement of a with $a + \|a\|$ if necessary, a may be assumed positive. Then $\text{Tr}\,\rho a = \sup_n \text{Tr}\,\rho^{(n)}a$, $\rho^{(n)}$, where the $\rho^{(n)}$ have the eigenvalues $\rho_1, \rho_2, \ldots, \rho_n, 0, 0, \ldots$. The changes of the orders of operation in what follows are justified for the $\rho^{(n)}$, and the suprema can also be interchanged:

(i) $\Rightarrow$: Let $\alpha_1 \geq \alpha_2 \geq \cdots$ be the decreasing sequence of eigenvalues of a and α_∞ be the upper boundary of $\sigma_{\text{ess}}(a)$ (to be understood in the sense analogous to (III: 3.5.21)). If $\rho = \sum \rho_i |i\rangle\langle i|$, then

$$\begin{aligned}\text{Tr}\,\rho a &= \sum \rho_i \langle i|a|i\rangle = (\rho_1 - \rho_2)\langle 1|a|1\rangle + (\rho_2 - \rho_3)[\langle 1|a|1\rangle + \langle 2|a|2\rangle] + \cdots \\ &\leq (\rho_1 - \rho_2)\alpha_1 + (\rho_2 - \rho_3)(\alpha_1 + \alpha_2) + \cdots = \sum \rho_i \alpha_i,\end{aligned}$$

and $\sup \text{Tr}\,U\rho U^{-1}a = \sum \rho_i \alpha_i$.

$$\begin{aligned}\sum_i \tilde{\rho}_i \alpha_i &= \tilde{\rho}_1(\alpha_1 - \alpha_2) + (\tilde{\rho}_1 + \tilde{\rho}_2)(\alpha_2 - \alpha_3) + \cdots + \alpha_\infty \\ &\leq \rho_1(\alpha_1 - \alpha_2) + (\rho_1 + \rho_2)(\alpha_2 - \alpha_3) + \cdots + \alpha_\infty = \sum \rho_i \alpha_i.\end{aligned}$$

$\Leftarrow$: Choose an n-dimensional projection for a and use the min–max principal.

The proof of (ii) is similar.

2.2 The Properties of Entropy

The information about a system in a mixed state is incomplete. The entropy is a measure of how far from maximal the information is.

In statistical physics, entropy is not an observable in the sense of an operator on Hilbert space, but rather a property of the state of the system, measuring the lack of our knowledge as expressed in the specification of the state. This section will consider what sorts of conditions single out a particular measure of this lack of knowledge and will see what conclusions can be drawn from it.

A primary requirement would be monotony with respect to the ordering introduced in the preceding section (we consider only normal states). In other words, a density matrix that is more mixed should have more entropy, which we denote S: $\tilde{\rho} \succeq \rho \Rightarrow S(\tilde{\rho}) \geq S(\rho)$. This leaves many possibilities open for the definition; every monotonic function of the trace of a concave function of ρ would satisfy this requirement (cf. (2.1.14)). A further reasonable requirement is the additivity of the entropies of independent systems. If their combination is represented on the tensor product of their Hilbert spaces, this means

$$S(\rho' \otimes \rho'') = S(\rho') + S(\rho''). \tag{2.2.1}$$

The two requirements together do not yet quite determine S uniquely. The whole one-parameter family of

α-Entropies (2.2.2)

$$S_\alpha(\rho) = \frac{1}{1-\alpha} \ln \operatorname{Tr} \rho^\alpha, \qquad \alpha \in \mathbb{R}^+ \setminus \{1\},$$

satisfy the general

Properties of Entropy (2.2.3)

(i) $0 \leq S_\alpha(\rho) \leq \ln \dim \mathscr{H}$;
(ii) $\tilde{\rho} \succeq \rho \Rightarrow S_\alpha(\tilde{\rho}) \geq S_\alpha(\rho)$;
(iii) $S_\alpha(\rho' \otimes \rho'') = S_\alpha(\rho') + S_\alpha(\rho'')$;
(iv) If $\rho = P/\dim P$, $P = P^2 = P^*$, then $S_\alpha(\rho) = \ln \dim P$.

(In particular, $S_\alpha(\rho) = 0$ iff ρ is a pure state, and $S_\alpha(\rho) = \ln \dim \mathscr{H}$ iff ρ is the chaotic state $1/\dim \mathscr{H}$.)

Proof

(i) If $\alpha > 1$, then $\sum_i \rho_i^\alpha \leq (\sum_i \rho_i)^\alpha = 1$, and if $\alpha < 1$, then $\sum_i \rho_i = 1 \leq (\sum_i \rho_i^\alpha)^{1/\alpha}$. This shows the left side of the inequality, and the right follows from (iv) and (ii).
(ii) The function ρ^α is concave for $\alpha < 1$ and convex for $\alpha > 1$. The logarithm is monotonic, and the $1 - \alpha$ accounts for the sign (see (2.1.18(iii)).
(iii) $\operatorname{Tr}(\rho' \otimes \rho'')^\alpha = \operatorname{Tr}[(\rho')^\alpha \otimes (\rho'')^\alpha] = \operatorname{Tr}(\rho')^\alpha \cdot \operatorname{Tr}(\rho'')^\alpha$.
(iv) If $n = \dim P$, then $S_\alpha(\rho) = (1/(1-\alpha)) \ln(n n^{-\alpha})$. □

The entropy can be fixed uniquely by a more stringent assumption of additivity (2.2.1), with which monotony emerges as a consequence rather than a separate axiom:

Characterization of the von Neumann Entropy (2.2.4)

The only entropy satisfying the following conditions is $S(\rho) = -\operatorname{Tr} \rho \ln \rho$

(i) *$S(\rho)$ is a continuous function of the eigenvalues of ρ;*

(ii) $S\begin{pmatrix} \frac{1}{2} & 0 \\ 0 & \frac{1}{2} \end{pmatrix} = \ln 2;$

(iii) *If*
$$\mathscr{H} = \bigoplus_{n=1}^{N} \mathscr{H}_n, \qquad \rho = \bigoplus_{n=1}^{N} p_n \rho_n, \qquad \sum_n p_n = 1,$$
$$0 \leq p_i \leq 1, \ \operatorname{Tr} \rho_n = 1,$$

then, regardless of the dimension of $\mathscr{H}_n$, $S(\rho) = \sum_{n=1}^{N} p_n S(\rho_n) + S(p)$, *where* p *is the diagonal matrix on* $\mathbb{C}^n$ *having eigenvalues* ρ_n.

Remark (2.2.5)

1. Since the representation should make no difference, S can only depend on the eigenvalues. It certainly does not seem unreasonable to demand continuity.
2. Condition (ii) is a normalization.
3. If all the $\mathscr{H}_n$ in condition (iii) have the same dimension and all ρ_n are equal, then $\mathscr{H} = \mathscr{H}_1 \otimes \mathbb{C}^n$, and (iii) reduces to (2.2.1). This generalization of (2.2.1), which makes possible an inductive proof, has the following interpretation: Suppose a system consists of two subsystems, one described by $\mathbb{C}^n$ and the other having several variants according to the position of the state vector of the first in $\mathbb{C}^n$. Then the entropy of the total system is just the sum of the entropy of the first subsystem and those of the second, averaged according to their probabilities.
4. The formula $S = -\mathrm{Tr}\, \rho \ln \rho$ can be justified in the spirit of Boltzmann as follows. Let the state corresponding to ρ be realized as a vector of a reducible representation of the algebra $\mathscr{A}$ of observables consisting of N identical representations. The ensemble described by ρ can be thought of as having been subjected to a sequence of N measurements, where ρ_i is N_i/N, N_i being the number of times the eigenvector e_i has been measured. The Hilbert space is $\mathscr{H} = \bigoplus_{j=1}^{N} \mathscr{H}_j$, where the spaces $\mathscr{H}_j$ are all identical and are spanned by $\{e_i\}$. The observables are represented as a direct sum of N identical representations. With the use of doubled indices, this can be written as $\mathscr{H}_j = \bigoplus_{i=1}^{\infty} e_{i,j}$. A ρ of rank r and with $\rho_i = N_i/N$, $i = 1, \ldots, r$, is represented by the vector

$$\frac{1}{\sqrt{N}}(e_{1,1} + e_{1,2} + \cdots + e_{1,N_1} + e_{2,N_1+1} + \cdots + e_{2,N_1+N_2} + \cdots + e_{r,N_1+N_2+\cdots+N_{r-1}+1} + \cdots + e_{r,N})$$

of $\mathscr{H}$. If the e_i are chosen from other spaces $\mathscr{H}_j$, the same state results, and there are clearly $W \equiv N!/\prod_i N_i!$ different vectors for the same ρ. If the numbers N_i are large enough, then $\ln W \cong N \ln N - \sum_i N_i \ln N_i = -N \sum_i \rho_i \ln \rho_i$, so $(1/N) \ln W \to -\mathrm{Tr}\, \rho \ln \rho$. Assuming that every vector of $\mathscr{H}$ is assigned the same probability, S turns out to be roughly the logarithm of the probability of the configuration, and there is an identification: the most mixed state = the state of greatest entropy = the most probable state.
5. $S(\rho) = \lim_{\alpha \to 1} S_\alpha(\rho)$, yet if the dimension is infinite, then $S(\rho)$ may become $+\infty$. However, Properties (2.2.3) remain valid in this limit, and apply to S as well.

6. A particular consequence of (2.2.3(ii)) is that $S(\alpha\rho + (1-\alpha)U\rho U^{-1}) \geq S(\rho)$. More generally, (2.1.7; 2) implies that the mapping $\rho \to S(\rho)$ is concave: $S(\alpha\rho_1 + (1-\alpha)\rho_2) \geq \alpha S(\rho_1) + (1-\alpha)S(\rho_2)$. This means that the entropy of a mixed state is greater than the constituent entropies weighted as in the mixing. If $\rho = \sum_n p_n \rho_n$, $0 \leq p_n \leq 1$, $\sum_n p_n = 1$, then the inequalities

$$\sum_n p_n S(\rho_n) \leq S(\rho) \leq \sum_n p_n S(\rho_n) + \sum_n p_n \ln \frac{1}{p_n}$$

necessarily follow (Problem 4). They are optimal in the sense that equality holds on the left if all ρ_n are equal, and on the right if all ρ_n have disjoint support, by (2.2.4(iii)).
7. Although by (2.2.3(iv)) all the S_α are the same with the chaotic state, with the canonical state $\rho = \exp(-\beta(H - F(\beta)))$, $\operatorname{Tr}\exp(-\beta H) = \exp(-\beta F(\beta))$, they are different (Problem 6).

Proof of (2.2.4)

We write $S(\rho_1, \rho_2, \ldots)$ for $S(\rho)$.

(a) Let $\mathscr{H} = \mathbb{C}^1$. Then $S(1) = 0$, because on $\mathbb{C}^2$, $S(\rho_1, \rho_2) = \rho_1 S(1) + \rho_2 S(1) + S(\rho_1, \rho_2)$.
(b) Let $\mathscr{H} = \mathbb{C}^n$, $f(n) \equiv S(1/n, 1/n, \ldots, 1/n)$, and let $n = m_1 m_2$. We write $\mathbb{C}^n = \mathbb{C}^{m_1} \oplus \mathbb{C}^{m_1} \oplus \cdots \oplus \mathbb{C}^{m_1}$ and use (iii) with

$$N = m_2, p_i = m_2^{-1}, \qquad \rho_i = \begin{pmatrix} 1/m_1 & & \\ & \ddots & \\ & & 1/m_1 \end{pmatrix},$$

$$f(m_1 m_2) = m_2 \frac{1}{m_2} f(m_1) + f(m_2) = f(m_1) + f(m_2).$$

The solution of this equation is $f(n) = C \ln n$, and the normalization (ii) makes $C = 1$. Other solutions are excluded by the continuity requirement (Problem 1).

(c)
$$f(m) = S\Bigl(\overbrace{\frac{1}{m}, \frac{1}{m}, \ldots, \frac{1}{m}}^{n}, \overbrace{\frac{1}{m}, \ldots, \frac{1}{m}}^{m-n}\Bigr)$$
$$= \frac{n}{m} f(n) + \frac{m-n}{m} f(m-n) + S\left(\frac{n}{m}, \frac{m-n}{m}\right),$$

so by step (b),

$$S\left(\frac{n}{m}, 1 - \frac{n}{m}\right) = -\frac{n}{m} \ln \frac{n}{m} - \left(1 - \frac{n}{m}\right) \ln\left(1 - \frac{n}{m}\right).$$

This holds initially only for integers n and m, and then by continuity holds generally, $S(\rho_1, \rho_2) = -\sum_{i=1}^{2} \rho_i \ln \rho_i$.

(d) The rest of the proof proceeds inductively: with $\mathbb{C}^{n+1} = \mathbb{C}^n \oplus \mathbb{C}$, $p_1 = 1 - \rho_n, p_2 = \rho_n$,

$$\begin{aligned} S(\rho_1, \rho_2, \ldots, \rho_{n-1}, \rho_n) &= (1 - \rho_n)S\left(\frac{\rho_1}{1 - \rho_n}, \ldots, \frac{\rho_{n-1}}{1 - \rho_n}\right) + \rho_n S(1) \\ &\quad + S(1 - \rho_n, \rho_n) \\ &= -\sum_{j=1}^{n-1} \rho_i \ln \frac{\rho_i}{1 - \rho_n} - \rho_n \ln \rho_n \\ &\quad -(1 - \rho_n)\ln(1 - \rho_n) = -\sum_{j=1}^{n} \rho_i \ln \rho_i. \end{aligned}$$ □

The Classical Entropy (2.2.6)

For a classical density $\rho_{\text{cl}}(\mathbf{x}, \mathbf{p})$ on phase space the entropy would be defined as $-\int d\Omega \rho_{\text{cl}} \ln \rho_{\text{cl}}$. This is not *a priori* positive-definite; for instance $\rho_{\text{cl}} = \chi(A)/\Omega(A)$ as in (2.1.6; 2) leads to $-\int d\Omega \chi \ln \chi = \ln \Omega(A)$, which is negative if $\Omega(A) < 1$. It is easy to see that this entropy also depends on the measure of volume in phase space. There are many ways to associate a density ρ_{cl} with a density matrix ρ or vice versa.

The most useful such expressions are obtained with a method of A. Wehrl, in which for a given density matrix ρ one calculates expectation values in coherent states, and, conversely, a classical density is used to mix coherent states. The coherent states $W(\mathbf{z})|u\rangle \equiv |\mathbf{z}\rangle$ of (III: 3.1.13) can be generalized for functions u that are even and normalized, but not necessarily Gaussian. The state $|\mathbf{z}\rangle$ has the wavefunction $\exp(i\mathbf{k} \cdot \mathbf{x})u(\mathbf{x} - \mathbf{q})$ if $\mathbf{z} = \mathbf{q} + i\mathbf{k} \in \mathbb{C}^{dN}$, which is the phase space for N particles in a physical space of dimension d. It is easy to check that $\mathbf{z} = \langle \mathbf{z}|\mathbf{x}|\mathbf{z}\rangle + i\langle \mathbf{z}|\mathbf{p}|\mathbf{z}\rangle$ still holds and that the states are complete, $\int d^{2Nd}z(2\pi)^{-Nd}|\mathbf{z}\rangle\langle\mathbf{z}| = \mathbf{1}$.

The Density Matrix and the Phase-Space Density (2.2.7)

If to an N-particle density matrix ρ we associate the phase-space density $\rho_{\text{cl}}(\mathbf{z}) = \langle \mathbf{z}|\rho|\mathbf{z}\rangle$, and to a classical density $f(\mathbf{z})$ on phase space we associate the density matrix $\rho_{\text{qu}} = \int d\Omega_z f(\mathbf{z})|\mathbf{z}\rangle\langle\mathbf{z}|$, $d\Omega_z^N = (2\pi)^{-Nd}d^{2N}z$, then

$$\begin{gathered} \rho \geq 0, \qquad \operatorname{Tr} \rho = 1 \Rightarrow 0 \leq \rho_{\text{cl}}(\mathbf{z}) \leq 1, \qquad \int d\Omega_z^N \rho_{\text{cl}}(\mathbf{z}) = 1, \\ f \geq 0, \qquad \int d\Omega_z f(z) = 1 \Rightarrow 0 \leq \rho_{\text{qu}} \leq 1, \qquad \operatorname{Tr} \rho_{\text{qu}} = 1. \end{gathered} \tag{2.2.8}$$

Proof

Positivity is trivial, and the connection between the trace and the phase-space integral follows from the n-dimensional version of a formula of (III: 3.1.14; 1):

$$\mathbf{1} = \int d\Omega_z^N |\mathbf{z}\rangle\langle\mathbf{z}| \Rightarrow \operatorname{Tr} a = \sum_i \langle i|a|i\rangle = \sum_i \int d\Omega_z^N \langle i|\mathbf{z}\rangle\langle\mathbf{z}|a|i\rangle$$
$$= \int d\Omega_z^N \langle\mathbf{z}|a|\mathbf{z}\rangle.$$

Conversely, $\operatorname{Tr} \int d\Omega_z f(\mathbf{z})|\mathbf{z}\rangle\langle\mathbf{z}| = \sum_i \int d\Omega_z^N f(\mathbf{z})|\langle\mathbf{z}|i\rangle|^2 = \int d\Omega_z f(\mathbf{z})$, since $\langle\mathbf{z}|\mathbf{z}\rangle = 1$. The denominator $(2\pi)^{dN}$ in $d\Omega_z^N$ reveals that the phase-space volume is measured in units of h rather than $\hbar = h/2\pi = 1$. □

Inequalities for the Classical and Quantum-Mechanical Entropies (2.2.9)

(i) $S(\rho) \le -\int d\Omega_z^N \rho_{\text{cl}}(\mathbf{z}) \ln \rho_{\text{cl}}(\mathbf{z}) \equiv S_{\text{cl}}(\rho)$;
(ii) $-\int d\Omega_z^N f(\mathbf{z}) \ln f(\mathbf{z}) \le S(\rho_{\text{qu}})$.

Remarks (2.2.10)

1. Inequality (i) implies that the ρ_{cl} of (2.2.7) always has more entropy than $S(\rho)$. This classical entropy is therefore always positive; the density ρ_{cl} defined in (2.2.7) can never be so concentrated as to make the classical entropy negative, and indeed $\rho_{\text{cl}} \le 1$.
2. It can also be shown that this classical entropy equals 1 if ρ is extremal, and otherwise it is greater than 1 [32].
3. If a quantum-mechanical density is associated with a classical density f by mixing the coherent states with f, then Inequality (ii) states that the quantum-mechanical entropy is greater than the classical entropy. The latter may even tend to $-\infty$, for instance if f tends to a delta function.
4. Inequality (ii) shows that the continuous analogue of the last inequality of (2.2.3; 6) is false: $S(|\mathbf{z}\rangle\langle\mathbf{z}|) = 0$, and in this case the inequality goes in the other direction, with the replacements $p_n \to f(\mathbf{z})$, $\sum_n \to \int d\Omega_z^N$:

$$-\int d\Omega_z^N f(\mathbf{z}) \ln f(\mathbf{z}) + \int d\Omega_z^N f(\mathbf{z}) S(|\mathbf{z}\rangle\langle\mathbf{z}|) \le S\left(\int d\Omega_z^N f(\mathbf{z})|\mathbf{z}\rangle\langle\mathbf{z}|\right).$$

5. If the particles are identical, states must be either symmetrized or antisymmetrized according to the statistics. For bosons this is accomplished most easily with the aid of the creation operator

$$a_{\mathbf{z}}^* \equiv a^*(\exp[i\mathbf{k}\cdot\mathbf{x}]u(\mathbf{q}-\mathbf{x})), \qquad |\mathbf{z}_1, \ldots, \mathbf{z}_N\rangle = \mathbf{a}_{z_1}^* \cdots \mathbf{a}_z^* |0\rangle,$$

with which

$$\mathbf{1} = \sum_{n=0}^{\infty} \frac{(2\pi)^{-nd}}{n!} \int d^{2d}\mathbf{z}_1 \cdots d^{2d}\mathbf{z}_n |\mathbf{z}_1, \ldots, \mathbf{z}_n\rangle\langle\mathbf{z}_1, \ldots, \mathbf{z}_n|.$$

So, with identical bosons, when the trace is taken the volume of the classical phase space has to be divided by $n!$. The states are not yet normalized to norm 1.

$$\langle\mathbf{z}'_1, \ldots, \mathbf{z}'_N|\mathbf{z}_1, \ldots, \mathbf{z}_N\rangle = \sum_P (\pm 1)^P \prod_{i=1}^{N} \langle\mathbf{z}'_i|\mathbf{z}_{P_i}\rangle \equiv \begin{matrix}\text{Per}\\ \text{Det}\end{matrix} (\langle\mathbf{z}'_i|\mathbf{z}_k\rangle),$$

where $P_1, \ldots, P_n$ is a permutation of $1, \ldots, n$, because the coherent states are not orthogonal:

$$\langle\mathbf{z}'|\mathbf{z}\rangle = \int d^d x \exp[i\mathbf{x}\cdot(\mathbf{k} - \mathbf{k}')]u^*(\mathbf{x} - \mathbf{q}')u(\mathbf{x} - \mathbf{q}).$$

These determinants and permanents crop up along with $d\Omega_z^N$ in the calculations of expectation values, making them more laborious.

6. Since these inequalities are valid for coherent states with a great degree of arbitrariness in u, they can be optimized by varying u.

Inequalities (2.2.9) will follow from a lemma of Berezin on the

Relationship between the Trace and the Phase-Space Integral (2.2.11)

Let K be a convex function and suppose $a^* = a$. Then

(i) $\text{Tr}\, K(a) \geq \int d\Omega_z^N K(\langle\mathbf{z}|a|\mathbf{z}\rangle)$;
(ii) $\int d\Omega_z^N K(f(\mathbf{z})) \geq \text{Tr}\, K(a)$, *where* $a = \int d\Omega_z^N f(\mathbf{z})|\mathbf{z}\rangle\langle\mathbf{z}|$, $K(a) \in C^1$, *and* f *is a measurable function* $\mathbb{C}^N \to \mathbb{R}$.

Proof

(i) As noted in the proof of Peierls's inequality, $\langle|K(a)|\rangle \geq K(\langle|a|\rangle)$ for expectation values in an arbitrary vector, so

$$\text{Tr}\, K(a) = \int d\Omega_z^N \langle\mathbf{z}|K(a)|\mathbf{z}\rangle \geq \int d\Omega_z^N K(\langle\mathbf{z}|a|\mathbf{z}\rangle).$$

(ii) If $|j\rangle$ denotes an eigenfunction of a, then

$$\begin{aligned}\text{Tr}\, K(a) &= \sum_j K(\langle j|a|j\rangle) = \sum_j K\left(\int d\Omega_z^N f(\mathbf{z})|\langle\mathbf{z}|j\rangle|^2\right)\\ &\leq \sum_j \int d\Omega_z^N |\langle\mathbf{z}|j\rangle|^2 K(f(\mathbf{z}))\\ &= \int d\Omega_z^N K(f(\mathbf{z})).\end{aligned}$$

□

Proof of (2.2.9)

The function $x \ln x$ is convex, and for the concave function $-x \ln x$ the inequalities for convex functions are reversed. □

The additivity of the entropy when $\rho = \rho_1 \otimes \rho_2$ generalizes to an inequality when ρ is not in the form of a product. To cover general ρ_1 and ρ_2 requires the

Definition of Partial Traces (2.2.12)

Let $\mathscr{H} = \mathscr{H}_1 \otimes \mathscr{H}_2$. The **partial traces** Tr_1 and Tr_2 are defined by $\mathrm{Tr}_{1,2}\, a = \sum_j \langle j|a|j\rangle \in \mathscr{B}(\mathscr{H}_{2,1})$ for any $a \in \mathscr{C}_1(\mathscr{H})$, where $\{|j\rangle\}$ is any complete orthonormal set in $\mathscr{H}_{1,2}$.

A consequence of this is the

Subadditivity of the Entropy (2.2.13)

Let $\rho_{1,2} = \mathrm{Tr}_{2,1}\, \rho$. Then $S(\rho) \leq S(\rho_1) + S(\rho_2)$.

Remarks (2.2.14)

1. If $\rho = \rho_1 \otimes \rho_2$, then $\rho_{1,2} = \mathrm{Tr}_{2,1}\, \rho$ and by (2.2.3(iii)) equality holds in (2.2.13).
2. The partial traces reproduce the reduced density matrices used in §1.1. At that time we noticed that the reduction entailed a loss of information. Inequality (2.2.13) indicates that there is less information in ρ_1 and ρ_2 than in the original ρ.
3. If $\alpha \neq 1$, then the α-entropies S_α (2.2.3) are not subadditive (Problem 2). It is consequently not necessarily true that $\rho_1 \otimes \rho_2 \succeq \rho$.
4. Subadditivity allows axiom (iii) of (2.2.4) to be replaced [7] with
   (iii (a)) $S(\rho) = S(V^* \rho V)$ for all isometries V; and
   (iii (b)) $S(\rho) \leq S(\rho_1) + S(\rho_2)$, equality holding iff $\rho = \rho_1 \otimes \rho_2$.

Proof

By Klein's inequality (2.1.7; 5), $\mathrm{Tr}\, a \ln a - \mathrm{Tr}\, a \ln b \geq \mathrm{Tr}\,(a - b)$. Put $a = \rho$ and $b = \rho_1 \otimes \rho_2$ and note that $\ln \rho_1 \otimes \rho_2 = \ln \rho_1 \otimes \mathbf{1} + \mathbf{1} \otimes \ln \rho_2$. □

Corollary (2.2.15)

Consider a sequence of ever larger systems on the tensor product $\mathscr{H}^n$, $n = 1, 2, 3, \ldots$. Suppose that the density matrices ρ_n are compatible so that when reduced to a subsystem they always become the density matrix of the smaller system: $\rho_m = \mathrm{Tr}_{n-m}\rho_n$, $m \leq n$. If $\sigma_n = -(1/n)\mathrm{Tr}\,\rho_n \ln \rho_n$, then $n\sigma_n \leq m\sigma_m + (n-m)\sigma_{n-m}$. In particular, $\sigma_{2n} \leq \sigma_n$, and hence the limits $\lim_{n\to\infty} \sigma_n = \inf_n \sigma_n$ must exist and be ≥ 0. Although the entropy itself does not tend to a limit as the size of the system gets arbitrarily large, the specific entropy does.

It will be asked by how far (2.2.13) misses equality. More precisely, it might be supposed that the entropy of a united system is always greater than that of any single one of its parts. Surprisingly, this is not necessarily so with quantum statistics; ρ could be a pure state, thus having entropy zero, while the ρ_i correspond to mixtures. This is the case that arose in the discussion of the time-evolution in §1.1; the additional information contained in ρ has to do with the correlations between the subsystems. The correlations are precisely pinned down in

Lemma (2.2.16)

Let ρ be pure; then ρ_1 and ρ_2 have the same spectrum with the same multiplicities, except possibly for an eigenvalue at 0.

Proof

See Problem 3. □

Corollary (2.2.17)

If ρ is pure, then $S(\rho_1) = S(\rho_2)$. Our information about the subsystems is correlated, so they possess the same amount of disorder.

In this case, $S(\rho) = S(\rho_1) - S(\rho_2)$; more generally there is a

Triangle Inequality (2.2.18)

$$|S(\rho_1) - S(\rho_2)| \leq S(\rho) \leq S(\rho_1) + S(\rho_2).$$

(Lieb and Araki [8]).

Remarks (2.2.19)

1. This inequality has no classical analogy; a counterexample is provided by a ρ with $S(\rho) < 0$ but $S(\rho_1) = S(\rho_2)$.
2. Even if the entropy of a subsystem can be greater than that of the whole system, the triangle inequality reveals that it can not exceed the sum of the total entropy and the entropy of the complementary subsystem.
3. Astonishingly, the classical entropy (2.2.9) of a quantum-mechanical density matrix is monotonic; it is always larger for the whole than for a part: $S_{\mathrm{cl}}(\rho) \geq S_{\mathrm{cl}}(\rho_1)$. (For the proof see Problem 5.)

Proof

According to Remark (2.1.6; 5), ρ may be regarded as a pure state ρ_{123} on a large Hilbert space $\mathscr{H}_1 \otimes \mathscr{H}_2 \otimes \mathscr{H}_3$, for which $\rho = \operatorname{Tr}_3 \rho_{123}$. Let $\rho_3 = \operatorname{Tr}_{12} \rho_{123}$, $\rho_{23} = \operatorname{Tr}_1 \rho_{123}$; then by Corollary (2.2.17), $S(\rho) = S(\rho_3)$, $S(\rho_1) = S(\rho_{23})$. Because of subadditivity, $S(\rho_1) = S(\rho_{23}) \leq S(\rho_3) + S(\rho_2)$ $= S(\rho) + S(\rho_2)$, and along with the same thing with 1 and 2 interchanged, this yields the left inequality of (2.2.18). □

An ideal measurement leaves the system in a pure state, reducing the entropy to 0. For this reason, $S(\rho)$ may be regarded as a measure of the amount of information to be gained by an ideal measurement. The difference $S(\rho) - S(\rho_1)$ specifies how much more information a measurement of the total system can yield than a measurement of a subsystem. Inequality (2.2.18) bounds this relative information gain by $S(\rho_2)$:

$$|S(\rho) - S(\rho_1)| \leq S(\rho_2).$$

With quantum statistics the difference can be either positive or negative. If ρ is pure, so that the greatest possible information about the total system is available, but ρ_1 is a mixture, then more information can be obtained by measuring the subsystem. On the other hand, there are some inequalities for this entropy difference that are analogous to those of the classical entropy:

Inequalities for the Entropy Difference (2.2.20)

Let ρ_{123} be given on $\mathscr{H}_1 \otimes \mathscr{H}_2 \otimes \mathscr{H}_3$, and $\rho_{12} = \operatorname{Tr}_3 \rho_{123}$, $\rho_1 = \operatorname{Tr}_2 \rho_{12}$, etc. Then

(i) $S(\rho_{12}) - S(\rho_1)$ *is concave in* ρ_{12};
(ii) $S(\rho_{13}) - S(\rho_1) + S(\rho_{23}) - S(\rho_2) \geq 0$ (Lieb and Ruskai [8]); *and*
(iii) $S(\rho_{123}) - S(\rho_2) \leq S(\rho_{12}) - S(\rho_2) + S(\rho_{32}) - S(\rho_2)$.

Remarks (2.2.21)

1. Proposition (i) implies that mixing increases the relative information gain. In particular, the relative information gain is a monotonic function in ρ_{12} with the ordering introduced in (2.1.9).
2. If Roman numerals are used to denote the systems corresponding to the Hilbert spaces $\mathscr{H}_i$, then Inequality (ii) implies that more information can be obtained by measuring I ∪ III and II ∪ III than I and II. If $\mathscr{H}_2$ is one-dimensional, so $S(\rho_2) = 0$ and $S(\rho_{23}) = S(\rho_3)$, then this proposition reduces to (2.2.18).

3. Inequality (iii) is subadditivity for the entropy difference. The information content of I ∪ II and III ∪ II relative to II is greater than that of I ∪ II ∪ III relative to II.

Proof

(i) Let $\rho_{12} = \alpha\rho'_{12} + (1-\alpha)\rho''_{12}$, $\rho_1 = \alpha\rho'_1 + (1-\alpha)\rho''_1$. Then

$$\begin{aligned}
-S(\rho_{12}) &+ \alpha S(\rho'_{12}) + (1-\alpha)S(\rho''_{12}) + S(\rho_1) - \alpha S(\rho'_1) - (1-\alpha)S(\rho''_1) \\
&= \alpha \operatorname{Tr}_{12} \rho'_{12}[\ln\rho_{12} - \ln\rho'_{12} - \ln\rho_1 + \ln\rho'_1] \\
&\quad + (1-\alpha) \operatorname{Tr}_{12} \rho''_{12}[\ln\rho_{12} - \ln\rho''_{12} - \ln\rho_1 + \ln\rho''_1] \\
&\equiv \alpha\Delta' + (1-\alpha)\Delta''.
\end{aligned}$$

If $a = -\beta H_0 - \ln \operatorname{Tr}\exp(-\beta H_0)$ and $b = -\beta V$, then because of (2.1.8; 3) and $\operatorname{Tr}\exp(a) = 1$, $\exp(\operatorname{Tr} b \exp(a)) \le \operatorname{Tr}\exp(a+b)$, so with $a = \ln\rho'_{12}$, $b = [\cdots]$, we find $\exp(\Delta) \le \operatorname{Tr}_{12}\exp(\ln\rho_{12} - \ln\rho_1 + \ln\rho'_1)$. Therefore, with Lieb's theorem (2.1.7; 8),

$$\begin{aligned}
\exp(\alpha\Delta' + (1-\alpha)\Delta'') &\le \alpha\exp(\Delta') + (1-\alpha)\exp(\Delta'') \\
&\le \alpha \operatorname{Tr}_{12}\exp(\ln\rho_{12} - \ln\rho_1 + \ln\rho'_1) \\
&\quad + (1-\alpha)\operatorname{Tr}_{12}\exp(\ln\rho_{12} - \ln\rho_1 + \ln\rho''_1) \\
&\le \operatorname{Tr}_{12}\exp(\ln\rho_{12} - \ln\rho_1 + \ln(\alpha\rho'_1 \\
&\quad + (1-\alpha)\ln\rho''_1)) = \operatorname{Tr}_{12}\exp(\ln\rho_{12}) = 1.
\end{aligned}$$

(ii) Since ρ_{ik} and ρ_i can be expressed linearly in ρ_{123}, part (i) makes the left side concave in ρ_{123}. The minimum is consequently attained when ρ_{123} is pure. But by Corollary (2.2.17), in this case $S(\rho_{13}) = S(\rho_2)$ and $S(\rho_{23}) = S(\rho_1)$, and the minimum is zero.

(iii) Choose a pure ρ_{1234} on $\mathscr{H}_{123} \otimes \mathscr{H}_4$, such that $\operatorname{Tr}_4\rho_{1234} = \rho_{123}$. Then by Corollary (2.2.17), $S(\rho_{123}) + S(\rho_2) - S(\rho_{12}) - S(\rho_{23}) = S(\rho_4) + S(\rho_2) - S(\rho_{12}) - S(\rho_{14})$ which is ≤ 0 by (ii). □

These general inequalities for density matrices reflect mixing properties of the entropy like those used in phenomenological thermodynamics, and thereby provide a deeper foundation for those classical rules.

Nearly as important as the entropy differences is the

Relative Entropy (2.2.22)

$$S(\sigma\,|\,\rho) \equiv \operatorname{Tr}\rho(\ln\rho - \ln\sigma), \qquad \rho, \sigma \ge 0, \qquad \operatorname{Tr}\rho = \operatorname{Tr}\sigma = 1,$$

for which

(i) $S(\sigma\,|\,\rho) \ge 0$;
(ii) *the function* $(\sigma, \rho) \to S(\sigma\,|\,\rho)$ *is strictly convex and lower semicontinuous*;
(iii) $S(\sigma\otimes\tau\,|\,\rho\otimes\tau) = S(\sigma\,|\,\rho)$ *for any density matrix* τ; *and*
(iv) $S(\sigma_1\,|\,\rho_1) \le S(\sigma\,|\,\rho)$ *for* $\mathscr{H} = \mathscr{H}_1 \otimes \mathscr{H}_2$, $(\sigma_1, \rho_1) = \operatorname{Tr}_2(\sigma, \rho)$.

Proof of the Properties of the Relative Entropy

(i) This was shown in the proof of subadditivity (2.2.13).

(ii) Convexity follows from (2.1.7; 8) when $\alpha \to 0$. The function is lower semicontinuous because $S(\sigma|\rho)$ can be written as the supremum of a set of continuous functions (Problem 7).

(iii)
$$\begin{aligned} S(\sigma \otimes \tau | \rho \otimes \tau) &= \mathrm{Tr}_{12}\, \rho \otimes \tau[(\ln \rho) \otimes \mathbf{1} + \mathbf{1} \otimes \ln \tau \\ &\quad - (\ln \sigma) \otimes \mathbf{1} - \mathbf{1} \otimes \ln \tau] \\ &= \mathrm{Tr}_1\, \rho(\ln \rho - \ln \sigma)\mathrm{Tr}_2\, \tau = S(\sigma|\rho). \end{aligned}$$

(iv) As in Problem (2.1.19; 1), write $\rho_1 \otimes \mathbf{1}/d_2 = \int d\mu U_2 \rho U_2^{-1}$, $d_2 = \dim \mathscr{H}_2$, and similarly for σ. By (iii) and (ii),

$$\begin{aligned} S(\sigma_1|\rho_1) &= S\left(\sigma_1 \otimes \frac{\mathbf{1}}{d_2}\middle|\rho_1 \otimes \frac{\mathbf{1}}{d_2}\right) \\ &= \int S\left(\int d\mu U_2 \sigma U_2^{-1} \middle| \int d\mu U_2 \rho U_2^{-1}\right) \\ &\leq \int d\mu S(U_2 \sigma U_2^{-1} | U_2 \rho U_2^{-1}) = S(\sigma|\rho). \end{aligned}$$

Since d_2 drops out of the expression, this proof for $d_2 < \infty$ extends to the infinite-dimensional case. □

Remarks (2.2.23)

1. If σ is the canonical density matrix $\sigma = \exp(-\beta H)/\exp(-\beta F)$, and the free energy is $F = -\beta^{-1} \ln \mathrm{Tr} \exp(-\beta H)$, then $S(\sigma|\rho) = \beta(\mathrm{Tr}\, \rho H - F) - S(\rho)$. If a free energy $F(\rho) \equiv \mathrm{Tr}\, \rho H - \beta^{-1} S(\rho)$ is ascribed to ρ, then $S(\sigma|\rho) = F(\rho) - F$. The relative entropy $S(\sigma|\rho)$ measures the difference from the canonical free energy $F(\sigma) = F$, which always lies lower because of (i).
2. By Property (ii), mixing and passing to limits bring the free energy closer to the canonical free energy.
3. Property (iii) states that the difference from the canonical free energy is the same for ρ_1 and ρ if there are two independent subsystems 1 and 2, where $\rho = \rho_1 \otimes \rho_2$, and ρ_2 is the canonical density matrix of system 2.
4. If a subsystem is weakly coupled, $H_{12} \cong H_1 \otimes \mathbf{1}_2 + \mathbf{1}_1 \otimes H_2$, i.e., $\exp(-\beta(H_1 - F_1)) \cong \mathrm{Tr}_1 \exp(-\beta(H_{12} - F_{12}))$, then its difference from its canonical free energy is always less than that of the whole system. The analogous argument for the entropy only leads to $S(\rho_1) \leq S(\rho) + \ln d_2$, which already follows from (2.2.5; 3).

A final matter to investigate is how sensitive S is to small changes in ρ.

Theorem (2.2.24)

The mapping $\mathscr{C}_1^+ \to \mathbb{R}^+ : \rho \to S(\rho)$ *is lower semicontinuous in the trace-norm topology of* $\mathscr{C}_1$.

Remarks (2.2.25)

1. The set $\mathscr{C}_1$ is topologized with the trace norm $\| \ \|_1$. If a sequence $\{\rho_N\}$ converges in this topology to ρ, then $S(\rho)$ is at most $\lim_{N\to\infty} S(\rho_N)$. However, we shall see in (2.4.19; 1) that for density matrices all topologies between the trace topology and the weak topology are equivalent.
2. Continuity does not occur, because in every $\| \ \|_1$-neighborhood of ρ there are density matrices with arbitrarily much entropy. This follows directly from concavity,
$$S\left(\frac{1}{N}\rho_N + \left(1 - \frac{1}{N}\right)\rho\right) \geq \frac{1}{N} S(\rho_N) + \left(1 - \frac{1}{N}\right)S(\rho).$$
Let $S(\rho) = 0$, and $S(\rho_N) = N^2$; then $S((1/N)\rho_N + (1 - 1/N)\rho) \geq N$, although
$$\left\|\frac{1}{N}\rho_N + \left(1 - \frac{1}{N}\right)\rho - \rho\right\|_1 \leq \frac{2}{N},$$
so the density matrices converge to ρ. The two terms in the expression $(1/N)\rho_N + (1 - (1/N))\rho$, however, can not be comparable in the sense of (2.1.9); that would contradict (2.1.10; 4), by which the limit of a sequence of equivalent density matrices can not be purer than the elements of the sequence.
3. The mappings $\mathscr{C}_1^+ \to \mathbb{R}^+ : \rho \to S_\alpha(\rho)$, $\alpha > 1$ are continuous (see below).
4. By lower semicontinuity the sets $S_n \equiv \{\rho : S(\rho) \leq n\}$ are closed, and by Remark 2 they are nowhere dense. This means that the set $\bigcup_n S_n$ of ρ's of finite entropy is of the first category, the topological analogue of a null set. In this sense the entropy is almost always $+\infty$.

Proof

Because $\mathrm{Tr}\, \rho^\alpha = \|\rho\|_\alpha^\alpha \leq \|\rho\|^{\alpha-1} \cdot \|\rho\|_1$, the mapping of $\mathscr{C}_1$ to $\mathbb{R}^+ : \rho \to S_\alpha(\rho)$ is continuous. As the supremum of a set of continuous functions, $S(\rho) = \sup_{\alpha>1} S_\alpha(\rho)$ is lower semicontinuous. □

The failure of $S(\rho)$ to be continuous does not diminish its usefulness. The density matrices ρ of very large S have their eigenvalues ρ_i spread so far apart that the average of the energy diverges.

The Continuity of the Entropy at Finite Energy (2.2.26)

Suppose that $H \geq 0$ *and* $\text{Tr}\exp(-\beta H) < \infty$ *for some* $\beta > 0$. *If the density matrices having* $\text{Tr}\,\rho H < \infty$ *are topologized with the norm* $\|\rho\|_H = \text{Tr}\,\rho(1 + H)$, *then* $S(\rho)$ *is a continuous mapping* $\mathscr{C}_H \to \mathbb{R}^+$, *where* $\mathscr{C}_H = \{\rho \in \mathscr{C}_1, \|\rho\|_H < \infty\}$.

Proof

According to Remark (2.2.23; 1), $S(\rho) = \beta(\text{Tr}(\rho H) - F) - S(\sigma|\rho)$, where $\sigma = \exp(-\beta H)/\exp(-\beta F)$. The function $\text{Tr}\,\rho H$ is continuous in the $\| \ \|_H$-topology, and $-S(\sigma|\rho)$ is upper semicontinuous, because the $\| \ \|_H$-topology is finer than the trace topology. Since $S(\rho)$ is lower semicontinuous in the trace-norm topology, it is also lower semicontinuous in $\| \ \|_H$, and hence continuous in $\| \ \|_H$. □

Problems (2.2.27)

1. (i) Show for the functions $f(n) \equiv S(1/n, \ldots, 1/n)$ that $\lim_{n\to\infty} [f(n) - f(n-1)] = 0$.
   (ii) Conclude from (i) that the only solution of the equation $f(mn) = f(m) + f(n)$ is of the form $f(n) = C \ln n$, supposing that S is continuous according to (2.2.4(i)).
2. For $\alpha \neq 1$, show that the α-entropies S_α of (2.2.2) are not subadditive.
3. Prove Lemma (2.2.16).
4. Show that $S(\sum_i \lambda_i \rho_i) \leq \sum_i \lambda_i S(\rho_i) - \sum_i \lambda_i \ln \lambda_i$, $\lambda_i > 0$, $\sum_i \lambda_i = 1$.
5. Show that $S_{\text{cl}}(\rho_1) \leq S_{\text{cl}}(\rho)$ if $\mathscr{H} = \mathscr{H}_1 \otimes \mathscr{H}_2$, where $\mathscr{H}_i$ are one-particle Hilbert spaces, particles 1 and 2 are distinguishable, $\rho_1 = \text{Tr}_2\,\rho$, and $S_{\text{cl}}(\rho)$ is defined as in (2.2.9).
6. Calculate $S_\alpha(\exp(-\beta[H - F(\beta)])$, where $\exp(-\beta F(\beta)) = \text{Tr}\exp(-\beta H)$.
7. Show that $S(\sigma|\rho)$ is lower semicontinuous. Hint: use
   (i) $S(\sigma|\rho) = \sup_{0<\lambda<1} S_\lambda(\sigma|\rho)$, $S_\lambda(\sigma|\rho) \equiv (1/\lambda)(S(\lambda\rho + (1-\lambda)\sigma) - \lambda S(\rho) - (1-\lambda)S(\sigma)) \geq 0$;
   (ii) if $a \geq 0$ then $\text{Tr}\,a = \sup_n P_n a$, $P_n \to \mathbf{1}$, is an increasing sequence of finite-dimensional projections; and
   (iii) the operator inequalities (III: 2.2.38; 11),
   to show that the function $s(x) \equiv -x \ln x$ is concave for operators, i.e., $s(\lambda a + (1-\lambda)b) \geq \lambda s(a) + (1-\lambda)s(b)$ for all $a, b \in \mathscr{B}(\mathscr{H})$.
8. Prove the formula for the identity operator in (2.2.10; 5).

Solutions (2.2.28)

1. (i) Let $d_n = f(n) - f(n-1)$ and $\delta_n = S(1/n, 1-(1/n))$. Because S is continuous, $\delta_n \to 0$.

$$f(n-1) = d_{n-1} + \cdots + d_2 \Rightarrow d_n = \delta_n + \frac{d_{n-1} + \cdots + d_2}{n},$$

$$\sum_{n=2}^{N} (n\, d_n + d_{n-1} + \cdots + d_2) = \sum_{n=2}^{N} n\, \delta_n,$$

$$\sum_{n=2}^{N} d_n = \frac{1}{N} \sum_{n=2}^{N} n\, \delta_n \Rightarrow d_N = \delta_N - \frac{1}{N(N-1)} \sum_{n=2}^{N-1} n\, \delta_n,$$

$$|d_N - \delta_N| \le \frac{1}{N-1} \sup_n \delta_n + \sup_{n \ge \sqrt{N}} \delta_n \quad \text{for all } N \ge 2, \text{ which} \Rightarrow \lim d_N = 0,$$

because $\sup_n \delta_n < \infty$ and $\lim_{n\to\infty} \delta_n = 0$.

(ii) It suffices to show that $\lim_{n\to\infty} f(n)/\ln n = f(n_0)/\ln n_0$ (for any fixed $n_0 \ge 2$); because $f(n^k) = kf(n)$ this implies that $f(n) = (f(n_0)/\ln n_0) \ln n$ for all $n \ge 2$. Define $g(n) = f(n) - (f(n_0)/\ln n_0) \ln n$; then it suffices to show that $g(n)/\ln n \to 0$. Let $n = n_1 n_0 + l_1$, with $0 \le l_1 \le n_0$. Because $g(n_0) = 0$, $g(n) = \sum_{n_1 n_0}^{n-1} \varepsilon_k + g(n_1 n_0)$, where $\varepsilon_k \equiv g(k+1) - g(k)) = \sum_k \varepsilon_k + g(n_1)$. Now let $n_1 = n_2 n_0 + l_2$; then $g(n) = g(n_2) + \sum_{j=1}^{2} \sum_{i=0}^{l_j - 1} \varepsilon_{n_j + i}$, etc. After $k_0 < \ln n/\ln n_0$ steps, $n_{k_0} < n_0$, and therefore $g(n) = \sum_k \varepsilon_k$. The sum has fewer than $n_0 k_0$ summands, and therefore $\lim g(n)/\ln n = 0$. since $\varepsilon_k \to 0$.

2. Let $\mathscr{H} = \mathbb{C}^2 \otimes \mathbb{C}^2$ and $\rho = (\rho_{ik,jl})$, where $\rho_{ik,jl} = \delta_{ij}\delta_{kl} r_{ik}$; $r_{11} = pq + \varepsilon$, $r_{12} = p(1-q) - \varepsilon$, $r_{21} = (1-p)q - \varepsilon$, $r_{22} = (1-p)(1-q) + \varepsilon$ with $0 < p,\ q < 1$, $p, q \ne \frac{1}{2}$. Since ρ is diagonal, this allows $S_\alpha(\rho)$ to be read off with no further ado: If $\varepsilon = 0$, then $\rho = \rho_1 \otimes \rho_2$,

$$\rho_1 = \mathrm{Tr}_2\, \rho = \begin{pmatrix} p & 0 \\ 0 & 1-p \end{pmatrix}, \qquad \rho_2 = \mathrm{Tr}_1\, \rho = \begin{pmatrix} q & 0 \\ 0 & 1-q \end{pmatrix}.$$

If $S_\alpha(\rho)$ were $\le S_\alpha(\rho_1) + S_\alpha(\rho_2)$, then the function $g(\varepsilon) \equiv (pq + \varepsilon)^\alpha + (p(1-q) - \varepsilon)^\alpha + ((1-p)q - \varepsilon)^\alpha + ((1-p)(1-q) + \varepsilon)^\alpha$ would have an extremum at $\varepsilon = 0$, but $g'(0) \ne 0$ if $\alpha \ne 1$.

3. Let $|x\rangle \in \mathscr{H}_1 \otimes \mathscr{H}_2$. $|x\rangle = \sum_{i,k} c_{ik} |i\rangle_1 \otimes |k\rangle_2$, where $\{|i\rangle_1\}$ and $\{|k\rangle_2\}$ are orthonormal sets in $\mathscr{H}_1$ and $\mathscr{H}_2$ respectively, and $\rho = |x\rangle\langle x|$.

$$\mathrm{Tr}_2 |x\rangle\langle x| = \mathrm{Tr}_2 \sum_{ijkl} c_{ik} c_{jl}^* |i\rangle_{11}\langle j| \otimes |k\rangle_{22}\langle l|$$

$$= \sum_{ijkl} c_{ik} c_{jl}^* |i\rangle_{11}\langle j| \delta_{kl} = \sum_{ijk} c_{ik} c_{jk}^* |i\rangle_{11}\langle j|,$$

which implies that the positive eigenvalues of $\mathrm{Tr}_2\, |x\rangle\langle x|$ are the same as those of the matrix CC^*, where $C = (c_{ij})$. A similar argument shows that the positive eigenvalues of $\mathrm{Tr}_1\, |x\rangle\langle x|$ are the same as those of C^*C and thus of CC^*.

4. Let $\lambda_i \rho_i = a_i$; then the proposition is equivalent to $S(\sum_i a_i) \le \sum_i S(a_i)$ for all $a_i \in \mathscr{C}_1^+$. Since $\ln x$ is monotonic as an operator function (III: 2.2.38; 11), if $a_k \ge 0$,

then $\ln a_i \leq \ln(\sum_j a_j)$, which implies $a_i^{1/2}(\ln a_i)a_i^{1/2} \leq a_i^{1/2}(\ln \sum_j a_j)a_i^{1/2}$, and therefore $\sum_i \operatorname{Tr}(a_i \ln a_i) \leq \operatorname{Tr}[(\sum_i a_i) \ln(\sum_i a_i)]$.

5. $\rho_1(\mathbf{z}_1) \equiv \langle \mathbf{z}_1 | \rho_1 | \mathbf{z}_1 \rangle = \sum_i \langle \mathbf{z}_1 \otimes e_i | \rho | \mathbf{z}_1 \otimes e_i \rangle \geq \langle \mathbf{z}_1 \otimes \mathbf{z}_2 | \rho | \mathbf{z}_1 \otimes \mathbf{z}_2 \rangle \equiv \rho(\mathbf{z}_1, \mathbf{z}_2)$, since $\{e_i\}$ may be chosen to be an arbitrary basis. Therefore

$$S_{\text{cl}}(\rho) - S_{\text{cl}}(\rho_1) = \int d\Omega_z^2 \rho(\mathbf{z}_1, \mathbf{z}_2) \ln\left(\frac{\rho(\mathbf{z}_1)}{\rho(\mathbf{z}_1, \mathbf{z}_2)}\right) \geq 0.$$

6. $$S_\alpha = \frac{1}{1-\alpha} \ln \operatorname{Tr} \exp(-\alpha\beta(H - F(\beta))) = \frac{1}{\alpha - 1}[F(\alpha\beta) - \alpha F(\beta)].$$

As $\alpha \to 1$, $S \to \partial F(\beta)/\partial\beta$

7. (i) The function λS_λ is concave in λ, because

$$S(\alpha(\lambda_1\sigma + (1 - \lambda_1)\rho) + (1 - \alpha)(\lambda_2\sigma + (1 - \lambda_2)\rho))$$
$$\geq \alpha S(\lambda_1\sigma + (1 - \lambda_1)\rho) + (1 - \alpha)S(\lambda_2\sigma + (1 - \lambda_2)\rho),$$

so

$$S(\sigma|\rho) = \frac{d}{d\lambda} \lambda S_\lambda|_{\lambda=0} = \sup_{0<\lambda<1} S_\lambda.$$

(ii) This is the normality of the trace.

(iii) The operator concavity of $-x \ln x = \int_0^\infty (1 - \alpha/(x + \alpha) - x/(1 + \alpha))d\alpha$ is equivalent to the operator convexity of $1/(x + 1)$, and it suffices to show convexity with $\alpha = \frac{1}{2}$:

$$\frac{1}{(A+B)/2+1} \leq \frac{1}{2(A+1)} + \frac{1}{2(B+1)} \Leftrightarrow \frac{4}{A+1+B+1} \leq \frac{1}{A+1} + \frac{1}{B+1}$$

$$\Leftrightarrow \frac{4}{(B+1)^{-1/2}(A+1)(B+1)^{-1/2}+1}$$
$$\leq (B+1)^{1/2}(A+1)^{-1}(B+1)^{1/2} + 1.$$

Since $4/(x + 1) \leq (1/x) + 1$ for all $x \in \mathbb{R}^+$, this is also valid for positive operators. Therefore, $(1/\lambda)[s(\lambda\sigma + (1 - \lambda)\rho) - \lambda s(\sigma) - (1 - \lambda)s(\rho)] \geq 0$, which implies $S(\sigma|\rho) = \sup_n \sup_{0<\lambda<1} (1/\lambda) \operatorname{Tr} P_n[s(\lambda\sigma + (1 - \lambda)\rho) - \lambda s(\sigma) - (1 - \lambda)s(\rho)]$, and $s(\rho)$ is continuous in finite dimensions. This also provides a new proof of the lower semicontinuity of $S(\rho)$.

8. The right side of the equation clearly leaves the number of particles invariant. Hence the formula is shown by

$$\langle f_1, \ldots, f_N | \int \frac{dz_1 \cdots dz_N}{N!(2\pi)^{dN}} |\mathbf{z}_1, \ldots, \mathbf{z}_N\rangle\langle\mathbf{z}_1, \ldots, \mathbf{z}_N | g_1, \ldots, g_N\rangle$$

$$= \sum_{P,Q} (\pm 1)^{P+Q} \int \frac{dz_1 \cdots dz_N}{N!(2\pi)^{dN}} \prod_{i=1}^{N} \langle f_{P_i} | \mathbf{z}_i \rangle \langle \mathbf{z}_i | g_{Q_i} \rangle$$

$$= \sum_{P,Q} (\pm 1)^{P+Q} \prod_i \langle f_{P_i} | g_{Q_i} \rangle \frac{1}{N!}$$

$$= \sum_{P'} (\pm 1)^{P'} \prod_i \langle f_i | g_{P'_i} \rangle = \langle f_1, \ldots, f_N | g_1, \ldots, g_N \rangle.$$

2.3 The Microcanonical Ensemble

Insight into the fundamental thermodynamic laws is gained by investigating the chaotic state below the energy surface.

Two trains of thought are usually followed to justify regarding the equilibrium state as predominating for macroscopic systems. Like Boltzmann, one can investigate the time-evolution of a system and show that most states tend to equilibrium. Alternatively, one can follow Gibbs and examine an ensemble of identical copies of the system and identify states of scanty information with equilibrium states. The set of problems connected with the first procedure is the subject of the next chapter, while in this section we shall study systems for which the only information concerns the energy. If it is known that the energy does not exceed some maximum value E_m, then, as remarked in (2.1.10; 2), the most mixed state containing no further information corresponds to the

Microcanonical Density Matrix (2.3.1)

$$\rho = \Theta(E_m - H)/\mathrm{Tr}\,\Theta(E_m - H), \qquad \Theta(x) = \begin{cases} 1 & \text{for } x \geq 0 \\ 0 & \text{for } x < 0 \end{cases},$$

where $E_m \geq \varepsilon_1 \equiv$ the lowest eigenvalue of H. Its

Entropy and Average Energy (2.3.2)

Are

$$S = \ln \mathrm{Tr}\,\Theta(E_m - H), \qquad E = \exp(-S)\mathrm{Tr}\,H\Theta(E_m - H).$$

Remarks (2.3.3)

1. The discontinuous function Θ of a self-adjoint operator is defined with the spectral representation of the operator.
2. It is assumed that H is bounded below and that $\sigma_{\mathrm{ess}}(H)$ is empty, so the traces in (2.3.2) are finite.
3. The entropy S is a discontinuous function of E_m, and has no well-defined inverse. On the other hand, E may be construed as a function of S, as shown in Figure 3. The function $E(S)$ increases monotonically.
4. By the min–max principle, $E(S)$ is also given by $E(S) = \exp(-S)\inf_{\mathscr{H}_n} \mathrm{Tr}_{\mathscr{H}_n} H$, where $\mathscr{H}_n$ is an n-dimensional subspace of $D(H)$ and $n = \exp(S)$. It is consequently a concave function of all parameters on which the dependence of H is concave.

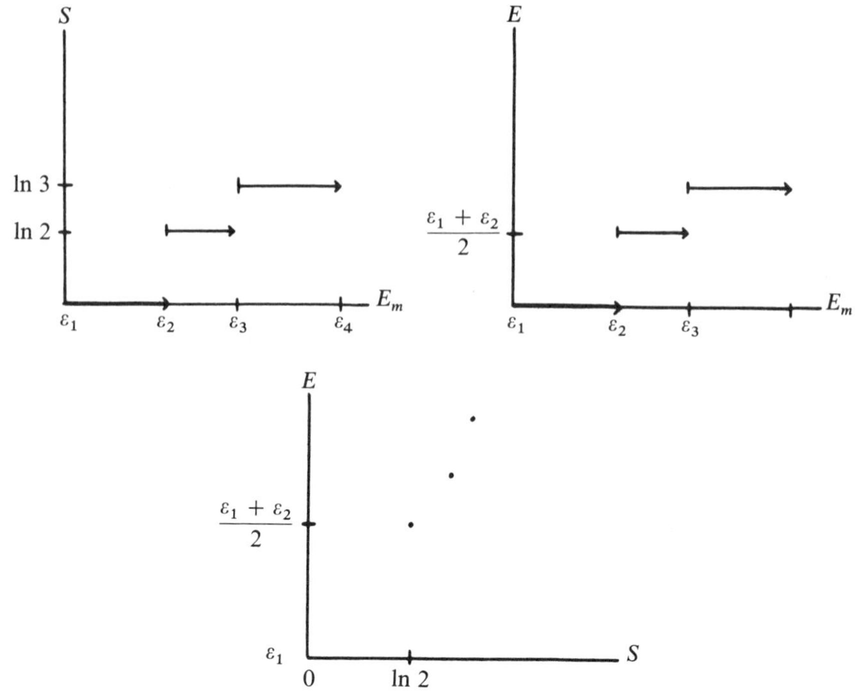

Figure 3 The thermodynamic functions for a finite system.

5. By Property (2.2.3(iv)), all α-entropies S_α lead to the same S (2.3.2), which can be identified as the entropy of phenonemological thermodynamics.
6. It will be seen shortly that in the systems under consideration here the density of states increases so rapidly with the energy that in the limit of an infinite system, any density matrix $\rho \sim \Theta(E - H) - \Theta(E(1 - \varepsilon) - H)$ yields the same entropy density for all $\varepsilon > 0$.

The further properties of $E(S)$ follow from the special form of the Hamiltonian,

$$H_N = \sum_{i=1}^{N} \frac{|\mathbf{p}_i|^2}{2m_i} + \sum_{i>j} v(\mathbf{x}_i - \mathbf{x}_j),$$

where v is assumed bounded relative to the kinetic energy. It will be most convenient to deal with the quadratic form associated with H_N (cf. (III: 2.5.18; 2)). The quadratic-form domain $Q(H_N)$ consists of functions ψ such that $\sum_i (1/2m_i) \int |\nabla_i \psi|^2 < \infty$ and with some other restrictions from the boundary conditions. The formula of Remark (2.3.3; 4) then holds with $\mathscr{H}_n \subset Q(H_N)$. The boundary conditions we shall choose are Dirichlet conditions on the surface of a volume $V \subset \mathbb{R}^3$, which mean specifically that: $\mathscr{H} \subset L^2(V^N)$

and $\psi|_{\partial(V^N)} = 0$. The Hilbert space $\mathscr{H}$ is $L^2(V^N)$ if the particles are distinguishable, and if they are identical bosons or fermions, then $\mathscr{H}$ must be restricted to functions of the appropriate symmetry. The energy can be treated as a function of S, V, and N, and its dependence on V is described by the following theorem.

Monotony of the Energy (2.3.4)

If $V' \supset V$, *then*

$$E(S, V', N) \leq E(S, V, N).$$

Proof

This follows from (2.3.3; 4) because $Q(H(V')) \supset Q(H(V))$, where $\supset$ is intended in the sense of the natural imbedding, i.e., functions ψ such that $\psi|_{\partial V} = 0$ are set to 0 in $V' \setminus V$. □

Subadditivity generalizes this monotony when particles in separated volumes do not repel one another.

Subadditivity of the Energy (2.3.5)

If $V_1 \cap V_2 = \emptyset$ *and* $v(\mathbf{x}_i - \mathbf{x}_j) \leq 0$ *for all* $\mathbf{x}_i \in V_1$, $\mathbf{x}_j \in V_2$, *then*

$$E(S_1 + S_2, V_1 \cup V_2, N_1 + N_2) \leq E_1(S_1, V_1, N_1) + E(S_2, V_2, N_2).$$

Proof

This again follows from (2.3.3; 4), since the right side results from taking the infimum over a subspace of $Q(H)$, which consists of tensor products of $\exp(S_1)$ vectors, for which N_1 particles lie within the volume V_1, with $\exp(S_2)$ vectors having N_2 particles within V_2. The tensor products have to be symmetrized or antisymmetrized if there are Bose or Fermi statistics. However, since symmetrization does not affect the expectation values of (2.3.5) when the functions have disjoint supports, (2.3.5) is independent of the statistics. □

The existence of $\lim_{V \to \infty} E/V$ can be derived from the subadditivity, though it is rather difficult to go beyond the restriction $v \leq 0$. This problem will have to be investigated later for each of the systems discussed in §1.2, and for now convergence will simply be assumed. The condition is satisfied trivially for free particles ($v = 0$). To draw conclusions like those of (2.2.15), assume that V is a cube, the volume of which will also be fearlessly denoted

$V \in \mathbb{R}^+$. If eight cubes are packed together as a single cube of double the side, then (2.3.5) implies

$$E(8S, 8V, 8N) \leq 8E(S, V, N). \tag{2.3.6}$$

Assuming in addition that there exists $A \in \mathbb{R}^+$ such that

$$H_N \geq -AN \quad \text{for all } N \in \mathbb{Z}^+, \tag{2.3.7}$$

the limit

$$\lim_{\mathbb{Z}^+ \ni \nu \to \infty} 8^{-\nu} E(8^\nu S, 8^\nu V, 8^\nu N) = \inf_\nu 8^{-\nu} E(8^\nu S, 8^\nu V, 8^\nu N)$$

exists. This allows the passage to an infinite system, for which the energy, entropy, and particle densities are defined by $E/V = \varepsilon, S/V = \sigma$, and $N/V = \rho$.

The Thermodynamic Limit of the Energy Density (2.3.8)

$$\varepsilon(\sigma, \rho) = \inf_{\mathbb{Z}^+ \ni \nu} 8^{-\nu} \rho E(8^\nu \sigma \rho^{-1}, 8^\nu \rho^{-1}, 8^\nu).$$

Remarks (2.3.9)

1. Equation (2.3.7) guarantees that $\varepsilon > -\infty$, so the infimum always exists; but (2.3.8) is only of interest when there is a well-defined limit, for only then is it certain that the thermodynamic properties do not depend on the exact number of particles. Even if the limit exists, as in the case of (2.3.6), it does not guarantee that the resulting ε is nontrivial. If, say, the particles can be distinguished (which does not invalidate the general conclusions), then classically,

$$\exp(S) = \int_{V^N} d^{3N}x \int d^{3N}p\Theta\left(E_m - \sum_{i=1}^N |\mathbf{p}_i|^2\right) = \pi^{3N/2} \frac{E_m^{3N/2} V^N}{(3N/2)!},$$

and

$$E = \frac{E_m}{1 + 2/3N}.$$

Therefore, as $N \to \infty$,

$$\frac{E}{V} = \frac{3}{2\pi e} \frac{\rho^{5/3}}{N^{2/3}} \exp(\tfrac{2}{3}\sigma\rho^{-1}) \to 0.$$

The familiar result obtains only with the replacement $\exp(S) \to (1/N!) \exp(S)$ to account for the particles being identical. A later calculation of $\varepsilon(\sigma, \rho)$ will reveal that (2.3.8) is then not without content.
2. Though the result has been derived only for cubes, the limit clearly exists for other shapes if they are not too different from cubes.

3. The effect of dilatations on the kinetic energy (cf. (III: 3.3.21; 8) and (III: 4.1.4)) of free particles implies, moreover, that

$$E(S, V, N) = \exp(2\tau)E(S, \exp(-3\tau)V, N).$$

Hence the one-parameter family of limits

$$\lim_{\nu \to \infty} 8^{-\nu(1-2\tau)} E(8^\nu S, 8^{\nu(1-3\tau)} V, 8^\nu N)$$

exist (cf. (1.2.1)). Ordinarily, the limit is taken with $\tau = 0$, and quantities proportional to N, like E, S, and V, are described as extensive, while N-independent quantities like ε, ρ, and σ are called intensive. The existence of some limit is important, for, whatever it may be like, it enables precise propositions to be formulated. In reality systems are large but still finite, but if a quantity converges as $N \to \infty$ the limit may be expected to be attained for practical purposes when, say, $N = 10^{24}$. Indeed, it will be shown in realistic situations that the limit is sometimes attained to $O(N^{-1/6})$, which is sufficient accuracy for macroscopic bodies. There are various ways to interpret the limit $N \to \infty$. As has been done here, the system may be thought of as becoming larger and larger, or, alternatively, the atoms may be imagined smaller and smaller with their number in the fixed volume of the container being increased at the same time.

Since monotony and convexity survive pointwise limits, there are the following

Properties of the Energy Density (2.3.10)

For the function $\mathbb{R}^+ \times \mathbb{R}^+ \to \mathbb{R}^+ : \sigma, \rho \to \varepsilon(\sigma, \rho)$,

(i) *ε increases monotonically in σ*:
(ii) *$\rho^{-1}\varepsilon(\alpha\rho, \rho)$ increases monotonically in ρ*;
(iii) *ε is convex in (σ, ρ)*;
(iv) *moreover, for free particles, $\varepsilon(\sigma, \rho) = \rho^{5/3} f(\sigma/\rho)$.*

Proof

Property (i) holds as remarked in (2.3.3; 3), and Property (ii) follows from Theorem (2.3.4). From subadditivity (2.3.5),

$$\varepsilon(\tfrac{1}{2}(\sigma_1 + \sigma_2), \tfrac{1}{2}(\rho_1 + \rho_2)) \leq \tfrac{1}{2}(\varepsilon(\sigma_1, \rho_1) + \varepsilon(\sigma_2, \rho_2)),$$

which implies (iii), and (iv) follows from (2.3.9; 3). □

Remarks (2.3.11)

1. Since $N \in \mathbb{Z}^+$, $S \in \ln \mathbb{Z}^+$, ε is at first defined only on the dense set for which $\sigma\rho^{-1}$ is a power of $(\ln z)/2$, $z \in \mathbb{Z}^+$. It extends continuously to $\mathbb{R}$, because monotony and concavity with the coefficient $\frac{1}{2}$ imply uniform

continuity. There are discontinuous functions that are concave with coefficient $\frac{1}{2}$, such as

$$f(x) = \begin{cases} x, & x \text{ rational}, \\ 0, & \text{otherwise}, \end{cases}$$

for which the equation $f(\alpha x) = \alpha f(x)$ holds for all rational α. However, this can not occur if the function is monotonic. The extension then in addition satisfies the inequality

$$\varepsilon(\alpha\sigma_1 + (1 - \alpha)\sigma_2, \alpha\rho_1 + (1 - \alpha)\rho_2) \leq \alpha\varepsilon(\sigma_1, \rho_1) + (1 - \alpha)\varepsilon(\sigma_2, \rho_2)$$
$$\text{for all } \alpha \in \mathbb{R}, 0 \leq \alpha \leq 1.$$

2. Subadditivity (2.3.5) is sufficient but not necessary for Property (iii); (2.3.5) may be violated if the interaction is partially repulsive, which is a necessary assumption or $H_N \geq -AN$ when the particles interact. However, if the potential goes to zero rapidly enough at infinity, the correction to (2.3.5) on any finite region is a surface effect, so the convexity of the energy density is still guaranteed in the thermodynamic limit. On the other hand, the special form (2.3.8) is crucial, and in §4.2 it will be seen that convexity (2.3.10(iii)) is violated in gravitating systems, although (2.3.5) is valid.
3. Since the limiting function is continuous, Dini's theorem ensures that the monotonic limit (2.3.8) is uniform on compact sets.
4. Let H be defined so that $\inf \varepsilon = 0$. Since ε is convex in σ, unless $\varepsilon \equiv 0$, there exists a σ_0 such that ε is strictly monotonic in σ for all $\sigma > \sigma_0$. There is consequently an inverse function $\sigma(\varepsilon, \rho)$ (see Figure 4), which is concave and monotonically increasing in ε.
5. As long as σ is strictly monotonically increasing in ε, the density matrices

$$\rho = \Theta(E_m - H)\exp(-S)$$

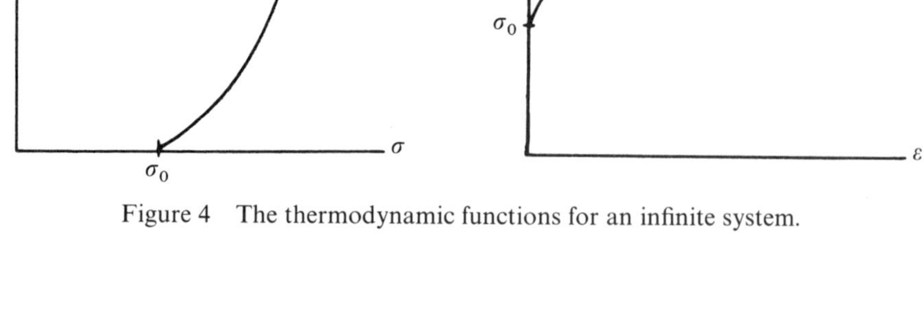

Figure 4 The thermodynamic functions for an infinite system.

and

$$\rho_\delta = (\Theta(E_m - H) - \Theta(E_m - V\delta - H))\exp(-S_\delta)$$

yield the same entropy densities in the limit $N \to \infty$:

$$\sigma_\delta = \lim_{V\to\infty} \frac{1}{V} \ln \operatorname{Tr}(\Theta(E_m - H) - \Theta(E_m - V\delta - H))$$

$$= \sigma(\varepsilon, \rho) + \lim_{V\to\infty} \frac{1}{V} \ln(1 - \exp[-V(\sigma(\varepsilon, \rho) - \sigma(\varepsilon - \delta, \rho))]) = \sigma.$$

This means that as $N \to \infty$ most of the states crowd just under the energy surface with arbitrarily high density.

6. For some systems $\sigma(\varepsilon)$ is constant for ε greater than some ε_1, in which case ρ and ρ_δ may have different entropies. Consider for example N spins in an external field ((1.1.3) with $\varepsilon = 0$). The density of states $(\partial/\partial E)\exp[S(E)]$ is invariant under $\sigma \to -\sigma$ and thus an even function in E. This makes $\operatorname{Tr} \rho_\delta$ a decreasing function of E_m when $E_m + \delta > 0$, which is impossible for $\operatorname{Tr} \rho$ (see Figure 5); Definition (2.3.1) rules negative temperatures out.
7. The number of energy levels below E_m is $\exp(N\sigma/\rho)$, which is immense for macroscopic bodies, $N \sim 10^{24}$. It would never be possible to isolate the energy levels completely—their widths are on the order of (macroscopic time)$^{-1}$, which is much larger than their spacing. Systems will later be idealized as infinite, having continuous energy spectra, which comes closer to reality than does the fiction of a discrete spectrum.

After this first exposure to these ideas, let us consider two systems the interaction between which is so weak that it can be neglected in comparison with other energies. They are to be considered as parts of a larger system with $\mathscr{H} = \mathscr{H}_1 \otimes \mathscr{H}_2$, $H = H_1 + H_2$. The question is how the energy and entropy are shared by the two subsystems. Even though H is a sum, the microcanonical density matrix (2.3.1) is not in the form of a product $\rho = \rho_1 \otimes \rho_2$, and we will have to see how the entropy of this state can nonetheless be additive for independent, macroscopic systems. Assume to

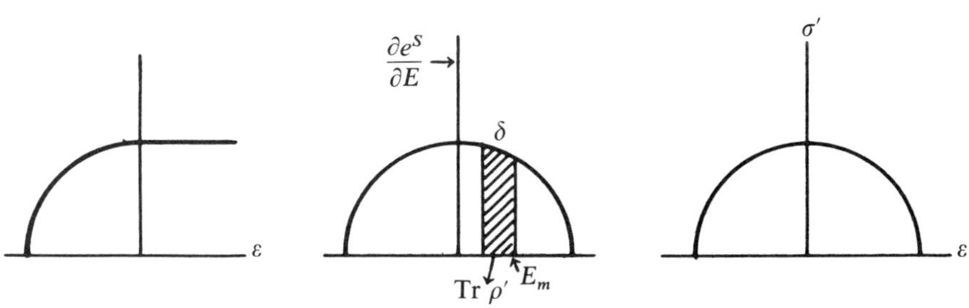

Figure 5 Inequivalence of the microcanonical ensembles for spins in a magnetic field.

this end that the systems are large and that the sequence (2.3.8) converges and has all the necessary kinds of continuity so that $\varepsilon = E/V$ can be regarded as a continuous variable for the purposes of integration and differentiation. For the problem at hand and other estimates we shall need

Lemma (2.3.12)

Let $\sigma(\varepsilon) \le 0$ and be concave on $[0, 1]$, and $\sigma(1) = 0$, $-\infty < \sigma(0) < 0$; this implies that σ is nondecreasing and that there exists an ε_0, $0 < \varepsilon_0 \le 1$, such that $\sigma' \equiv \sigma'(\varepsilon_0) > 0$. Then

$$\frac{1 - \exp(-V|\sigma(0)|)}{V|\sigma(0)|} \le \int_0^1 d\varepsilon \exp(V\sigma(\varepsilon)) \le 1 - \varepsilon_0 + \frac{1 - \exp(-V\varepsilon_0\sigma')}{V\sigma'}.$$

Proof.

By assumption (see Figure 6),

$$(1 - \varepsilon)\sigma(0) \le \sigma(\varepsilon) \le \begin{cases} 0 & \text{for } \varepsilon_0 \le \varepsilon \le 1 \\ -(\varepsilon_0 - \varepsilon)\sigma' & \text{for } 0 \le \varepsilon \le \varepsilon_0. \end{cases}$$ □

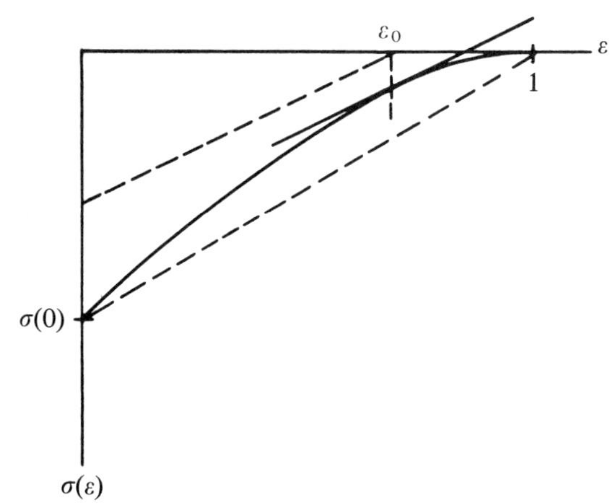

Figure 6 Bounds for the concave function $\sigma(\varepsilon)$.

Corollaries (2.3.13)

1. If σ is concave but not necessarily negative, then the formula

$$\int_a^b V\,d\varepsilon \exp(V\sigma(\varepsilon)) = \exp(V\bar{\sigma}) \int_a^b d\varepsilon V \exp(V(\sigma(\varepsilon) - \bar{\sigma})) \text{ with } \bar{\sigma} = \max_{a \le \varepsilon \le b} \sigma(\varepsilon)$$

can be used instead, since $-\infty < \bar{\sigma} < \infty$ unless $\sigma \equiv \pm\infty$. By an application of the lemma, possibly after subdivision of the region of integration,

$$\lim_{V\to\infty} \frac{1}{V} \ln \int_a^b d\varepsilon V \exp(V(\sigma(\varepsilon) - \bar{\sigma})) = 0.$$

Thus only the maximum value of σ contributes in the infinite limit:

$$\lim_{V\to\infty} \frac{1}{V} \ln \int_a^b V d\varepsilon \exp(V\sigma(\varepsilon)) = \sup_{a\le\varepsilon\le b} \sigma(\varepsilon) = \bar{\sigma}(\varepsilon_1) = \sup_{a\le\varepsilon\le b} \bar{\sigma}(\varepsilon).$$

2. Remark (2.3.11; 5) leads one to expect that E_m and E may become equal for large systems. More precisely, if σ is concave in ε, $d\sigma/d\varepsilon > 0$, $\lim_{V\to\infty} (E - E_m)/V = 0$. This follows because E may be written as

$$E = \exp(-S)\mathrm{Tr}\, H\Theta(E_m - H) = \int_0^{E_m} dE'E' \frac{\partial}{\partial E'} \mathrm{Tr}\, \Theta(E' - H)\exp(-S)$$
$$= E_m - \int_0^{E_m} dE'\, \mathrm{Tr}\, \Theta(E' - H)\exp(-S).$$

With $\varepsilon_0 = 1$ and $E' = \varepsilon V$ the lemma now implies that the last integral is $O(1)$, whereas $E_m \sim V$.

3. We next calculate $\exp(S(E)) = \mathrm{Tr}\, \Theta(E - H_1 - H_2)$, $H_i \ge 0$, as $V = V_1 + V_2 \to \infty$ with V_i/V fixed. Because of the assumption of subadditivity,

$$\sigma_{1,V_1}(\varepsilon) \equiv \frac{1}{V_1} \ln \mathrm{Tr}_1\, \Theta(V_1\varepsilon - H_1)$$

is concave in ε and increases monotonically to $\sigma_1(\varepsilon)$. Let $E_2[n]$ denote the ordered sequence of eigenvalues of H_2. If the entropies are considered as functions of the maximum energy, which leads to the same function in the limit $V \to \infty$ because of Corollary 2, then n may be identified with $\exp S$, and $E_2(S_2) \equiv E_2[\exp(S_2)]$ becomes the function introduced in (2.3.3; 3). With $E = \varepsilon V$,

$$\sigma(\varepsilon) = \lim_{V\to\infty} \frac{1}{V} \ln \mathrm{Tr}\, \Theta(E - H_1 - H_2)$$
$$= \lim_{V\to\infty} \frac{1}{V} \ln \sum_{n=1}^{\exp(S_2(E))} \exp(S_1(E - E_2[n])).$$

Now regard n as a continuous variable, and interpolate $E_2[n]$ linearly. Since the integrand decreases monotonically, the sum $\sum_{n=1}^{\exp(S_2(E))} \cdots$ lies between $\int_0^{\exp(S_2(E))} dn \cdots$ and $\int_1^{\exp(S_2(E))+1} dn \cdots$, and the evaluation of the error is unnecessary, since $\exp(S_2(E)) \sim \exp(10^{23})$. With the variables $\sigma_2 = (1/V_2)\ln n$, $\sigma(\varepsilon)$ can be written as

$$\lim_{V\to\infty} \frac{1}{V} \ln \int_0^{\sigma_2(\varepsilon)} V_2\, d\sigma_2 \exp\left[V_1\sigma_{1,V_1}\left(\frac{V}{V_1}\varepsilon - \frac{V_2}{V_1}\varepsilon_{2,V_2}(\sigma_2)\right) + V_2\sigma_2 \right].$$

Now note that $\sigma_2 \to a - b\varepsilon_2(\sigma_2)$ is concave if $b \geq 0$, σ_{1,V_1} is concave and increasing, and that (concave, increasing) $\circ$ concave $=$ concave. This allows the lemma to be applied, to show

$$\sigma(\varepsilon) = \lim_{V\to\infty} \sup_{0\leq\sigma_2\leq\sigma_2(\varepsilon)} \left[\frac{V_1}{V}\sigma_{1,V_1}\left(\frac{V}{V_1}\varepsilon - \frac{V_2}{V_1}\varepsilon_{2,V_2}(\sigma_2)\right) + \frac{V_2}{V}\sigma_2\right]$$

$$= \sup_{0\leq\sigma_2\leq\sigma_2(\varepsilon)} \left[\frac{V_1}{V}\sigma_1\left(\frac{V}{V_1}\varepsilon - \frac{V_2}{V_1}\varepsilon_2(\sigma_2)\right) + \frac{V_2}{V}\sigma_2\right].$$

The interchange of the limit $V \to \infty$ and the supremum is justified because $\varepsilon_{2,V_2}(\sigma_2)$ increases monotonically in σ_2 for all V_2, and since $\sigma_{1,V_1}(\varepsilon)$ likewise increases in ε, it decreases in σ_2, and consequently the first term in the brackets [ ] converges uniformly on compact sets to

$$\sigma_2 \to \sigma_1\left(\frac{V}{V_1}\varepsilon - \frac{V_2}{V_1}\varepsilon_2(\sigma_2)\right).$$

Although the concavity of σ is preserved in the limit $V \to \infty$, strict concavity, which is needed to guarantee that the maximum is attained at only one point, may break down. A lack of strict concavity means that there is a phase transition, and will be examined in detail later. If, however, $\sigma_i(\varepsilon_i)$ are strictly concave and continuously differentiable, then the result of Corollary 3 can be improved upon and the additivity of the entropies demonstrated.

Equilibrium Condition (2.3.14)

Let $\sigma_i(\varepsilon_i) = \lim_{V_i\to\infty}(1/V_i)\ln \operatorname{Tr}\Theta(V_i\varepsilon_i - H_i)$ *be strictly concave and continuously differentiable,* $\lim_{\varepsilon\to 0}\sigma'(\varepsilon) = \infty$ *and* $\lim_{V\to\infty} V_i/V \equiv \alpha_i, \alpha_1 + \alpha_2 = 1$. *Then*

$$\lim_{V\to\infty}\frac{1}{V}\ln \operatorname{Tr}\Theta(V\varepsilon - H_1 - H_2) \equiv \sigma(\varepsilon) = \alpha_1\sigma_1(\varepsilon_1) + \alpha_2\sigma_2(\varepsilon_2),$$

where ε_i *are determined uniquely by*

$$\alpha_1\varepsilon_1 + \alpha_2\varepsilon_2 = \varepsilon, \qquad \frac{\partial}{\partial\varepsilon_1}\sigma_1(\varepsilon_1) = \frac{\partial}{\partial\varepsilon_2}\sigma_2(\varepsilon_2).$$

Remarks (2.3.15)

1. The energy densities can equally well be regarded as functions of the entropy densities, which reformulates the equilibrium condition as

$$\frac{\partial}{\partial\sigma_1}\varepsilon_1(\sigma_1) = \frac{\partial}{\partial\sigma_2}\varepsilon_2(\sigma_2) \quad \text{and} \quad \alpha_1\sigma_1 + \alpha_2\sigma_2 = \sigma.$$

2. Convexity of $\varepsilon(\sigma)$ is equivalent to concavity of $\sigma(\varepsilon)$, which is equivalent to the number of states below E_m not increasing faster than exponentially with the energy. This is not a general property of quantum-mechanical systems, and has to be checked in individual cases. A simple counter-example is the hydrogen atom, for which $E_n \sim -1/n^2$, $\exp(S(E_n)) \sim n^3$, where n is the principal quantum number, and therefore

$$E \sim -\exp(-\tfrac{2}{3}S), \qquad \frac{\partial E}{\partial S} \sim \tfrac{2}{3}\exp(-\tfrac{2}{3}S) > 0, \qquad \frac{\partial^2 E}{\partial S^2} \sim -\tfrac{4}{9}\exp(-\tfrac{2}{3}S) < 0.$$

In such cases there may be many solutions of the equilibrium condition (see Figure 7).
3. Condition (2.3.14) implies that the energy is apportioned between the two systems so as to maximize the total entropy. From the point of view of $\varepsilon(\sigma)$ this means distributing entropy so as to minimize the total energy. As a consequence, the subadditivity inequality (2.3.5) becomes an equality in the limit $V \to \infty$.
4. If $\varepsilon_i(\sigma) \in \mathbb{C}^2$, then at the minimum, $\varepsilon_1''/\alpha_1 + \varepsilon_2''/\alpha_2 \geq 0$, where $\varepsilon'' = \partial^2\varepsilon/\partial\sigma^2$. Then by Problem 4, at the minimum, $1/\varepsilon'' = \alpha_1/\varepsilon_1'' + \alpha_2/\varepsilon_2''$.

If the total system consists of a system immersed in a thermal reservoir, then the system of interest is not affected by the fine details of the reservoir, but only by $\partial\sigma_2/\partial\varepsilon_2$, which not only determines $\partial\sigma_1/\partial\varepsilon_1$, but also equals $\partial\sigma/\partial\varepsilon$, because

$$\frac{d}{d\varepsilon}\left(\alpha_1\sigma_1(\varepsilon_1(\varepsilon)) + \alpha_2\sigma_2\left(\frac{\varepsilon}{\alpha_2} - \frac{\alpha_1}{\alpha_2}\,\varepsilon_1(\varepsilon)\right)\right)$$
$$= \sigma_2'(\varepsilon_2(\varepsilon)) + \alpha_1\frac{d\varepsilon_1}{d\varepsilon}(\sigma_1'(\varepsilon_1(\varepsilon)) - \sigma_2'(\varepsilon_2(\varepsilon))),$$

where

$$\varepsilon_2(\varepsilon) \equiv \frac{\varepsilon}{\alpha_2} - \frac{\alpha_1}{\alpha_2}\,\varepsilon_1(\varepsilon),$$

and the latter term vanishes because of (2.3.14). This is the justification for

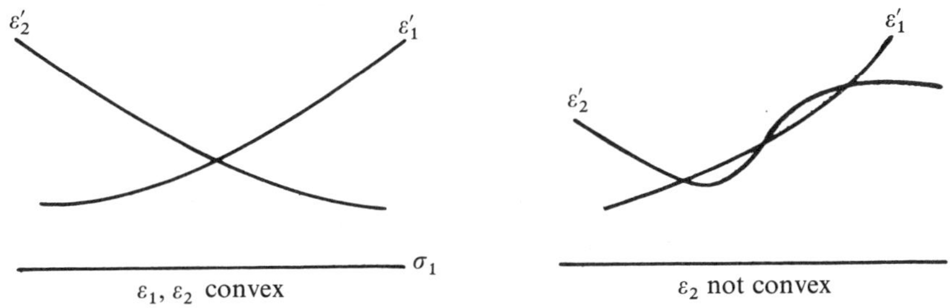

Figure 7 Uniqueness of the equilibrium temperature.

Definition (2.3.16)

The **temperature** is

$$T = \frac{\partial \varepsilon}{\partial \sigma}.$$

Remarks (2.3.17)

1. The temperature has the dimension of energy in units where Boltzmann's constant k is set to 1.
2. The temperature is always positive with the microcanonical ρ (2.3.1), but ρ_δ gives the spin system of (2.3.11; 6) a negative temperature at $E > 0$.
3. The concavity of σ means that the specific heat at constant volume,

$$V^{-1}C_N \equiv c_V = \frac{d\varepsilon}{dT} = \frac{d\varepsilon}{d\sigma}\left(\frac{dT}{d\sigma}\right)^{-1} = \frac{T}{d^2\varepsilon/d\sigma^2}$$

is positive. In particular, by Remark (2.3.15; 4), the heat capacity (at constant volume) $C_V = V \cdot 1/\varepsilon''$ of the total system is the sum of the heat capacities $V_i \cdot 1/\varepsilon_i''$ of the subsystems. The condition of stability $\varepsilon_1''/\alpha_1 + \varepsilon_2''/\alpha_2 \geq 0$ implies that two systems of negative specific heat can not coexist in equilibrium. Heat transferred from the hotter system to the colder one would make the hot one hotter and the cold one colder. Large temperature fluctuations would arise, making the situation unstable. If only subsystem 1 has negative specific heat, while that of subsystem 2 is positive, then the heat capacities must satisfy $|C_1| > C_2$: The transfer of heat from 1 to 2 would warm subsystem 1 less than 2, so 2 would immediately cool off by transferring heat back to 1, making the temperature equilibrium between the subsystems stable. This means that the temperature of a system of negative specific heat should be taken with a small thermometer, and never with a large thermal reservoir.

Now allow the wall between the subsystems to be slowly moveable. The energy as a function of V acts as a potential energy for the wall, just as the electron energy acted as the potential for the atomic nuclei in the Born–Oppenheimer approximation in volume III. Stable equilibrium occurs when the total volume V is apportioned so as to minimize the energy. Let $V_2 = V - V_1$, and look for

$$E(S, V, N_1 + N_2) = \inf_{\substack{0 \leq S_1 \leq S \\ 0 \leq V_1 \leq V}} (E_1(S_1, V_1, N_1) + E_2(S - S_1, V - V_1, N_2)). \tag{2.3.18}$$

In the cases of interest here, E depends differentiably on V even for finite systems, and $E \to \infty$ if $V \to 0$. Hence the infimum is attained within the interval $0 < V_1 < V$, and is determined by the

Equilibrium Condition (2.3.19)

For E of (2.3.18), the equilibrium volume V_1 satisfies

$$\frac{\partial E_1}{\partial V_1} = \left.\frac{\partial E_2}{\partial V_2}\right|_{V_2 = V - V_1}.$$

Remarks (2.3.20)

1. Because the energy is monotonic (2.3.4), with the boundary conditions $\psi_{|\partial V} = 0$, it follows that $\partial E/\partial V < 0$, and so (2.3.19) definitely has a solution V_1. At that minimum,

$$\frac{\partial E}{\partial V} = \frac{\partial E_1}{\partial V_1} = \left.\frac{\partial E_2}{\partial V_2}\right|_{V_2 = V - V_1},$$

   and

$$\frac{\partial^2 E_1}{\partial V_1^2} + \frac{\partial^2 E_2}{\partial V_2^2} \geq 0, \qquad \left(\frac{\partial^2 E}{\partial V^2}\right)^{-1} = \left(\frac{\partial^2 E_1}{\partial V_1^2}\right)^{-1} + \left(\frac{\partial^2 E_2}{\partial V_2^2}\right)^{-1}.$$

2. With other boundary conditions it may not be true that $\partial E/\partial V < 0$. For example, if a hydrogen atom is confined to a sphere on the surface of which $d\psi_{|\partial V} = 0$, then $E = E_\infty - \alpha V^{-1/3}$, so $\partial E/\partial V > 0$. This kind of boundary condition can be approximately realized physically with a very strong δ' potential. The lesson of this is that it is necessary to verify the hope that in infinite systems the pressure (see (2.3.21)) satisfies $P \equiv -\partial E/\partial V \geq 0$. It is not guaranteed that $\partial^2 E/\partial V^2 \geq 0$ even with the boundary condition $\psi_{|\partial V} = 0$, which makes the proof of the convexity of $\varepsilon(\sigma, \rho)$ all the more important for real matter.
3. Since $\partial E/\partial V|_S = -\partial E/\partial S|_V \, \partial S/\partial V|_E$, another interpretation of (2.3.19) is that the condition $\partial(S_1(E_1, V_1) + S_2(E_2, V - V_1))/\partial V_1 = 0$ determines V_1; that is, the volumes arrange themselves to maximize the total entropy.

   Analogous to (2.3.16) is

Definition (2.3.21)

The **pressure** is $P \equiv -\partial E/\partial V$. In the limit $V \to \infty$ it becomes

$$P = -\varepsilon + \rho\frac{\partial \varepsilon}{\partial \rho} + \sigma\frac{\partial \varepsilon}{\partial \sigma} = T\left(\sigma - \varepsilon\frac{\partial \sigma}{\partial \varepsilon} - \rho\frac{\partial \sigma}{\partial \rho}\right).$$

Remarks (2.3.22)

1. For realistic systems it can be shown how the pressure defined in (2.3.21) arises from the forces exerted by the system on the wall [9].
2. The equilibrium condition states that the pressures of the two subsystems are equal, with the same value as the total system has.

3. Remark (2.3.20; 1) implies for the compressibility

$$\kappa = -\left[V\frac{\partial P}{\partial V}\right]^{-1} \xrightarrow{V\to\infty} \left[\rho^2\frac{\partial^2\varepsilon}{\partial\rho^2} + 2\rho\sigma\frac{\partial^2\varepsilon}{\partial\rho\partial\sigma} + \sigma^2\frac{\partial^2\varepsilon}{\partial\sigma^2}\right]^{-1}$$

that

$$\kappa = \frac{V_1}{V}\kappa_1 + \frac{V_2}{V}\kappa_2.$$

4. For the systems to be stable against displacements of their interface, their volumes and compressibilities must be related by $(\kappa_1 V_1)^{-1} + (\kappa_2 V_2)^{-1} \geq 0$. For reasons like those of (2.3.17; 3) it is not possible for two systems of negative compressibility to coexist, because the pressure of one system would increase with its volume and force that of the other one down. If only subsystem 1 has negative compressibility, then a necessary condition for stable equilibrium is $V_1 \geq V_2 \cdot \kappa_2/|\kappa_1|$. The increase of pressure in subsystem 1 when it expands is then less than that of 2 when it contracts. If V_1 is large enough in comparison with V_2, then subsystem 2 undergoes a large relative compression and exerts more pressure back on 1 than 1 exerts on 2. The volumes adjust in the other direction and stable equilibrium is established.

Consider finally what happens to the particle configuration if the subsystems can exchange particles to maximize the entropy. Formally, this means that the Hilbert space is

$$\mathscr{H} = \bigoplus_{N_1=1}^{N} \mathscr{H}_{N_1,V_1} \otimes \mathscr{H}_{N_2,V_2},$$

and the quantity to be calculated is

$$\operatorname{Tr}\Theta(E - H) = \sum_{N_1=0}^{N} \exp(S(N_1))\exp(S(N - N_1)). \tag{2.3.23}$$

In the limit $V \to \infty$, $N \to \infty$, $V_i/V \to \alpha_i$, $N_i/V_i \to \rho_i$, if S is concave in N, then arguments like those made earlier yield

$$\sigma(\rho) = \sup_{\alpha_1\rho_1+\alpha_2\rho_2=\rho} (\alpha_1\sigma_1(\rho_1) + \alpha_2\sigma_2(\rho_2)). \tag{2.3.24}$$

If the functions $\sigma_i(\rho_i)$ are nice, we obtain the

Equilibrium Condition (2.3.25)

Let $\sigma_i(\rho_i)$ be strictly concave and continuously differentiable. Then $\sigma(\rho) = \alpha_1\sigma_1(\rho_1) + \alpha_2\sigma_2(\rho_2)$, where ρ_i are determined uniquely by the conditions

$$\alpha_1\rho_1 + \alpha_2\rho_2 = \rho \quad \text{and} \quad \frac{\partial\sigma_1}{\partial\rho_1} = \frac{\partial\sigma_2}{\partial\rho_2}.$$

Remarks (2.3.26)

1. For a given ε and a given ρ, the six variables ε_i, ρ_i, α_i satisfy the three equations $\alpha_1\varepsilon_1 + \alpha_2\varepsilon_2 = \varepsilon$, $\alpha_1\rho_1 + \alpha_2\rho_2 = \rho$, $\alpha_1 + \alpha_2 = 1$. The three variations δE, δV, and δN corresponding to the equilibrium conditions are not independent, because $S(E, V, N)$ is of the special form $V\sigma(E/V, N/V)$, and there is one equation too few to fix six variables. Suppose for simplicity that the two subsystems are identical, $\sigma_1 = \sigma_2 = \sigma$; then because of the concavity, the maximum of $\alpha_1\sigma(\varepsilon_1, \rho_1) + \alpha_2\sigma(\varepsilon_2, \rho_2)$ is assumed when $\varepsilon_1 = \varepsilon_2 = \varepsilon$, $\rho_1 = \rho_2 = \rho$, and $\alpha_1 = 1 - \alpha_2$ is not determined by (2.3.25) and can be specified arbitrarily. Equality of the temperatures and the chemical potentials (see (2.3.27)) suffices to guarantee that the pressures are equal. After the onset of equilibrium, the wall allowing the exchange of energy and particles no longer exerts any force, and can be placed anywhere.
2. It is still possible to minimize the energy instead of maximizing the entropy. But this does not furnish a new stability condition, since if $\partial\varepsilon/\partial\sigma > 0$ the concavity of $(\varepsilon, \rho) \to \sigma(\varepsilon, \rho)$ is equivalent to the convexity of $(\sigma, \rho) \to \varepsilon(\sigma, \rho)$ (Problem 2). Besides $c_V > 0$ and $\kappa > 0$, this requires that

$$\frac{\partial^2 E}{\partial S^2}\frac{\partial^2 E}{\partial V^2} > \left(\frac{\partial^2 E}{\partial S\,\partial V}\right)^2,$$

or, in terms of the adiabatic expansivity

$$\alpha = \frac{1}{V}\frac{\partial V}{\partial T}\bigg|_S, \qquad \alpha^2 > c_V \kappa/T.$$

This amounts physically to the requirement of stability under a simultaneous change in the entropy and volume, related by

$$\delta S \sim \frac{\partial^2 E}{\partial V^2}, \qquad \delta V \sim -\frac{\partial^2 E}{\partial S\,\partial V}.$$

The equilibrium condition (2.3.25) requires the chemical potentials of the subsystems to be equal, if they are defined as with (2.3.26; 2) by minimizing the energy:

Definition (2.3.27)

The **chemical potential** is

$$\mu = \frac{\partial \varepsilon}{\partial \rho} = -\frac{\partial \varepsilon}{\partial \sigma}\bigg|_\rho \frac{\partial \sigma}{\partial \rho}\bigg|_\varepsilon.$$

Remarks (2.3.28)

1. The intuitive meaning of the temperature is the amount of energy it would take to raise the system from the quantum number n to en ($e = 2.718 \cdots$). Analogously, the chemical potential is the energy increase when a particle is added to the system without changing V or S.
2. Although T and P are always positive with the assumptions and boundary conditions that have been postulated, μ can in general have either sign. Because the density of states increases with N, the e^Sth eigenvalue may decrease with N even if $H \geq 0$.

In phenomenological thermodynamics entropy increases if the energy, volume, or particle number increases, according to the relationship $T\,dS = dE + P\,dV - \mu\,dN$. As we have seen, some of these differentials are well defined only in the thermodynamic limit, and are then considered as intensive properties. For future convenience, we collect the

Interrelationships among the Thermodynamic Properties (2.3.29)

$$T = \frac{\partial \varepsilon}{\partial \sigma}, \qquad \mu = \frac{\partial \varepsilon}{\partial \rho} = -T\,\frac{\partial \sigma}{\partial \rho},$$

$$P = -\varepsilon + \sigma\frac{\partial \varepsilon}{\partial \sigma} + \rho\frac{\partial \varepsilon}{\partial \rho} = T\left(\sigma - \varepsilon\frac{\partial \sigma}{\partial \varepsilon} - \rho\frac{\partial \sigma}{\partial \rho}\right),$$

$$c_V = T\left[\frac{\partial^2 \varepsilon}{\partial \sigma^2}\right]^{-1}, \qquad \kappa = \left[\sigma^2\frac{\partial^2 \varepsilon}{\partial \sigma^2} + 2\rho\sigma\frac{\partial^2 \varepsilon}{\partial \rho \partial \sigma} + \rho^2\frac{\partial^2 \varepsilon}{\partial \rho^2}\right]^{-1}.$$

Gloss

The sense of the partial derivatives is that, of the two variables on which a function has been regarded as depending, the one not written explicitly is to be held fixed. In any doubtful case the fixed argument will be indicated explicitly.

Remark (2.3.30)

Without knowledge of the Hamiltonian nothing can be said about the values the thermodynamic functions can assume. In (2.3.11; 6) there was an example in which $\varepsilon(\sigma)$ was even bounded above. If the function $\varepsilon(\sigma)$ is convex and asymptotically linear, then there is a maximum temperature. This is quite possibly the case realized in Nature, and $T_{\max} = 140$ MeV. In a model to be investigated shortly (2.3.32; 2), the function $\varepsilon(\sigma)$ has a kink, so T skips over certain values. It depends on the system whether the minimum entropy σ_0 defined in (2.3.11; 4) equals zero as postulated in the third law of thermodynamics. For instance, with a system consisting of N spins without energy $\otimes$ a system with entropy $N\sigma$, the total entropy divided by N equals $\sigma + \ln 2$, and when $\sigma \to 0$ the total entropy is the ln 2 left over. It is true that the ground state of this system is degenerate, but it is also easy to find examples with nondegenerate ground states for which the third law fails, simply by taking the previous Hamiltonian $\oplus$ a one-dimensional system with a lower energy level. The resulting ground state is simple, but that has no effect on what happens as $N \to \infty$.

It has been seen that the concavity of the function $\sigma(\varepsilon, \rho)$ is at the root of thermodynamic stability. Concavity is jeopardized when σ is maximized with respect to all of its parameters—the supremum of a set of concave functions is not necessarily concave, in contrast to the infimum. However, there is a useful

Lemma on the Envelope of a Set of Concave Functions (2.3.31)

If $\sigma(\varepsilon, \alpha)$ is jointly concave in ε and α, then $\bar{\sigma}(\varepsilon) = \sup_\alpha \sigma(\varepsilon, \alpha)$ is concave in ε.

Picture of the Proof

Think of the silhouette of a concave mountain slope and of a mountain with hollows.

Formal Proof if $\boldsymbol{\sigma(\epsilon, \alpha) \in C^2(K)}$

With this assumption, the maximum is attained at a point $\alpha(\varepsilon)$, $\bar{\sigma}(\varepsilon) = \sigma(\varepsilon, \alpha(\varepsilon))$, and

$$\sigma_{,\alpha}(\varepsilon, \alpha(\varepsilon)) = 0 \Rightarrow \sigma_{,\alpha\varepsilon} + \frac{d\alpha(\varepsilon)}{d\varepsilon}\sigma_{,\alpha\alpha} = 0.$$

Then

$$\frac{d^2\bar{\sigma}}{d\varepsilon^2} = \sigma_{,\varepsilon\varepsilon} + \sigma_{,\varepsilon\alpha}\frac{d\alpha(\varepsilon)}{d\varepsilon} = \frac{\sigma_{,\varepsilon\varepsilon}\sigma_{,\alpha\alpha} - (\sigma_{,\varepsilon\alpha})^2}{\sigma_{,\alpha\alpha}}.$$

Since $\sigma_{,\alpha\alpha} \leq 0$ and $\sigma_{,\varepsilon\varepsilon}\sigma_{,\alpha\alpha} - (\sigma_{,\varepsilon\alpha})^2 \geq 0$, $d^2\bar{\sigma}/d\varepsilon^2 \leq 0$. If $\sigma_{,\alpha\alpha} = 0$, it follows that $\sigma_{,\alpha\varepsilon}(\varepsilon, \alpha(\varepsilon)) = 0$, and therefore $\bar{\sigma}_{,\varepsilon\varepsilon} = \sigma_{,\varepsilon\varepsilon} \leq 0$. (For the proof without the assumption that $\sigma(\varepsilon, \alpha) \in C^2$, see Problem 3.) □

If the entropy is maximized with respect to parameters in the absence of joint concavity, then thermodynamic stability may be lost, and it will be necessary to reconsider the foregoing assumptions.

Examples (2.3.32)

1. Model of a star

   Consider N classical particles in a container V and attracting each other pairwise only within some $V_0 \subset V$. Suppose the potentials are constant in V_0 and $\sim N^{-1}$, to ensure that E be extensive.

$$H_N = \sum_{i=1}^{N} |\mathbf{p}_i|^2 - \frac{1}{N} \sum_{i,j=1}^{N} \chi_{V_0}(\mathbf{x}_i)\chi_{V_0}(\mathbf{x}_j),$$

$$\chi_{V_0}(\mathbf{x}) = \begin{cases} 1 & \text{for } \mathbf{x} \in V_0 \\ 0 & \text{otherwise.} \end{cases}$$

   With indistinguishable particles, the volume of phase space below the energy surface,

$$\exp(S(E, V, N))$$

$$= \frac{1}{N!} \int d^{3N}p\, d^{3N}x \Theta \left(E - \sum_{i=1}^{N} |\mathbf{p}_i|^2 + \frac{1}{N} \sum_{i,j=1}^{N} \chi_{V_0}(\mathbf{x}_i)\chi_{V_0}(\mathbf{x}_j)\right)$$

$$= \frac{\pi^{3N/2}}{N!(3N/2)!} \int_{(\cdots)>0} d^{3N}x \left(E + \frac{1}{N} \sum_{i,j=1}^{N} \chi_{V_0}(\mathbf{x}_i)\chi_{V_0}(\mathbf{x}_j)\right)^{3N/2},$$

   can be calculated exactly, because the integrand is piecewise constant. Let N_0 be the number of the $\mathbf{x}_i$ in V_0. Then

$$\exp(S) = \frac{V_0^N \pi^{3N/2}}{(3N/2)!} \sum_{-NE \leq N_0^2 \leq N^2} \left(\frac{V}{V_0} - 1\right)^{N-N_0} \frac{(E + N_0^2/N)^{3N/2}}{N_0!(N - N_0)!}$$

$$\equiv \sum_{N_0=1}^{N} \exp(S(E, V, N; N_0)).$$

   Only the dependence on E matters, so let $E = \varepsilon \cdot N$, $\rho = N/V = 1$, $N_0/N \equiv \alpha$, $(\max(0, -\varepsilon))^{1/2} \leq \alpha \leq 1$. Then it remains to evaluate

$$\sigma(\varepsilon) = \sup_{\alpha} \lim_{N \to \infty} \frac{1}{N} S(N\varepsilon, N, N; \alpha N) \equiv \sup_{\alpha} \sigma(\varepsilon, \alpha),$$

   and with the help of Stirling's formula,

$$\sigma(\varepsilon, \alpha) = \tfrac{3}{2} \ln(\varepsilon + \alpha^2) - \alpha \ln \alpha - (1 - \alpha)\ln(1 - \alpha) + F(1 - \alpha) + \text{constant},$$

$$F = \ln\left(\frac{V}{V_0} - 1\right). \tag{2.3.33}$$

   A calculation of the derivatives yields

$$\sigma_{,\varepsilon} = \frac{\frac{3}{2}}{\varepsilon + \alpha^2}, \qquad \sigma_{,\alpha} = \frac{3\alpha}{\varepsilon + \alpha^2} + \ln\left(\frac{1}{\alpha} - 1\right) - F,$$

$$\sigma_{,\varepsilon\varepsilon} = \frac{-\frac{3}{2}}{(\varepsilon + \alpha^2)^2}, \qquad \sigma_{,\varepsilon\alpha} = \frac{-3\alpha}{(\varepsilon + \alpha^2)^2}, \qquad \sigma_{,\alpha\alpha} = \frac{3\varepsilon - 3\alpha^2}{(\varepsilon + \alpha^2)^2} - \frac{1}{\alpha(1 - \alpha)}.$$

The maximum is achieved on the curve

$$\varepsilon(\alpha) = -\alpha^2 + \frac{3\alpha}{F - \ln(1/\alpha - 1)},$$

and the ranges of values of the variables are such that $\varepsilon + \alpha^2 \geq 0$, so only the branch of $F > \ln(1/\alpha - 1)$ comes into consideration. Because

$$\sigma_{,\alpha\alpha} = -\frac{(\varepsilon - \varepsilon_1(\alpha))(\varepsilon - \varepsilon_2(\alpha))}{(\varepsilon + \alpha^2)^2\alpha(1 - \alpha)},$$

$$\varepsilon_{1,2} = \frac{3\alpha}{2}\left(1 - \tfrac{5}{3}\alpha \pm \sqrt{1 - \alpha}\sqrt{1 - 11\alpha/3}\right),$$

$\sigma(\varepsilon, \alpha)$ is concave in α except when $\varepsilon_2 < \varepsilon < \varepsilon_1$. The sign of $d\varepsilon/d\alpha = -\sigma_{,\alpha\alpha}/\sigma_{,\alpha\varepsilon}$ changes in the interval $\varepsilon_2 < \varepsilon < \varepsilon_1$, so three values of α belong to a single ε, and the maximum needed is the greater of the two. Joint concavity requires that

$$\sigma_{,\varepsilon\varepsilon}\sigma_{,\alpha\alpha} - (\sigma_{,\alpha\varepsilon})^2 = \frac{3(\varepsilon - 3\alpha + 4\alpha^2)}{2(\varepsilon + \alpha^2)^3\alpha(1 - \alpha)} \geq 0$$

and implies $\varepsilon \geq 3\alpha - 4\alpha^2$. If $\varepsilon(\alpha)$ lies in this range of values, then the system has positive specific heat, and otherwise not (see Figure 8). Indeed,

$$\tfrac{3}{2}T = \varepsilon + \alpha^2 = \frac{3\alpha}{F - \ln(1/\alpha - 1)}$$

behaves as a function of ε as shown in Figure 9. The physical significance is that if energy is removed, the temperature falls until a certain fraction of the particles reside in V_0, which causes the system to start heating back up. If most of the particles are eventually in V_0, then they behave normally again. The system can be thought of as a normal system with

$$\sigma(\varepsilon, \rho) = \rho(\tfrac{3}{2}\ln\varepsilon - \tfrac{5}{2}\ln\rho)$$

put into contact with a peculiar system with

$$\sigma(\varepsilon, \rho) = \rho(\tfrac{3}{2}\ln(\varepsilon + \rho^2) - \tfrac{5}{2}\ln\rho) - F\rho.$$

If the energy is apportioned between them according to

$$\sigma(\varepsilon, \alpha) = \sup_{\varepsilon_1}(\tfrac{3}{2}(\alpha\ln(\varepsilon_1 + \alpha^2) + (1 - \alpha)\ln(\varepsilon - \varepsilon_1)) - \alpha F$$
$$- \tfrac{5}{2}(\alpha\ln\alpha + (1 - \alpha)\ln(1 - \alpha))),$$

then the entropy becomes exactly that of (2.3.33).

2. Model of a Ferromagnet
This problem is quantum-mechanical, but its analysis soon begins to resemble that of Example 1, for which reason we shall boldly plunge on

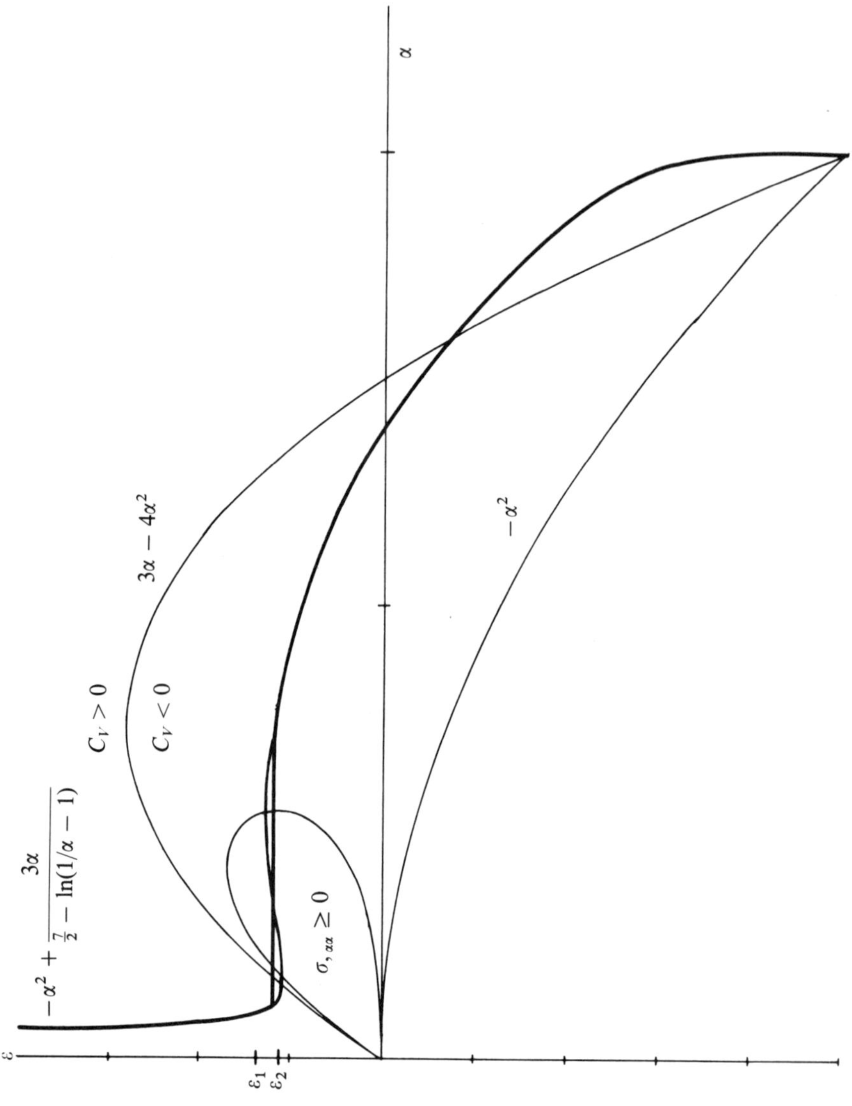

Figure 8 $\varepsilon(\alpha)$ in Example (2.3.32; 1).

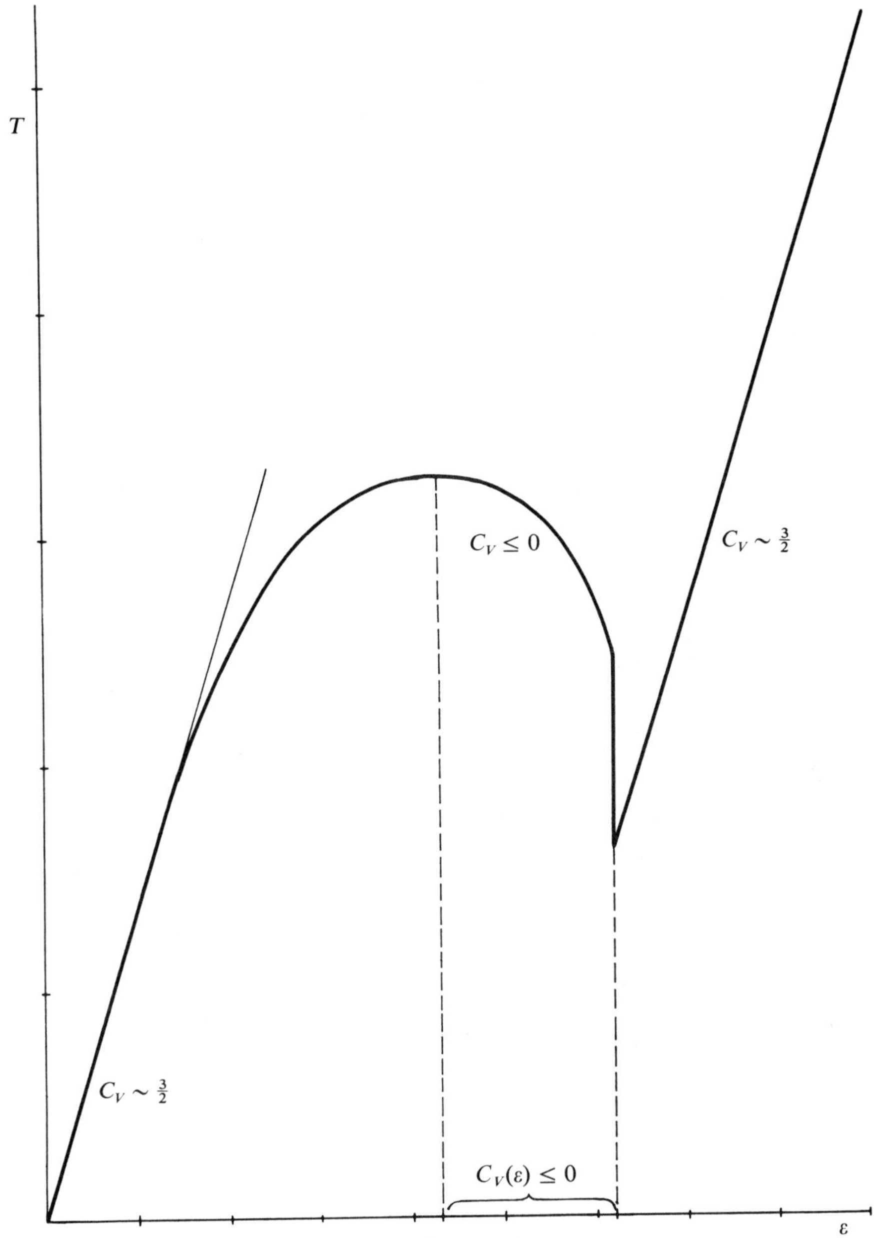

Figure 9 $T(\varepsilon)$ in Example (2.3.32; 1).

to the estimates without wasting time about epsilontic details. The Hamiltonian of (1.1.3) is modified to

$$H = B\sum_{j=1}^{N} \sigma_j^{(z)} - \frac{1}{N}\sum_{i,j=1}^{N} \sigma_i \cdot \sigma_j,$$

which contains a magnetic field in the z-direction and a spin–spin interaction favoring parallel spins. The strength of the interaction is the same for all pairs and must be $\sim 1/N$ for H to be $\sim N$. The mean magnetization $\mathbf{M}_N$ can be introduced as before, $H/N = BM_N^{(z)} - \mathbf{M}_N \cdot \mathbf{M}_N$, and it was shown in (III, §3.2) that the two parts of H can be diagonalized simultaneously. If the eigenvalues of $M_N^{(z)}$ are m_z and those of $\mathbf{M}_N \cdot \mathbf{M}_N$ are $m(m + 2/N)$, $0 \le m \le 1$, $-m \le m_z \le m$, then m_z and m are always multiples of $1/N$ spaced $2/N$ apart. To calculate Tr $\Theta(E - H)$ it is also necessary to find the multiplicities of the eigenvalues: If $m = 1$, then all spins must be parallel, and for one of these vectors, $m_z = 1$. There are now N vectors with $m_z = 1 - 2/N$, corresponding to the N possible ways to flip one spin. One of those possibilities has $m = 1$ (apply M^- to the previous vector) and the others must have $m = 1 - 2/N$. The general rule is that of the $\binom{N}{r}$ vectors with $m_z = 1 - 2r/N$, $\binom{N}{r-1}$ of them have $m > 1 - 2r/N$, and the remaining

$$\binom{N}{r} - \binom{N}{r-1} = \frac{N!(N - 2r + 1)}{r!(N - r + 1)!}$$

have $m = 1 - 2r/N$. This means that the number of vectors with the eigenvalues (m, m_z) is

$$\frac{N!(Nm + 1)}{((N/2)(1 - m))!((N/2)(1 + m) + 1)!} \sim \sqrt{\frac{2}{\pi(1 - m^2)N}}\,\frac{2m}{m + 1}$$
$$\times \exp\left\{N\left[\ln 2 - \left(\frac{1 + m}{2}\right)\ln(1 + m) - \left(\frac{1 - m}{2}\right)\ln(1 - m)\right]\right\}.$$

The last step used Stirling's formula $x! \sim (x/e)^x\sqrt{2\pi x}$, which is justified only for $m < 1$ even when $N \gg 1$, but in the limit being taken the contributions from the boundaries of the summation region are inconsequential. Since the integrand is a continuous function, as $N \to \infty$ the sum $\sum_{m=0}^{1}\sum_{m_z=-m}^{m}\cdots$ can be replaced with the integral $(N/2)^2\int_0^1 dm \int_{-m}^{m} dm_z \cdots$, and with $\varepsilon = E/N$ this leaves

$$\exp(S(\varepsilon)) = N^{3/2}\int_0^1 \frac{dm}{m + 1}\sqrt{\frac{m^2}{2\pi(1 - m^2)}}$$
$$\times \exp\left\{N\left[\ln 2 - \left(\frac{1 + m}{2}\right)\ln(1 + m) - \left(\frac{1 - m}{2}\right)\ln(1 - m)\right]\right\}$$
$$\times \int_{-m}^{m} dm_z \Theta(\varepsilon + m^2 - Bm_z). \tag{2.3.34}$$

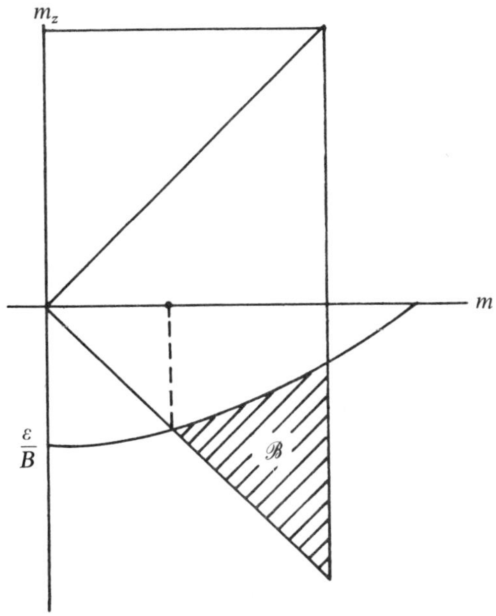

Figure 10 The region of integration in the $m - m_z$-plane.

Therefore the domain $\mathscr{B}$ of integration is $\{(m, m_z): 0 \leq m \leq 1, -m \leq m_z \leq m\} \cap \{(m, m_z): m_z \leq (\varepsilon + m^2)/B\}$. The entropy S is obviously even in B, so we may restrict consideration to $B \geq 0$ (see Figure 10).

Since the exponential function decreases rapidly with m, the appropriate generalization of Lemma (2.3.12) makes $\sigma = \lim_{N\to\infty} S/N$ sensitive only to $m_0 \equiv \inf_{m, m_z \in \mathscr{B}} m$ (the exponent in (2.3.34) decreases monotonically in m):

$$m_0 = \Theta(-\varepsilon)\left(\sqrt{\frac{B^2}{4} - \varepsilon} - \frac{B}{2}\right),$$
$$\sigma = \ln 2 - \tfrac{1}{2}\Theta(-\varepsilon)[(1 + m_0)\ln(1 + m_0) + (1 - m_0)\ln(1 - m_0)], \tag{2.3.35}$$

if $\varepsilon \geq -1 - B$, and is otherwise 0. Since σ is concave but decreasing in m_0, the concavity in ε remains to be verified:

$$T^{-1} = \frac{d\sigma}{d\varepsilon} = \frac{\Theta(-\varepsilon)}{4(B^2/4 - \varepsilon)^{1/2}} \ln\frac{1 + m_0}{1 - m_0} \geq 0,$$
$$-\frac{1}{T^2 c} = \frac{d^2\sigma}{d\varepsilon^2} = -\frac{\Theta(-\varepsilon)}{8}$$
$$\times\left[-\left(\frac{B^2}{4} - \varepsilon\right)^{-3/2} \ln\frac{1 + m_0}{1 - m_0} + \frac{2}{(B^2/4 - \varepsilon)(1 - m_0^2)}\right]. \tag{2.3.36}$$

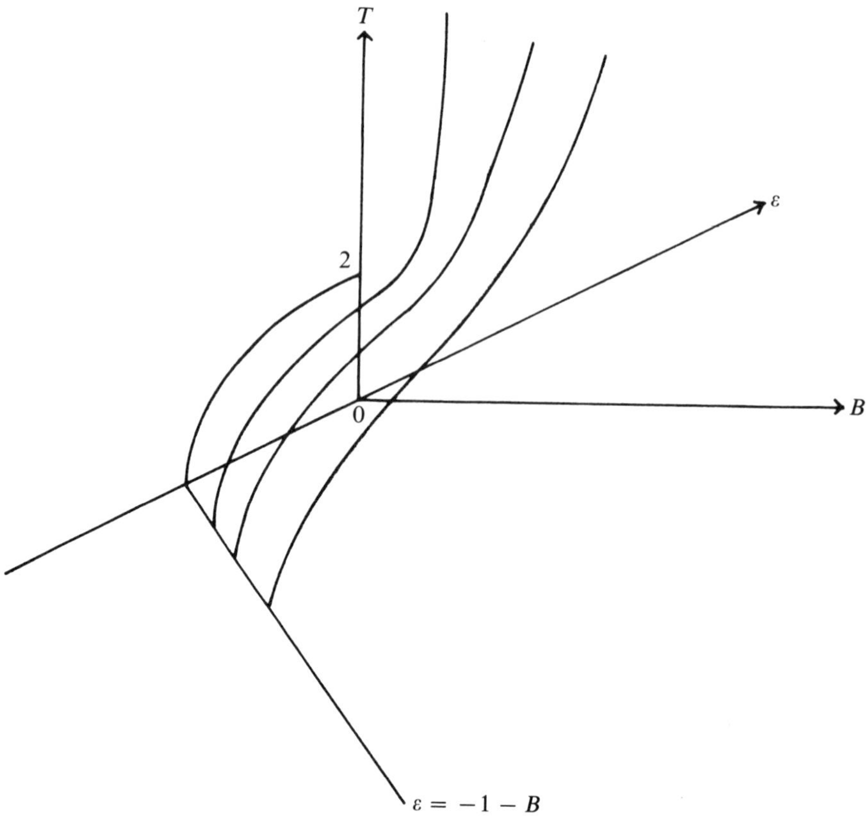

Figure 11 The surface of states in $T - \varepsilon - B$-space.

In a lucky break, the positive term in the brackets $[\cdots]$ is always greater than the negative one, and c_V is always positive. If $-1 > -1 - B \leq \varepsilon \leq 0$, then T increases continuously from 0 to ∞. The heat capacity c_V increases from 0 to a maximum value and then falls back to 0. If $B = 0$, then T reaches the value 2 for $\varepsilon = 0$, at which c_V has risen to $\frac{3}{2}$. Afterwards, T jumps up to ∞ and c_V falls back to 0 (see Figure 11).

Thus if $B = 0$ and $T < 2$, the thermal motion is no longer strong enough to counter the ordering tendency of H, and a spontaneous magnetization m_0 appears. As no direction is preferred, the thermal expectation value $|\text{Tr}\, \rho \mathbf{M}|$ remains 0. We shall learn later that as $N \to \infty$, the GNS representations of the σ's constructed with ρ become integrals over all directions of thermal representations (1.4.7). If $B > 0$, then $\text{Tr}\, \rho \mathbf{M}$ points in the z-direction, and m_0 grows smoothly from 0 to 1 as T decreases.

The interactions in these examples could have been replaced with average fields. This is typical of forces of long range like gravity. If the long-range forces neutralize each other—for instance if they are electric—then the system is basically the sum of its parts, i.e., it can be decomposed into parts in such a way that the entropy, energy, volume, and particle number are all additive. In that case the maximum entropy is concave.

Thermodynamic Stability of Decomposable Systems (2.3.37)

For an arbitrary function σ,

$$\bar{\sigma}(\varepsilon, \rho) \equiv \sup_n \sup_{K_n} \sum_{i=1}^{n} \alpha_i \sigma(\varepsilon_i, \rho_i),$$

where

$$K_n = \left\{(\alpha_i), (\varepsilon_i), (\rho_i) \,\middle|\, \sum_{i=1}^{n} \alpha_i = 1, \sum_{i=1}^{n} \alpha_i \varepsilon_i = \varepsilon, \sum_{i=1}^{n} \alpha_i \rho_i = \rho\right\},$$

is jointly concave in its two variables.

Proof

Let $\varepsilon = \gamma\varepsilon' + (1 - \gamma)\varepsilon''$, $\rho = \gamma\rho' + (1 - \gamma)\rho''$. Divide (α_i) into (α_i') and (α_i''), and take the supremum over $K_{n'}$ and $K_{n''}$:

$$K_{n'} \equiv \left\{(\alpha_i'), (\varepsilon_i'), (\rho_i') \,\middle|\, \sum_{i=1}^{n'} \alpha_i' = \gamma, \sum_{i=1}^{n'} \alpha_i' \varepsilon_i' = \varepsilon', \sum_{i=1}^{n'} \alpha_i' \rho_i' = \rho'\right\},$$

$$K_{n''} \equiv \left\{(\alpha_i''), (\varepsilon_i''), (\rho_i'') \,\middle|\, \sum_{i=1}^{n''} \alpha_i'' = 1 - \gamma, \sum_{i=1}^{n''} \alpha_i'' \varepsilon_i'' = \varepsilon'', \sum_{i=1}^{n''} \alpha_i'' \rho_i'' = \rho''\right\}.$$

Since this is only a particular division,

$$\begin{aligned}\bar{\sigma}(\varepsilon, \rho) &\geq \sup_{n', n''} \sup_{K_{n'}, K_{n''}} \left(\sum_i \alpha_i' \sigma(\varepsilon_i', \rho_i') + \sum_i \alpha_i'' \sigma(\varepsilon_i'', \rho_i'')\right) \\ &= \gamma\bar{\sigma}(\varepsilon', \rho') + (1 - \gamma)\bar{\sigma}(\varepsilon'', \rho'').\end{aligned}$$

□

Remark (2.3.38)

The construction (2.3.37) gives the concave envelope of σ, but nothing guarantees that $\bar{\sigma}$ is strictly concave. If σ is linear, then $\bar{\sigma} = \sigma$, and σ is of the form of Example (2.3.32; 1). The convex part of the curve gets bridged by a straight line, as shown in Figure 12.

The function $\bar{\sigma}$ is simply $\alpha\sigma(\varepsilon_1) + (1 - \alpha)\sigma(\varepsilon_2)$ in the intervening region where $\varepsilon = \alpha\varepsilon_1 + (1 - \alpha)\varepsilon_2$ for fixed ε_1 and ε_2. An interpretation is that the system consists of two phases in this region, having energies ε_1 and ε_2, and the temperature remains constant as the total energy varies, while the proportions of the phases present change. This suggests a

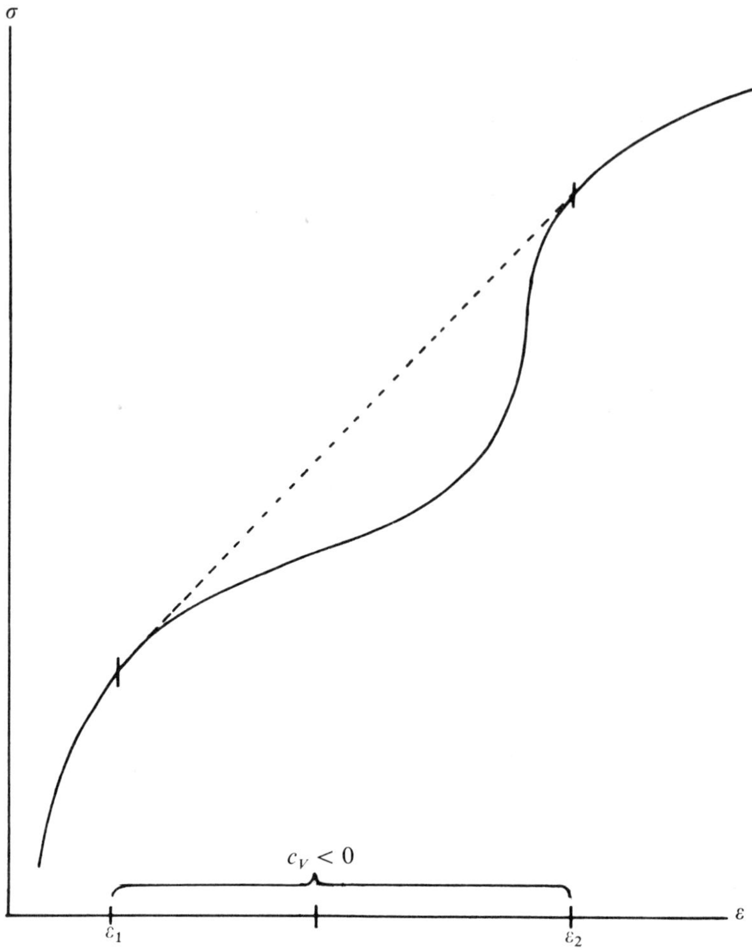

Figure 12 The region of negative specific heat.

Rough Definition of the Thermodynamic Phases (2.3.39)

The extreme points of the concave function $\sigma(\varepsilon, \rho)$ correspond to pure phases, and in the regions of coexistence of more than one phase the function σ is not strictly concave.

Examples (2.3.40)

1. If the graph of $\sigma(\varepsilon, \rho)$ shows a belt-like region the curvature of which vanishes in only one direction, then two phases coexist in its interior. The sides of the belt correspond to pure phases and the end to a critical point (see Figure 13):

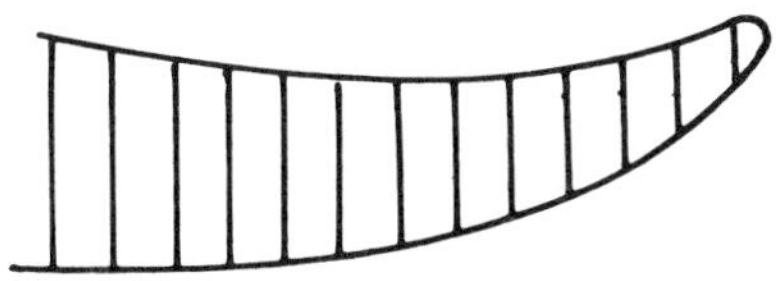

Figure 13 The region of coexistence of two phases.

2. In the usual solid–liquid–gas phase diagram, the triple point occurs in a region at which the curvature of $\sigma(\varepsilon, \rho)$ vanishes in both directions (Figure 14):

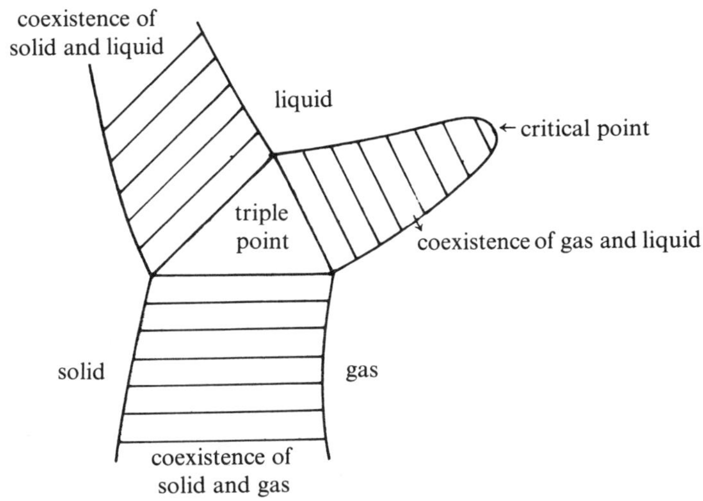

Figure 14 Regions of coexistence.

Remarks (2.3.41)

1. The sum, in the sense of (2.3.37), of many copies of Example (2.3.32; 1) produces a concave $\bar{\sigma}$, since the convex part lies below the phase-transition line. Some concave pieces of the curve are also bridged over, and are known as metastable phases, which arise in superheated stars and supercooled gases. They have positive specific heats and are locally stable (see Figure 15):
2. Gibbs's phenomenological phase rule states that whenever a material has two coexisting phases, there is always a one-parameter family of coexisting phases described by $T(\alpha)$ and $\mu(\alpha)$. Three coexisting phases can only exist at discrete values of (T, μ). This is exactly what went on in (2.3.40; 1) and (2.3.40; 2), where the parts that are flat in one direction

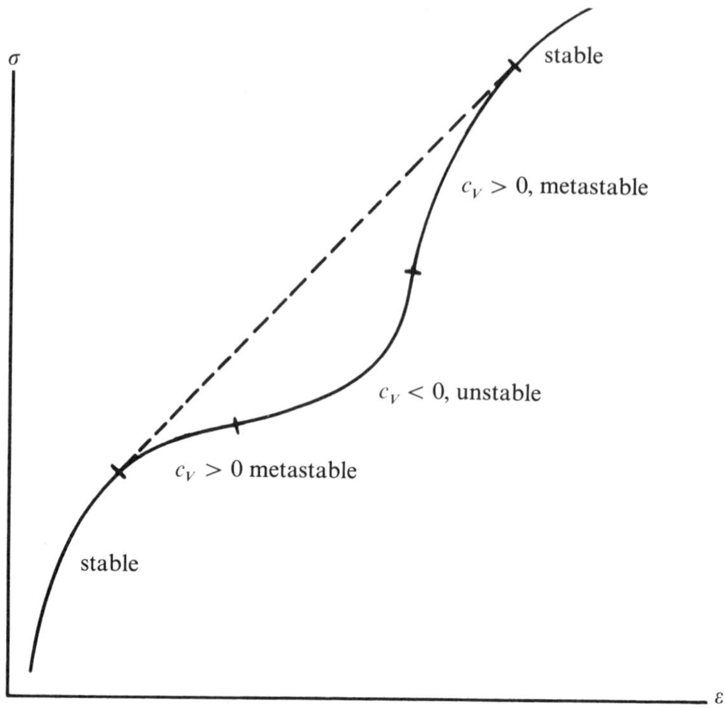

Figure 15 Stability of the regions of (2.3.32; 1).

are two-dimensional, but is not a consequence of concavity alone; for instance the function $\sigma = -\varepsilon^p \rho^{-q}$, $p > q + 1 > 1$, has a straight line segment only if $\varepsilon = 0$, but is nonetheless concave in (ε, ρ).

3. A quadruple point of a substance would be a flat rectangle in the energy surface. The nonexistence of quadruple points does not follow from concavity, but amounts to the assumption that the flat pieces of the energy surface form a simplex. If they do not form a simplex, then the ratio of the phases in the mixture is not even necessarily determined by ε and ρ:

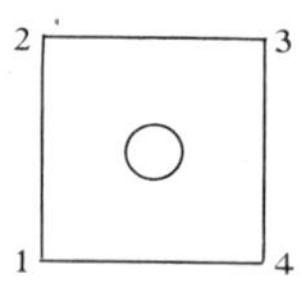

$$\bigcirc = \tfrac{1}{2}(1 + 3) = \tfrac{1}{2}(2 + 4).$$

At this point we have no arguments that would show that quadruple points do not occur, and in fact it is easy to construct models with quadruple points by taking the sum of two independent systems each of which has a phase transition. We shall have to take the issue up anew in (3.2.12; 2).

Problems (2.3.42)

1. Show that if $\sigma(\varepsilon, \rho)$ is concave, then $(E, N, V) \to S(E, N, V)$ is concave.
2. Show that for $\varepsilon_{,\sigma} > 0$, $\sigma(\varepsilon, \rho)$ is concave iff $\varepsilon(\sigma, \rho)$ is convex.
3. Without assuming differentiability, show that if $\sigma(\varepsilon, \alpha)$ is concave, then $\bar{\sigma}(\varepsilon) = \sup_\alpha \sigma(\varepsilon, \alpha)$ is concave.
4. Prove the relationship $V/\varepsilon'' = V_1/\varepsilon_1'' + V_2/\varepsilon_2''$ of (2.3.15; 4).

Solutions (2.3.43)

1. For simplicity assume that σ is twice differentiable. Then

$$D^2 S = \frac{1}{V}\begin{vmatrix} \sigma_{,\varepsilon\varepsilon} & \sigma_{,\varepsilon\rho} & -\varepsilon\sigma_{,\varepsilon\varepsilon} - \rho\sigma_{,\varepsilon\rho} \\ \sigma_{,\varepsilon\rho} & \sigma_{,\rho\rho} & -\varepsilon\sigma_{,\varepsilon\rho} - \rho\sigma_{,\rho\rho} \\ -\varepsilon\sigma_{,\varepsilon\varepsilon} - \rho\sigma_{,\varepsilon\rho} & -\varepsilon\sigma_{,\varepsilon\rho} - \rho\sigma_{,\rho\rho} & \varepsilon^2\rho_{,\varepsilon\varepsilon} + 2\varepsilon\rho\sigma_{,\varepsilon\rho} + \rho^2\sigma_{,\rho\rho} \end{vmatrix}.$$

Observe that the concavity of S is equivalent to $D^2 S \leq 0$, which means that $D^2\sigma \leq 0$ and $\det D^2 S \leq 0$. However, $D^2 S = 0$ because the mapping $\lambda \to S(\lambda E, \lambda N, \lambda V)$ is affine.

2. The function σ is concave iff the concave hull $\bar{\Gamma} = \{(x, y, z) = \sum \lambda_i(x_i, y_i, z_i), (x_i, y_i, z_i) \in \Gamma, 0 \leq \lambda_i \leq 1, \sum_i \lambda_i = 1\}$ of the graph $\Gamma = \{(x, \varepsilon, \rho): x = \sigma(\varepsilon, \rho)\}$ lies completely below Γ. However, looked at from the other side, Γ is also the graph of the inverse function $\varepsilon(\sigma, \rho)$, except that "below" becomes "above" and vice versa.

3. Let $\varepsilon = \gamma\varepsilon_1 + (1 - \gamma)\varepsilon_2$, and choose $\alpha_{1,2}$ so that $\sup_\alpha \sigma(\varepsilon_i, \alpha) = \sigma(\varepsilon_i, \alpha_j)$, $i = 1, 2$, or at least comes arbitrarily close to equality.

$$\begin{aligned}\sup_\alpha \sigma(\varepsilon, \alpha) &\geq \sigma(\gamma\varepsilon_1 + (1 - \gamma)\varepsilon_2, \gamma\alpha_1 + (1 - \gamma)\alpha_2) \\ &\geq \gamma\sigma(\varepsilon_1, \alpha_1) + (1 - \gamma)\sigma(\varepsilon_2, \alpha_2) \\ &= \gamma\bar{\sigma}(\varepsilon_1) + (1 - \gamma)\bar{\sigma}(\varepsilon_2).\end{aligned}$$

4. $$\varepsilon_1'(\sigma_1) = \varepsilon_2'\left(\frac{V\sigma - V_1\sigma_1}{V_2}\right) \Rightarrow \sigma_1'\varepsilon_1'' = \left(\frac{V}{V_2} - \frac{V_1}{V_2}\sigma_1'\right)\varepsilon_2'' \Rightarrow \sigma_1' = \frac{V}{V_2}\frac{\varepsilon_2''}{\varepsilon_1'' + \varepsilon_2'' V_1/V_2},$$

$$\varepsilon'(\sigma) = \varepsilon_1'(\sigma_1) \Rightarrow \varepsilon'' = \sigma_1'\varepsilon_1'' = \frac{V}{V_2}\varepsilon_2''\frac{\varepsilon_1''}{\varepsilon_1'' + \varepsilon_2'' V_1/V_2} \Rightarrow \frac{V}{\varepsilon''} = \frac{V_1}{\varepsilon_1''} + \frac{V_2}{\varepsilon_2''}.$$

2.4 The Canonical Ensemble

The Maxwell–Boltzmann distribution arises from the state of a system in contact with a thermal reservoir. If the system is large, this state is indistinguishable from that of the microcanonical ensemble.

In the preceding section it was shown that the entropy of two large subsystems without interaction is additive. The entropy was always defined with the microcanonical density matrix (2.3.1), but when the density matrix is restricted to a subsystem,

$$\rho_1 = \frac{\mathrm{Tr}_2\, \Theta(E - H_1 - H_2)}{\mathrm{Tr}\, \Theta(E - H_1 - H_2)} \equiv \exp(S(E - H_1))/\mathrm{Tr}_1 \exp(S(E - H_1)), \tag{2.4.1}$$

it appears quite different. It will now be shown that ρ_1 does not depend on the nature of the second system if it is infinitely large (a thermal reservoir). We shall also find out that this so-called canonical density matrix is equivalent to the microcanonical density matrix if the system is large. The convergence of ρ_1 as the second subsystem becomes infinitely large is described by

Lemma (2.4.2)

Suppose that the concave, increasing functions $(1/V)S(E) \equiv \sigma_V(E/V)$ *and their derivatives converge uniformly on some neighborhood of* $\varepsilon = E/V$ *to a function* $\sigma(\varepsilon) \in C^1$ *and to* $\sigma'(\varepsilon)$. *Then as* $V \to \infty$,

$$\rho_V \equiv \frac{\exp[V\sigma_V((E - H_1)/V)]}{\mathrm{Tr}\exp[V\sigma_V((E - H_1)/V)]} \to \frac{\exp(-H_1\sigma'(\varepsilon))}{\mathrm{Tr}\exp(-H_1\sigma'(\varepsilon))}$$

in the trace norm, provided that $\exp(-H_1\sigma'(\varepsilon))$ *is of the trace class* $\mathscr{C}_1$.

Remarks (2.4.3)

1. As in (2.3.13; 2), E and E_m can be identified.
2. *A priori*, $S(E)$ has been defined only for discrete values. We assume that it can be interpolated with a concave, strictly increasing, continuously differentiable function.
3. The facts $\sigma_{\mathrm{ess}}(H) = \varnothing$ and $H \geq 0$ do not suffice to make $\exp(-\beta H) \in \mathscr{C}_1$; $\mathrm{Sp}(H)$ could be $\mathbb{Z}^+$ and the eigenvalues $n \in \mathbb{Z}^+$ could have multiplicity n^n. More assumptions are needed than (2.3.3; 2).
4. The significance of the lemma is that temperature is the only property of a reservoir in the infinitely large limit that enters into the reduced density matrix. The reduced density matrix has the canonical form regardless of the structure of the reservoir, when the energy of interaction can be neglected.

Proof of (2.4.2)

With $\mathrm{Tr}_1\ \Theta(E_1 - H_1) = \exp(S_1(E_1))$, $\mathrm{Tr}\ \Theta(E - H_1 - H_2) = \int dE_1 \exp(S(E - E_1) + S_1(E_1))S_1'(E_1)$, ρ_V can be written as

$$\rho_V = \frac{\exp\{V[\sigma_V(\varepsilon - (H_1/V)) - \sigma_V(\varepsilon)]\}}{\int dE_1 \exp\{S_1(E_1) + \ln S_1'(E_1) + V[\sigma_V(\varepsilon - (E_1/V)) - \sigma_V(\varepsilon)]\}}.$$

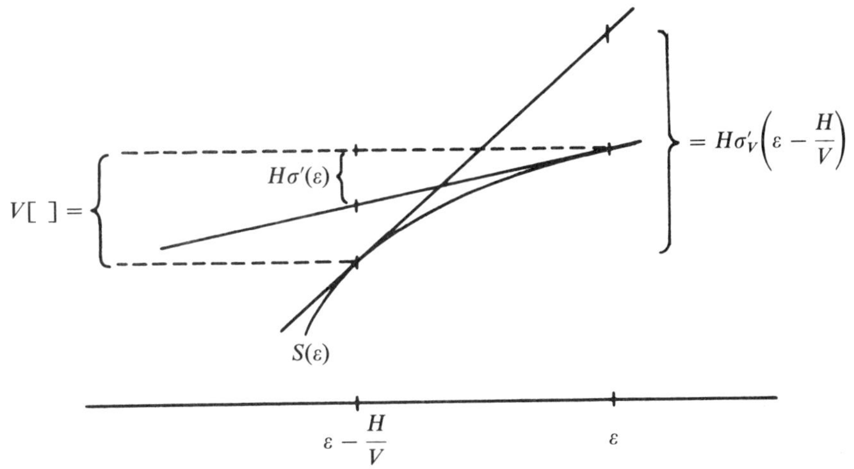

Figure 16 Estimating the slope of $S(\varepsilon)$.

Because of concavity, if $H_1 \geq 0$, then $H_1\sigma_V'(\varepsilon) \leq V[\sigma_V(\varepsilon) - \sigma_V(\varepsilon - (H_1/V)]$ $\leq H_1\sigma_V'(\varepsilon - (H_1/V))$ (see Figure 16).

The assumption that σ' converges uniformly then makes $V[\sigma_V(\varepsilon - (H_1/V)) - \sigma_V(\varepsilon)]$ converge uniformly to $-H_1\sigma'(\varepsilon)$ on compact sets in $\mathrm{Sp}(H_1)$. Moreover, there exist V' and β such that for all $V > V'$, there is an operator inequality, $\exp[V(\sigma_V(\varepsilon - (H_1/V)) - \sigma_V(\varepsilon))]$ $\leq \exp(-\beta H_1)$. In the spectral representation of H_1, $\exp[V(\sigma_V(\varepsilon - (H_1/V) - \sigma_V(\varepsilon))] \to \exp(-H_1\sigma'(\varepsilon))$ in the strong topology, by the Lebesgue dominated convergence theorem. If the operator on the right belongs to $\mathscr{C}_1$, then by the dominated convergence theorem again,

$$\begin{aligned}\operatorname{Tr}\exp[-H_1\sigma'(\varepsilon)] &= \int dE_1 \exp[S_1(E_1) + \ln S_1'(E_1) - E_1\sigma'(\varepsilon)] \\ &= \lim_{V\to\infty} \int dE_1 \exp\left\{S_1(E_1) + \ln S_1'(E_1)\right. \\ &\qquad \left. + V\left[\sigma_V\left(\varepsilon - \frac{E_1}{V}\right) - \sigma_V(\varepsilon)\right]\right\}.\end{aligned}$$

The proof is completed by appealing to the theorem (Problem 1) that strong convergence of density matrices implies convergence in the trace norm. □

Corollaries (2.4.4)

1. Since ρ_V converges in the sense of the strong topology of $\mathscr{B}(\mathscr{H})^*$ (cf. (2.1.2)), $\operatorname{Tr}\rho_V a \to \operatorname{Tr} a \exp[-\beta(H_1 - F)]$ for all $a \in \mathscr{B}(\mathscr{H}_1)$, where $\beta \equiv \sigma'(\varepsilon)$, $\exp(-\beta F) = \operatorname{Tr}\exp(-\beta H_1)$.
2. Because of Theorem (2.2.24), $S(\exp[-\beta(H_1 - F)]) \leq \underline{\lim}_{V\to\infty} S(\rho_V)$.

Recall that the microcanonical state is the most mixed state below E_m. The canonical state instead satisfies

The Maximum Principle for the Canonical Entropy (2.4.5)

Let $\rho = \exp(-\beta H)/\mathrm{Tr}\exp(-\beta H)$*and let* $\bar{\rho}$ *be any density matrix such that* $\mathrm{Tr}\,\bar{\rho}H = \mathrm{Tr}\,\rho H$*. Then* $S(\rho) \geq S(\bar{\rho})$.

Remarks (2.4.6)

1. Proposition (2.4.5) states that with a given average energy, the canonical state has the greatest possible entropy. The proposition does not work for all α-entropies S_α, so it can not be improved to the statement that $\rho \succeq \bar{\rho}$.
2. According to inequality (2.1.7; 2), since $x \to -x \ln x$ is strictly concave, S is a strictly concave function on the convex set of density matrices ρ such that $\mathrm{Tr}\,\rho H = E$. This means that the maximum is unique, and there can not even be local maxima elsewhere.
3. Not all $S_\alpha(\rho)$ are equal with the canonical ρ: $S_\alpha = \beta(F(\alpha\beta) - F(\beta))/(\alpha - 1)$.
4. This maximum principle is sometimes invoked as the motivation for the canonical density matrix, without appealing to the microcanonical state.
5. The free energy satisfies the inequality $F(\bar{\rho}) \geq F(\rho)$ without the assumption that $\mathrm{Tr}\,\bar{\rho}H = \mathrm{Tr}\,\rho H$.

Proof

Proposition (2.4.5) follows directly from Remark (2.2.23; 1). □

The canonical **partition function** $Z \equiv \mathrm{Tr}\exp(-\beta H)$ is easier to work with than the microcanonical partition function, because it does not involve discontinuous functions; if the dimension is finite, it is even an entire function of β. If the dimension is infinite, then $\exp(-\beta H)$ is required to belong to $\mathscr{C}_1$, so the spectrum of H must be bounded below and extend to $+\infty$. This, however, means that $\exp(-\beta H) \notin \mathscr{C}_1$ for $\beta < 0$, so the most that can be hoped for is analyticity in $\mathbb{C}^+ \equiv \{x + iy : x > 0\}$. For the cases of interest, there is in fact a proposition on

The Analyticity of the Partition Function of Finite Systems (2.4.7)

Let $\exp(-\beta H_0) \in \mathscr{C}_1$ *for all* $\beta > 0$ *and suppose* V *is* ε*-bounded with respect to* H_0 (cf. (III: 3.4.1)). *Then the mapping* $\mathbb{C} \times \mathbb{C}^+ \to \mathbb{C} : (\alpha, \beta) \to \mathrm{Tr}\exp[-\beta(H_0 + \alpha v)]$ *is analytic, and* $(\partial/\partial\alpha)\,\mathrm{Tr}\exp[-\beta(H_0 + \alpha v)]_{|\alpha=0} = -\mathrm{Tr}\,\beta v \exp[-\beta H_0]$.

Remarks (2.4.8)

1. Since the operator $H_0 + \alpha v$ is not normal when α is nonreal, the exponential function has to be defined. This can be done as in (2.1.8; 7) or by integrating the resolvent,
$$\exp[-\beta(H_0 + \alpha v)] = \int_C \frac{dz}{2\pi i} \frac{\exp(-\beta z)}{(H_0 + \alpha v - z)},$$
in which the integration contour runs through the region of analyticity (cf. (III: 3.5.13)) so that the integral converges in norm.
2. The next task is to make sense of $\operatorname{Tr}\exp[-\beta(H_0 + av)]$ and show that it belongs to $\mathscr{C}_1$ for $(\alpha, \beta) \in \mathbb{C} \times \mathbb{C}^+$. If $\alpha, \beta \in \mathbb{R} \times \mathbb{R}^+$, then this follows from $H_0 + \alpha v \geq H_0/2 - C(\alpha)$, $\exp(-\beta H_0) \in \mathscr{C}_1$, and the observation that if $0 < a < b \in \mathscr{C}_1$ then $a \in \mathscr{C}_1$. If α and β are complex, then Corollary (2.1.8; 7) can be appealed to for $|\operatorname{Tr}\exp(\alpha a + \beta b)| \leq \operatorname{Tr}|\exp(\alpha a)||\exp(\beta b)|$, with $\exp(a)$ and $\exp(b)$ Hermitian, and in particular $|\operatorname{Tr}\exp[-aH_0 - bv + i(cH_0 + dv)] \leq \operatorname{Tr}\exp(-aH_0 - bv)$ for all real a, b, c, and d.
3. The proposition implies that the free energy $F = -T\ln z$ can have singularities only at the zeros of Z. If $(\alpha, \beta) \in \mathbb{R} \times \mathbb{R}^+$ then $Z > 0$, so F is analytic in a neighborhood of $\mathbb{R} \times \mathbb{R}^+$. In addition, Corollary (2.1.8; 3) states that $-\ln Z$ is concave in $(\beta, \alpha\beta) \in \mathbb{R} \times \mathbb{R}^+$, so F is concave in $(T, \alpha/T)$ (cf. (III: 3.5.24)). The equation $\partial F/\partial\alpha = \langle v \rangle$ generalizes the Feynman–Hellmann formula (III: 3.5.19; 2).

Proof

See Problem 2. □

Since the exponential function is convex, the free energy can be bounded in terms of phase-space integrals by means of (2.2.11), and the upper bound of (2.2.11) can be improved upon with Corollary (2.1.8: 7).

The Connection with the Classical Free Energy (2.4.9)

Let

$$H = \sum_{i=1}^{N} |\mathbf{p}_i|^2 + v(\mathbf{x}), \qquad \exp(-\beta F) = \operatorname{Tr}\exp(-\beta H) < \infty,$$

and

$$\exp[-\beta F_{cl}(v)] = \int d^{3N}x \frac{d^{3N}p}{(2\pi)^{3N}} \exp\left[-\beta\left(\sum_{i=1}^{N} |\mathbf{p}_i|^2 + v(\mathbf{x})\right)\right].$$

Then

$$F_{\text{cl}}(v) \le F \le \inf_u F_{\text{cl}}(v_u),$$

where

$$v_u(\mathbf{x}) = \int d^{3N}x' v(\mathbf{x}') |u(\mathbf{x} - \mathbf{x}')|^2 + \int d^{3N}x |\nabla u(\mathbf{x})|^2.$$

Remarks (2.4.10)

1. The function $v(\mathbf{x})$ contains the interaction between the particles, as well as a possible external field. It must even account for the box confining the system, as the Hilbert space is $L^2(\mathbb{R}^{3N})$.
2. The proposition shows that quantum effects can only increase the free energy, either with a kinetic zero-point energy or a smeared-out effective potential.
3. The particles have been assumed distinguishable; the modifications needed for indistinguishable particles will be discussed below.
4. Countless attempts at expansions in $\hbar$ have been made in the literature, but the results are not conclusive because rigorous bounds on the higher-order contributions have not been obtained.
5. If $\hbar$ is not set to 1, the dimensionless volume in phase space becomes $d^{3N}x\, d^{3N}p h^{-3N}$, rather than $d^{3N}x\, d^{3N}p\hbar^{-3N}$.

Proof

The lower bound for F. By Corollary (2.1.8; 7),

$$\begin{aligned} \operatorname{Tr} \exp[-\beta(H_0 + v)] &\le \operatorname{Tr} \exp(-\beta H_0) \exp(-\beta v) \\ &= \int d^{3N}x \langle \mathbf{x} | \exp(-\beta H_0) | \mathbf{x} \rangle \exp(-\beta v(\mathbf{x})), \end{aligned}$$

and it was observed in (III: 3.3.3) that $\exp(-\beta H_0)$ has the integral kernel

$$K(\mathbf{x}, \mathbf{x}) = \left(\frac{1}{4\pi\beta}\right)^{3N/2} = \int \frac{d^{3N}p}{(2\pi)^{3N}} \exp\left(-\beta \sum_{k=1}^{N} |\mathbf{p}_k|^2\right).$$

The upper bound for F follows immediately from (2.2.11), for $\langle z \| \mathbf{p} |^2 | z \rangle = (\operatorname{Im} z^2) + \int dx |\nabla u|^2$. □

Example (2.4.11)

The one-dimensional harmonic oscillator; $u(x) = \exp(-bx^2/2)/\sqrt[4]{\pi}$, $H = p^2 + \omega^2 x^2$,

$$\operatorname{Tr} \exp(-\beta H) = \sum_{n=0}^{\infty} \exp[-\beta\omega(2n + 1)] = \frac{\exp(-\omega\beta)}{1 - \exp(-2\omega\beta)},$$

$$v_u = \omega^2\left(x^2 + \frac{1}{2b}\right) + \frac{b}{2},$$

which has the minimum $\omega^2 x^2 + \omega$ when $b = \omega$. Since

$$\int_{-\infty}^{\infty} \frac{dp\,dx}{2\pi} \exp[-\beta(p^2 + \omega^2 x^2)] = \frac{1}{2\omega\beta},$$

the bounds (2.4.9) yield the inequalities

$$\frac{\exp(-\alpha/2)}{\alpha} \leq \frac{\exp(-\alpha/2)}{1 - \exp(-\alpha)} \leq \frac{1}{\alpha}, \qquad \alpha \equiv 2\omega\beta \in \mathbb{R}^+.$$

The interest in the bounds (2.4.9) is mainly academic, since the particles in real physics are either fermions or bosons. In addition to multiplying the volume element of the phase-space integral by $1/N!$, the generalization for indistinguishable particles entails an effective interaction that vanishes as $mT \to \infty$, and is repulsive for fermions and attractive for bosons.

Bounds on F for Indistinguishable Particles (2.4.12)

Suppose that

$$H = \frac{1}{2m} \sum_{i=1}^{N} |\mathbf{p}_i|^2 + v(\mathbf{x}_1, \ldots, \mathbf{x}_N),$$

$$\exp[-\beta F_{cl}(H)] = \frac{1}{(2\pi)^{3N} N!} \int d^{3N}x\, d^{3N}p \exp[-\beta H(\mathbf{p}_1, \ldots, \mathbf{p}_N, \mathbf{x}_1, \ldots, \mathbf{x}_N)],$$

and that $F_B(H)$ and $F_F(H)$ equal $-T \ln \operatorname{Tr} \exp(-\beta H)$, where the trace is taken over the symmetric (resp. antisymmetric) tensor product of the one-particle spaces. Then

$$F_{cl}(H) \leq F_F(H) \leq F_{cl}(h + v_F),$$
$$F_{cl}(H + v_B) \leq F_B(H) \leq F_{cl}(h),$$

where the function $h(\mathbf{p}_i, \mathbf{x}_i)$ is the expectation value of H in the symmetrized (resp. antisymmetrized) states of (2.2.10; 5):

$$h(\mathbf{z}_1, \ldots, \mathbf{z}_N) = \frac{\langle \mathbf{z}_1, \ldots, \mathbf{z}_N | H | \mathbf{z}_1, \ldots, \mathbf{z}_N \rangle}{\langle \mathbf{z}_1, \ldots, \mathbf{z}_N | \mathbf{z}_1, \ldots, \mathbf{z}_N \rangle}, \qquad \mathbf{z}_i = \mathbf{x}_i + i\mathbf{p}_i.$$

If the coherent states are chosen with $u(\mathbf{x}) = \exp(-mT|\mathbf{x}|^2/2)$, then the effective potentials are

$$v_F = \begin{cases} T \ln 2 \sum_{i \neq k} \exp(-mT|\mathbf{x}_i - \mathbf{x}_k|^2) & \text{if } \sup_j \sum_{i \neq j} \exp\left(\dfrac{-mT|\mathbf{x}_i - \mathbf{x}_j|^2}{2}\right) \leq \dfrac{1}{2}, \\ \infty & \text{otherwise;} \end{cases}$$

and

$$v_B = -T \sum_{i,k} \exp\left(\frac{-mT|\mathbf{x}_i - \mathbf{x}_k|^2}{2}\right).$$

Proof

The lower bounds. For one particle in x-space (see (III: 3.3.3)),

$$\langle \mathbf{x}| \exp\left(\frac{-\beta|\mathbf{p}|^2}{2m}\right)|\mathbf{x}'\rangle = \left(\frac{mT}{2\pi}\right)^{3/2} \exp\left(\frac{-mT|\mathbf{x}-\mathbf{x}'|^2}{2}\right),$$

so in the properly symmetrized or antisymmetrized basis, if there are N particles, then

$$\langle \mathbf{x}_1, \ldots, \mathbf{x}_N|\exp\left(\frac{-\beta\sum_i|\mathbf{p}_i|^2}{2m}\right)|\mathbf{x}_1, \ldots, \mathbf{x}_N\rangle$$

$$= \frac{1}{N!}\left(\frac{mT}{2\pi}\right)^{3N/2} \sum_P (\pm 1)^P \exp\left(\frac{-mT\sum_i|\mathbf{x}_i - \mathbf{x}_{P_i}|^2}{2}\right).$$

The sum over permutations amounts to just a permanent or determinant of the form $\langle \mathbf{z}_1, \ldots, \mathbf{z}_N|\mathbf{z}_1, \ldots, \mathbf{z}_N\rangle$, by (2.2.10; 5). It is therefore ≥ 1 or, respectively, ≤ 1, since the length of a vector is increased or, respectively, decreased when acted upon by a_f^* with $\|f\| = 1$:

$$\|a_f^*|\ \rangle\|^2 = \langle|a_f a_f^*|\rangle = \langle\ |\ \rangle \pm \langle|a_f^* a_f|\rangle \gtreqless \langle\ |\ \rangle.$$

For fermions, $\mathrm{Det}(\langle \mathbf{z}_i|\mathbf{z}_k\rangle) \leq 1$, whereas for bosons the permanent has an upper bound from Problem 4, $\mathrm{Per}(\langle \mathbf{z}_i|\mathbf{z}_k\rangle) \leq \exp[\sum_{i,k}|\langle \mathbf{z}_i|\mathbf{z}_k\rangle|]$. The rest of the proof is similar to that of the lower bound of (2.4.10):

$$\mathrm{Tr}\exp[-\beta(H_0 + v)] \leq \mathrm{Tr}\exp(-\beta H_0)\exp(-\beta v)$$

$$= \frac{1}{N!(2\pi)^{3N}} \int d^{3N}x\, d^{3N}p \exp[-\beta(H_0(\mathbf{p}_1, \ldots, \mathbf{p}_N) + v(\mathbf{x}_1, \ldots, \mathbf{x}_N))]$$

$$\times \begin{matrix}\mathrm{Per}\\ \mathrm{Det}\end{matrix}\left(\exp\left(-\frac{m}{2}|\mathbf{x}_i - \mathbf{x}_j|^2 T\right)\right)$$

$$\leq \frac{1}{N!(2\pi)^{3N}} \int d^{3N}x\, d^{3N}p \exp\left[-\beta\left(H_0(\mathbf{p}_1, \ldots, \mathbf{p}_N) + v(\mathbf{x}_1, \ldots, \mathbf{x}_N)\right.\right.$$

$$\left.\left. - T\begin{Bmatrix}\exp(-\sum_{i,j}(mT/2)|\mathbf{x}_i - \mathbf{x}_j|^2)\\ 0\end{Bmatrix}\right)\right].$$

The upper bounds. Since the symmetrized and antisymmetrized coherent states are not normalized,

$$\langle \mathbf{z}_1, \ldots, \mathbf{z}_N|\mathbf{z}_1, \ldots, \mathbf{z}_N\rangle \equiv n(z) = \begin{matrix}\mathrm{Per}\\ \mathrm{Det}\end{matrix}(\langle \mathbf{z}_i|\mathbf{z}_k\rangle) \gtreqless 1,$$

the normalization has to be accounted for in (2.2.11(i)):

$$\mathrm{Tr}\, k(a) \geq \int d\Omega_z n(z) k\left(\frac{\langle z|a|z\rangle}{n(z)}\right).$$

For bosons the inequality follows now from $n(z) \geq 1$. For fermions, with $u(\mathbf{x}) = \exp(-mT|\mathbf{x}|^2/4)$, it is necessary to estimate $\mathrm{Det}(1 + K)$, where

$$K_{ij} = \begin{cases} \exp(-(mT/2)|\mathbf{x}_i - \mathbf{x}_j|^2), & i \neq j \\ 0, & i = j. \end{cases}$$

Since

$$\|K\| \leq \sup_j \sum_{i \neq j} \exp\left(-\frac{mT}{2}|\mathbf{x}_i - \mathbf{x}_j|^2\right),$$

we find

$$\begin{aligned} \ln \mathrm{Det}(\langle \mathbf{z}_i | \mathbf{z}_j \rangle) &= \ln \mathrm{Det}(1 + K) = \mathrm{Tr} \ln(1 + K) \\ &= \sum_{n=2}^{\infty} \mathrm{Tr}\, K^n \frac{(-1)^{n+1}}{n} \leq \mathrm{Tr}\, K^2 \sum_{n=0}^{\infty} \frac{\|K\|^n}{n+2} \\ &\leq \ln \frac{1}{1 - \|K\|} \mathrm{Tr}\, K^2 \\ &\leq \begin{cases} \ln 2\, \mathrm{Tr}\, K^2 & \text{for } \|K\| \leq \frac{1}{2} \\ \infty & \text{otherwise.} \end{cases} \end{aligned}$$

Finally,

$$\mathrm{Tr}\, K^2 = \sum_{i \neq j} \exp[-mT|\mathbf{x}_i - \mathbf{x}_j|^2]. \qquad \square$$

Remarks (2.4.13)

1. If $\min_{i,j}|\mathbf{x}_i - \mathbf{x}_j| \equiv b > 0$, then $\|K\| \cong b^{-3} \int_b^\infty dr r^2 \exp(-r^2 mT/2) \cong \exp(-mTb^2/2)$, so v_F can be replaced with a hard-core potential with a radius depending on T and energy $\sim N$.
2. The ranges of the potentials v_B and v_F are approximately the thermal wavelength, i.e., the wavelength of a particle with kinetic energy $3T/2$, so when the particles are about this close together, as in a degenerate quantum gas, the bounds spread wide apart.

In closing, let us study the limit $N \to \infty$ in the framework of the canonical ensemble. Not only the reservoir but also the subsystem will be made infinite at the same time, and we wish to know whether the free energy density F/V tends to a limit φ. This should be the case whenever this limit exists microcanonically. Then the issue is how to recover the microcanonical quantities from knowledge of φ:

Theorem (2.4.14)

Suppose that, with $H \geq 0$, $\sigma_V(\varepsilon, \rho) = (1/V) \ln \mathrm{Tr}\, \Theta(V\varepsilon - H)$ converges uniformly on compact sets to a concave function $\sigma(\varepsilon, \rho)$ and is bounded above *by a function* $s(\varepsilon, \rho)$ *such that* $0 = s(\varepsilon_0, \rho) = \lim_{\varepsilon \to \infty} s(\varepsilon, \rho)/\varepsilon$, *when* V *is big enough. Writing as usual* $\beta = 1/T$, *then*

$$\lim_{V \to \infty} \left(-\frac{T}{V} \ln \operatorname{Tr} \exp(-\beta H) \right) = \inf_{\varepsilon} (\varepsilon - T\sigma(\varepsilon, \rho)) \equiv \varphi(T, \rho).$$

Remarks (2.4.15)

1. Since σ is concave, it has a right derivative,

$$\sigma' \equiv \lim_{\delta \downarrow 0} (\sigma(\varepsilon + \delta, \rho) - \varepsilon(\varepsilon, \rho)) \frac{1}{\delta}.$$

   The infimum is attained at the point $\varepsilon(T, \rho)$ for which $\sigma'(\varepsilon(T, \rho), \rho) = 1/T$ (see Figure 17). If σ' has a discontinuity, jumping over the value $1/T$, then $\varepsilon(T, \rho)$ is the point at which the jump takes place. The usual thermodynamic relationship $\varphi(T, \rho) = \varepsilon(T, \rho) - T\sigma(\varepsilon(T, \rho), \rho)$ holds for the free energy.
2. The function $\beta\varphi$ is a Legendre transform $(\mathscr{L}(\sigma))(\beta) = \inf_\varepsilon (\beta\varepsilon - \sigma(\varepsilon))$. The transformation $\mathscr{L}$ has the following properties:
   (i) $\mathscr{L} \circ \mathscr{L}$ produces the concave envelope of any function so $\mathscr{L} \circ \mathscr{L} = 1$ on concave functions;
   (ii) $\mathscr{L}$ maps a linear piece of a concave function to the point of a corner and vice versa;
   (iii) $\mathscr{L}$ maps the set of strictly concave, continuously differentiable functions into itself. By Property (i),

$$\sigma(\varepsilon) = \inf_{\beta} (\beta\varepsilon - \mathscr{L}(\sigma)(\beta)) = \inf_{T} \frac{\varepsilon - \varphi(T)}{T}.$$

3. If $\sigma(\varepsilon)$ is strictly concave and continuously differentiable, then by Problem 3 the limit $V \to \infty$ and the derivative by β can be taken in either order.

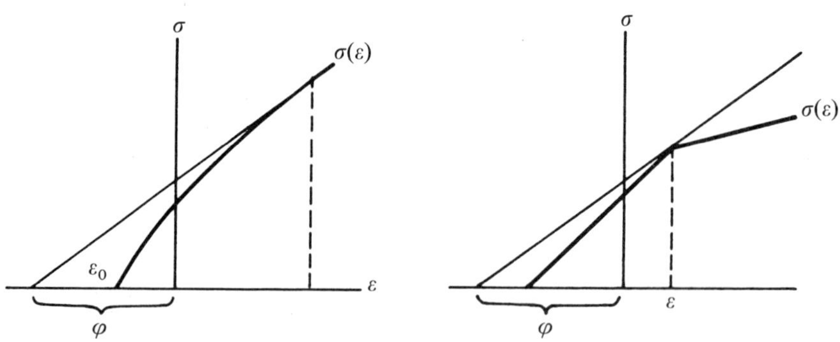

Figure 17 The geometric meaning of the free energy.

The energy and entropy densities calculated with the canonical density matrix are

$$\lim_{V\to\infty} \operatorname{Tr} \frac{H}{V} \frac{\exp(-\beta H)}{\operatorname{Tr}\exp(-\beta H)} = -\lim_{V\to\infty} \frac{\partial}{\partial\beta}\frac{1}{V} \ln \operatorname{Tr}\exp(-\beta H)$$

$$= -T^2 \frac{\partial}{\partial T}\frac{\varphi}{T} = \varphi + T\sigma$$

and

$$\lim_{V\to\infty} \frac{T}{V} S\left(\frac{\exp(-\beta H)}{\operatorname{Tr}\exp(-\beta H)}\right) = \varepsilon - \varphi,$$

which are obviously identical to the microcanonical energy and entropy densities. This fact is known as the **equivalence of the ensembles**.

4. The concavity of σ in ε is a necessary condition for the ensembles to be equivalent, since the specific heat in the canonical ensemble,

$$\frac{\partial\varepsilon}{\partial T} = \frac{\beta^2}{V}\frac{\partial^2}{\partial\beta^2} \ln \operatorname{Tr}\exp(-\beta H)$$

is automatically positive by Corollary (2.1.8; 3).

5. The bounding function s is necessary to ensure that

$$\lim_{V\to\infty} \sup_{\varepsilon}(T\sigma_V(\varepsilon, \rho) - \varepsilon) = \sup_{\varepsilon}(T\sigma(\varepsilon, \rho) - \varepsilon);$$

without it, $T\sigma_V(\varepsilon) - \varepsilon = 1 - (1 - \varepsilon/V)^2$ is a counterexample.

(The assumption that $H \geq 0$ is a normalization.)

Proof

$$\operatorname{Tr}\exp(-\beta H) = \int_0^\infty dE \exp(-\beta E) \frac{\partial}{\partial E} \operatorname{Tr}\Theta(E - H)$$

$$= \beta \int_0^\infty dE \exp[-\beta E + S(E)]$$

$$= \beta V \exp[-\beta V \varphi_V(T, \rho)]$$

$$\times \int_0^\infty d\varepsilon \exp[-\beta V(\varepsilon - T\sigma_V(\varepsilon) - \varphi_V)],$$

where

$$\varphi_V(T, \rho) = \inf_{\varepsilon}(\varepsilon - T\sigma_V(\varepsilon, \rho)).$$

If V is taken large enough, then the infimum lies between 0 and ε_0: $\varepsilon_0 - T\sigma(\varepsilon_0, \rho) = 0$. By assumption the functions σ_V converge uniformly on this compact interval, so $\varphi_V(T, \rho) \to \varphi(T, \rho)$. A modification of Lemma (2.3.12) shows that the contribution of the integral to φ is negligible in this limit. This step uses the assumption to ensure that for all $T > 0$ the exponent is dominated by $-\beta E$ for large E, so that the dominated convergence lemma applies. □

Several general properties of the Legendre transform of σ can be deduced from those of the microcanonical energy density (2.3.10), and are listed below:

Properties of the Free Energy Density (2.4.16)

1. As the infimum of a set of linear functions, $\varphi(T, \rho)$ is concave in T. If $H \geq 0$, then $\varphi(T, \rho) \leq 0$, and $\varphi(0, \rho) = 0$.
2. The function $\varphi(T, \rho)$ is convex in ρ, because $f(x, y)$ being convex in (x, y) implies that $\inf_x f(x, y)$ is convex in y (see (2.3.31)).
3. $\rho^{-1}\varphi(T, \rho)$ is an increasing function of ρ, since $\mathrm{Tr}\exp(-\beta H)$ is an increasing function of V when N and β are fixed.
4. $T^{-1}\varphi(T, \rho)$ is a decreasing function of T, since for $H \geq 0$, $\exp(-\beta H)$ is a decreasing function of β.

Remark (2.4.17)

Although convexity survives the thermodynamic limit, the analyticity (2.4.8; 3) of F is less hardy. The zeros of Z may approach the real axis as the system is made infinite, causing discontinuities in the derivatives of φ. Example (2.3.32; 2) can be modified to a degenerate BCS model, with

$$H = B\sum_{j=1}^{N} \sigma_j^{(z)} - \frac{1}{N}\sum_{i,j=1}^{N} (\boldsymbol{\sigma}_i \cdot \boldsymbol{\sigma}_j - \sigma_i^{(z)}\sigma_j^{(z)}).$$

This Hamiltonian has the eigenvalues $N(Bm_z - m(m + 2/N) + m_z^2)$, and, as in (2.3.34),

$$\varphi(T, B) = \inf_{0 \leq |m_z| \leq m \leq 1} \left(-m^2 + \left(m_z + \frac{B}{2}\right)^2 - \frac{B^2}{4} - T\sigma(m)\right),$$

$$\sigma(m) = \ln 2 - \frac{1+m}{2}\ln(1+m) - \frac{1-m}{2}\ln(1-m).$$

The infimum with respect to m_z is attained at $\max\{-B/2, -m\}$, assuming $B \geq 0$. If $m_z = -B/2$, then setting the derivative by m to zero leads to the equation

$$m(T) = \tanh\left(\frac{2m(T)}{T}\right).$$

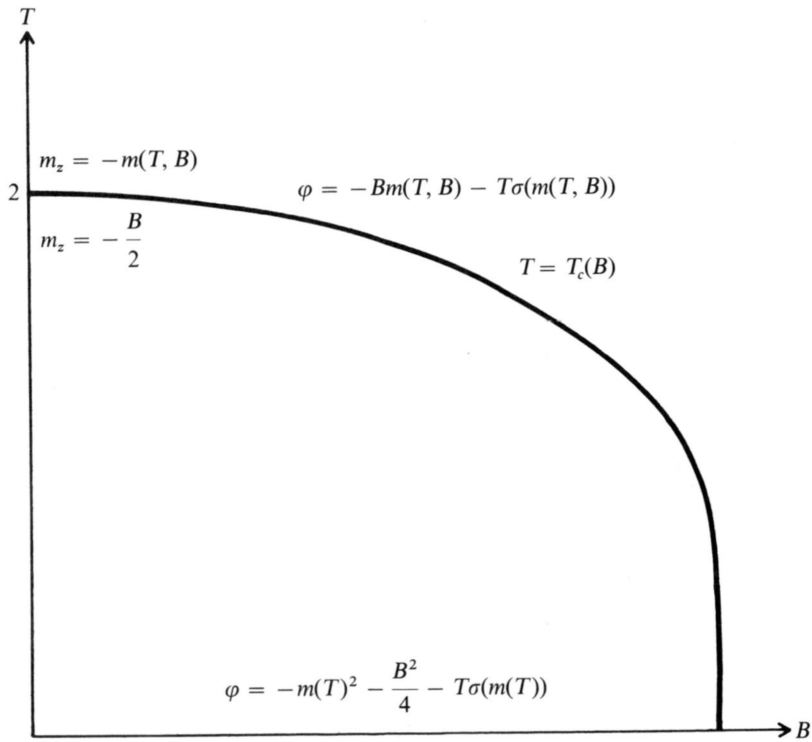

Figure 18 The free energy in Example (2.2.32; 2).

If $m_z = -m$, then instead of this, the minimizing value is $m(T, B) = \tanh(B/T)$. The two different possibilities give critical temperatures

$$T_c(B) \equiv \begin{cases} B/\operatorname{arctanh}(B/2) & \text{if } 0 < B < 2, \\ 0 & \text{if } 2 \leq B. \end{cases}$$

Figure 18 depicts $\varphi(T, B)$. The values of m and m_z are continuous at the transition point, but their derivatives are not. The function φ remains continuous along with its first derivatives—the derivatives by m and m_z vanish—but the second derivatives of $\varphi(T, B)$ are discontinuous at $T = T_c(B)$. Such properties as the specific heat display the discontinuity characteristic of a phase transition.

Problems (2.4.18)

1. Let ρ_n and ρ be density matrices for which $\rho_n \to \rho$. Show that $\operatorname{Tr} |\rho_n - \rho| \to 0$. (Hint: use the following lemma: *If ρ is a density matrix and Q a projection such that $\operatorname{Tr} \rho Q < \varepsilon$, then for all $a \in \mathscr{B}(\mathscr{H})$, $|\operatorname{Tr} \rho Q a| < \|a\| \sqrt{\varepsilon}$.*)

2. Prove (2.4.7) by applying Hartogs's theorem: *If $f(z_1, z_2)$ is separately analytic in z_1 and z_2, then it is jointly analytic.* Also observe that the trace is a continuous mapping $\mathscr{C}_1 \to \mathbb{C}$, where $\mathscr{C}_1$ has the norm $\|\cdot\|_1$.

3. Suppose $\varphi_V(\varepsilon)$ is a sequence of concave functions converging pointwise to $\varphi(\varepsilon)$. Let $\varphi'_{V,r}(\varepsilon)$ and $\varphi'_{V,l}$ denote the right and left derivatives of $\varphi_V(\varepsilon)$, and likewise for $\varphi'_r(\varepsilon)$ and $\varphi'_l(\varepsilon)$. Show that for all ε,

$$\varphi'_r(\varepsilon) \leq \liminf_{V \to \infty} \varphi'_{V,r}(\varepsilon) \leq \limsup_{V \to \infty} \varphi'_{V,l}(\varepsilon) \leq \varphi'_l(\varepsilon),$$

and that if φ_V and φ are differentiable at the point ε, then $\lim \varphi'_V(\varepsilon) = \varphi'(\varepsilon)$.

4. Show that $|\mathrm{Per}\langle \mathbf{z}_i | \mathbf{z}_k \rangle| \leq \exp \sum_{i,k} |\langle \mathbf{z}_i | \mathbf{z}_k \rangle|$.

5. Find a function of x and y that is convex in each variable separately but not jointly convex.

Solutions (2.4.19)

1. Lemma: $\rho = \sum c_i |x_i)(x_i|$, where $c_i \geq 0$, $\sum_i c_i = 1$, and $\{x_i\}$ is an orthonormal basis. Then

$$\mathrm{Tr}\, \rho Q = \sum_i c_i (x_i | Q x_i) = \sum_i c_i \|Q x_i\|^2 < \varepsilon,$$

$$|\mathrm{Tr}\, \rho Q a| = |\sum c_i (Q x_i | a x_i)| \leq \|a\| \cdot \sum_i c_i \|Q x_i\| < \|a\| \sqrt{\varepsilon},$$

since by the Cauchy–Schwarz inequality,

$$\sum_i c_i \|Q x_i\| = \sum_i \sqrt{c_i} \|Q x_i\| \sqrt{c_i} \leq \left(\sum_i c_i \|Q x_i\|^2\right)^{1/2} \cdot \left(\sum_i c_i\right)^{1/2} = \left(\sum_i c_i \|Q x_i\|^2\right)^{1/2}.$$

Proof of the proposition: For any finite-rank operator a, $\mathrm{Tr}\, \rho_n a \to \mathrm{Tr}\, \rho a$, and $\mathrm{Tr}\, \rho_n (\mathbf{1} - a) \to \mathrm{Tr}\, \rho (\mathbf{1} - a)$. Now let P be the projection onto the first N eigenvalues of ρ and choose N such that $\mathrm{Tr}\, \rho(\mathbf{1} - P) < \varepsilon$. Then

$$\begin{aligned}\mathrm{Tr}(\rho_n - \rho) a = \mathrm{Tr}\, \rho_n (\mathbf{1} - P) a + \mathrm{Tr}(\mathbf{1} - P) \rho_n P a + \mathrm{Tr}(\rho_n - \rho) P a P \\ + \mathrm{Tr}(P \rho P - \rho) a.\end{aligned}$$

$$\mathrm{Tr}(\rho_n - \rho) P a P < \varepsilon \|P a P\| < \varepsilon \|a\|$$

for sufficiently large n, since all topologies are equivalent on the finite-dimensional space $P\mathscr{B}(\mathscr{H})P$, and $\mathrm{Tr}(\rho_n - \rho) P a P \to 0$. $|\mathrm{Tr}(P \rho P - \rho) a| \leq \|a\| \mathrm{Tr}(\mathbf{1} - P) \rho < \|a\| \cdot \varepsilon$. $\mathrm{Tr}\, \rho_n (\mathbf{1} - P) \to \mathrm{Tr}\, \rho (\mathbf{1} - P) < \varepsilon$, which implies that for n large enough, $\mathrm{Tr}\, \rho_n (\mathbf{1} - P) < 2\varepsilon$. Hence, by the lemma,

$$|\mathrm{Tr}\, \rho_n (\mathbf{1} - P) a| < \sqrt{2\varepsilon}\, \|a\|,$$

$$|\mathrm{Tr}(\mathbf{1} - P) \rho_n P a| = |\mathrm{Tr}\, \rho_n (\mathbf{1} - P) a^* P| \leq \sqrt{2\varepsilon}\, \|a^* P\| \leq \sqrt{2\varepsilon}\, \|a\|.$$

Consequently,

$$|\mathrm{Tr}(\rho_n - \rho) a| < (2\varepsilon + 2\sqrt{2\varepsilon})\, \|a\|,$$

$$\mathrm{Tr}\, |\rho_n - \rho| = \sup_{\|a\| \leq 1} |\mathrm{Tr}(\rho_n - \rho) a| < 2\varepsilon + 2\sqrt{2\varepsilon}.$$

2. $U(\alpha, \beta) \equiv \exp[-\beta(H_0 + \alpha v)] \in \mathscr{C}_1$

(i) Analyticity (=complex differentiability) in β:

$$\left\| \frac{U(\alpha, \beta + \beta') - U(\alpha, \beta)}{\beta'} + (H_0 + \alpha v)U(\alpha, \beta) \right\|_1$$
$$\leq \left\| \left(\frac{U(\alpha, \beta') - 1}{\beta'} + (H_0 + \alpha v) \right) U\left(\alpha, \frac{\beta}{2}\right) \right\| \left\| U\left(\alpha, \frac{\beta}{2}\right) \right\|_1 \to 0$$

as $\beta' \to 0$, since U is a $\|\cdot\|$-convergent integral of $\|\cdot\|$-analytic functions and therefore a $\|\cdot\|$-analytic mapping, $\mathbb{C} \times \mathbb{C}^+ \to \mathscr{B}$.

(ii) Analyticity in α:

$$U(\alpha + \alpha', \beta) - U(\alpha, \beta) = -\beta\alpha' \int_0^1 d\tau U(\alpha + \alpha', \beta(1 - \tau)) v U(\alpha, \tau\beta),$$

$$\|U(\alpha + \alpha', \beta(1 - \tau)) v U(\alpha, \tau\beta)\|_1 \leq \|U(\alpha + \alpha', \beta(1 - \tau))\|$$
$$\times \left\| vU\left(a, \frac{\tau\beta}{2}\right) \right\| \left\| U\left(\alpha, \frac{\tau\beta}{2}\right) \right\|_1 \leq \text{constant},$$

when $\frac{1}{2} \leq \tau \leq 1$. If $0 \leq \tau \leq \frac{1}{2}$, then the first factor has to be divided up. This shows that the mapping $\mathbb{C} \times \mathbb{C}^+ \to \mathscr{B}_1 : (\alpha, \beta) \to U(\alpha, \beta)$ is analytic, and therefore the mapping $\mathbb{C} \times \mathbb{C}^+ \to \mathbb{C}: (\alpha, \beta) \to \operatorname{Tr} U(\alpha, \beta)$ is analytic, because the trace is continuous and linear $\mathscr{B}_1 \to \mathbb{C}$, and thus also analytic.

3. Concavity yields $(1/\varepsilon')(\varphi_V(\varepsilon + \varepsilon') - \varphi_V(\varepsilon)) \leq \varphi'_{V,r}(\varepsilon) \leq \varphi'_{V,l}(\varepsilon) \leq (1/\varepsilon')(\varphi_V(\varepsilon - \varepsilon') - \varphi_V(\varepsilon))$ for all $\varepsilon' > 0$, and the statement follows from this with the limits $\lim_{\varepsilon' \to 0} \lim_{V \to \infty}$.

4. $$\operatorname{Per}\langle \mathbf{z}_i | \mathbf{z}_k \rangle \leq \operatorname{Per}|\langle \mathbf{z}_i | \mathbf{z}_k \rangle| = \sum_P \prod_{i=1}^N |\langle \mathbf{z}_i | \mathbf{z}_{P_i} \rangle| \leq \prod_{(i,j)} (1 + |\langle \mathbf{z}_i | \mathbf{z}_j \rangle|)$$
$$\leq \exp\left(\sum_{i,j} |\langle \mathbf{z}_i | \mathbf{z}_j \rangle| \right).$$

5. $f(x, y) = -xy$. The Hessian matrix $\begin{pmatrix} 0 & -1 \\ -1 & 0 \end{pmatrix}$ is not positive.

2.5 The Grand Canonical Ensemble

The thermodynamic functions are easier to calculate explicitly if the constraint of a fixed number N of particles is dropped. It is physically realistic for a system coupled to a reservoir of particles.

This section will investigate the situation of a system with a reservoir with which it can exchange particles as well as heat. As in (2.3.23), the underlying Hilbert space is taken as

$$\bigoplus_{N_1=0}^{N} \mathscr{H}_{N_1, V_1} \otimes \mathscr{H}_{N - N_1, V_2},$$

and the Hamiltonian is

$$H = \bigoplus_{N_1=0}^{N} (H_1(V_1, N_1) + H_2(V_2, N - N_1)).$$

We consider the limit as $N \to \infty$ and $V_2 \to \infty$, and begin by collecting the immediate generalizations of some of the results of §2.4. Proofs will not be given, as they entail only slight modifications of the earlier ones.

Convergence of the Reduced Density Matrix (2.5.1)

Suppose that the concave, increasing functions

$$\frac{1}{V_2} \operatorname{Tr}_2 \Theta(E_2 - H_2) = \frac{1}{V_2} S_2(E_2, V_2, N_2) \equiv \sigma_{V_2}\left(\frac{E_2}{V_2}, \frac{N_2}{V_2}\right)$$

and their derivatives converge uniformly on a neighborhood of $\varepsilon = E_2/V_2$ *and* $\rho = N_2/V_2$ *to* $\sigma(\varepsilon, \rho)$, $\partial\sigma/\partial\varepsilon$, *and* $\partial\sigma/\partial\rho$. *Then with* $V = V_1 + V_2$, $N = N_1 + N_2$,

$$\lim_{V_2 \to \infty} \frac{\operatorname{Tr}_2 \Theta(E - H)}{\operatorname{Tr} \Theta(E - H)} \to \frac{\exp\left[-H_1(V_1, N_1)\dfrac{\partial\sigma}{\partial\varepsilon} - N_1 \dfrac{\partial\sigma}{\partial\rho}\right]}{\operatorname{Tr}_1 \exp\left[-H_1(V_1, N_1)\dfrac{\partial\sigma}{\partial\varepsilon} - N_1 \dfrac{\partial\sigma}{\partial\rho}\right]} = \rho_{GC}$$

in the trace norm.

Remarks (2.5.2)

1. The symbol Tr_2 denotes the trace in the second factor of

$$\bigoplus_{N_1=0}^{N} \mathscr{H}_{N_1, V_1} \otimes \mathscr{H}_{N - N_1, V_2},$$

so in the limit $N \to \infty$, $H_1(N_1, V_1)$ operates on $\sum_{N_1=0}^{\infty} \mathscr{H}_{N_1, V_1}$. This operator on the Hilbert space of an indefinite number of particles is most conveniently written in terms of the field operators (1.3.2).
2. The values of μ for which $\exp[-\beta(H - \mu N)] \in \mathscr{C}_1$ depend on the problem. If, for instance,

$$-\ln \operatorname{Tr}_{|\mathscr{H}_{N_1}} \exp[-\beta H_1(N_1)] > -c N_1,$$

then the trace exists whenever $\operatorname{Re} \beta\mu < -c$.

Many of the results of §2.4 may be reformulated for the grand canonical ensemble merely be replacing H with $H - \mu N$. An example is

The Principle of Maximum Entropy (2.5.3)

Let $\bar{\rho}$ be a density matrix such that $\mathrm{Tr}\,\bar{\rho}H = \mathrm{Tr}\,\rho_{GC}H$, $\mathrm{Tr}\,\bar{\rho}N = \mathrm{Tr}\,\rho_{GC}N$. *Then* $S(\rho_{GC}) \geq S(\bar{\rho})$.

If system 1 is now taken infinitely large, presupposing the extensivity following from $H > -N_c$, then T/V times the logarithm of the grand canonical partition function has a limit, which may be identified as the pressure, with reference to (2.3.29).

The Thermodynamic Limit (2.5.4)

If the assumptions of (2.4.14) *are satisfied. then*

$$\begin{aligned}\lim_{V\to\infty} \frac{T}{V} \ln \mathrm{Tr} \exp[-\beta(H - \mu N)] &= \lim_{V\to\infty} \frac{T}{V} \ln \sum_{N=0}^{\infty} \exp\left[-\beta V\left(\varphi_V\left(T, \frac{N}{V}\right) - \mu N\right)\right] \\ &= \sup(\mu\rho - \varphi(T, \rho)) = P(T, \mu).\end{aligned}$$

Remarks (2.5.5)

1. The supremum is attained where the right derivative
$$\lim_{\delta\downarrow 0} (\varphi(T, \rho + \delta) - \varphi(T, \rho))\delta^{-1} = \mu,$$
unless μ is on an endpoint of the interval on which $P(T, \mu)$ is defined. This means that with (2.3.29), μ can be identified with
$$\left.\frac{\partial \varepsilon}{\partial \rho}\right|_{\sigma} = \left.\frac{\partial \varepsilon}{\partial \rho}\right|_{T} - T\left.\frac{\partial \sigma}{\partial \rho}\right|_{T} = \left.\frac{\partial \varphi}{\partial \rho}\right|_{T}.$$
Because
$$\mu\rho - \varphi = \rho\left.\frac{\partial \varepsilon}{\partial \rho}\right|_{\sigma} + \sigma\left.\frac{\partial \varepsilon}{\partial \sigma}\right|_{\rho} - \varepsilon = P,$$
the grand canonical partition function turns out to be $\exp(PV/T)$. We shall also speak of P as the pressure when the system is finite, although it does not exactly agree with the definition as the force per area on the wall.
2. As before, the ensembles are equivalent, on account of the identities
$$\rho = \left.\frac{\partial P}{\partial \mu}\right|_{T}, \qquad \varepsilon = T\left.\frac{\partial P}{\partial T}\right|_{\mu/T} - P = \mu\rho - T\left.\frac{\partial \varphi}{\partial T}\right|_{\rho} - \mu\rho + \varphi = \varphi + T\sigma,$$
$$T\sigma = \varepsilon - \mu\rho + P.$$
Observe that the grand canonical averages of N/V and H_N/V approach ρ and ε, and that the entropy density of ρ_{GC} equals σ.

Properties of the Pressure (2.5.6)

1. The function $(T, \mu) \to P$ is convex, since it is the supremum of convex functions.
2. The pressure increases with μ, since it is the supremum of increasing functions.
3. If $H - \mu N \geq 0$, then $T^{-1}P$ is an increasing function of T, since $\exp[-\beta(H - \mu N)]$ is a decreasing function of β.

The grand canonical ensemble is particularly useful for identical particles, and allows the thermodynamic functions of bosons or fermions interacting with an external field to be evaluated more explicitly. For this purpose, we write the Hamiltonian and the particle number in terms of the field operators (1.3.2) and our orthogonal basis $\{f_m\}$, as

$$\begin{aligned} H &= \sum_{m,n} a_m^* a_n \left[\int d^3x \nabla f_m^*(\mathbf{x}) \cdot \nabla f_n(\mathbf{x}) + f_m^*(\mathbf{x}) f_n(\mathbf{x}) v(\mathbf{x}) \right] \\ &\equiv \sum_{m,n} a_m^* a_n \langle f_m | h | f_n \rangle, \\ N &= \sum_m a_m^* a_m, \end{aligned} \tag{2.5.7}$$

where $h = |\mathbf{p}|^2 + v(\mathbf{x})$ is the one-particle Hamiltonian, and a_m stands for $a(f_m)$. If h has pure-point spectrum with eigenvalues ε_m, and f_m are taken as the eigenvectors associated with ε_m, then

$$\operatorname{Tr} \exp[-\beta(H - \mu N)] = \operatorname{Tr} \exp\left[-\beta \sum_m a_m^* a_m (\varepsilon_m - \mu) \right]. \tag{2.5.8}$$

Taking the trace leads to easily computed sums, since a^*a has the eigenvalues 0 and 1 for fermions and 0, 1, 2, ..., for bosons. In these cases, P_F and P_B become

$$P_F(z) = -P_B(-z) = \frac{T}{V} \sum_m \ln(1 + z \exp(-\beta \varepsilon_m)), \tag{2.5.9}$$

where $z \equiv \exp(\beta\mu)$ is known as the **fugacity**. When written in terms of the one-particle Hamiltonian $h = |\mathbf{p}|^2 + V(\mathbf{x})$ and the trace tr on the one-particle space $L^2(\mathbb{R}^3)$,

The Pressure of Fermions or Bosons in an External Field (2.5.10)

becomes

$$P_F(T, z) = \frac{T}{V} \operatorname{tr} \ln(1 + z \exp(-\beta h)) = -P_B(T, -z).$$

Remarks (2.5.11)

1. In the limit $z \to 0$, $P_F(T, z) = P_B(T, z) = z(T/V) \sum_m \exp(-\beta\varepsilon_m)$, which corresponds to very dilute matter, for which both Bose and Fermi statistics become the same (Boltzmann statistics).
2. If $h \geq 0$ and $\exp(-\beta h) \in \mathscr{C}_1$, then the singularities of $\exp(P)$ occur where $z = -\exp(\beta\varepsilon_m) < -1$, $m = 0, 1, 2, \ldots$. The function $\exp(P)$ is analytic in z until the singularities are reached, i.e., the power series in z converges. The analytic function $P_F(T, z)$ describes all three kinds of statistics. Fermi statistics correspond to $z = \exp(\mu/T) > 0$, Boltzmann statistics to $z \to 0$, and Bose statistics to $-\exp(\varepsilon_0) < z < 0$ (see Figure 19).

It is easy to calculate expectation values as well as the partition function:

$$\langle a_m^* a_{m'} \rangle \equiv \text{Tr}\, a_m^* a_{m'} \exp[-\beta(H - \mu N + PV)] = \frac{\delta_{mm'}}{\exp[\beta(\varepsilon_m - \mu)] \pm 1}. \tag{2.5.12}$$

Since every one-particle vector $|f\rangle \in L^2(\mathbb{R}^3)$ can be expanded in eigenvectors of h, and when restricted to $L^2(\mathbb{R}^3)$, $a_f^* a_f$ equals $P_f = |f\rangle\langle f|$, the information about the one-particle observables is contained in the

Effective One-Particle Density Matrix (2.5.13)

One-particle expectation values are given by $\rho_1 = (\exp[\beta(h - \mu)] \pm 1)^{-1}$ with the formula $\langle a_f^* a_f \rangle = \text{Tr}\, \rho_1 P_f = \langle f | \rho_1 | f \rangle$. The density matrix ρ_1 has the properties

$$\text{Tr}\, \rho_1 = \overline{N},$$

and

$$0 \leq \rho_1 \leq \begin{cases} 1 & \text{for fermions} \\ \overline{N} & \text{for bosons.} \end{cases}$$

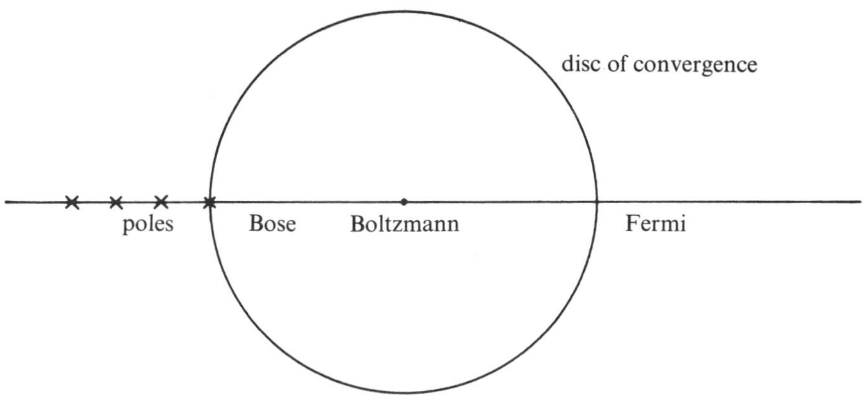

Figure 19 Singularities of P in the complex z-plane.

Remarks (2.5.14)

1. The number $\bar{N}$ is defined by $\langle \sum_m a_m^* a_m \rangle = \text{tr}[\exp(\beta(h - \mu)) \pm 1]^{-1}$. If it is preferred to deal with these more understandable variables of the canonical ensemble, then this can be taken as the equation determining μ.
2. Similarly, $\langle H \rangle = \text{tr}\ \rho_1 h$, etc.
3. If a reduced density matrix on the one-particle phase space is defined with coherent states (cf. (2.2.7) with (2.2.10; 5)), $\rho(\mathbf{x}, \mathbf{p}) = \langle a_z^* a_z \rangle = \langle z | \rho_1 | z \rangle$, then the properties of ρ_1 generalize as

$$\int \frac{d^3x\, d^3p}{(2\pi)^3} \rho(\mathbf{x}, \mathbf{p}) = \bar{N},$$

and

$$0 \leq \rho(\mathbf{x}, \mathbf{p}) \leq \begin{cases} 1 & \text{for fermions} \\ \bar{N} & \text{for bosons.} \end{cases}$$

This shows that the exclusion principle of fermions has the effect of reducing the maximum value $\bar{N}$ of $\rho(z)$ allowed in quantum mechanics to 1.

As well as the one-particle observables, global properties like $\langle H \rangle$ and P can be calculated with ρ_1, and even the many-particle entropy can be expressed in terms of ρ_1:

The Effective One-Particle Entropy (2.5.15)

$$S(\rho_{GC}) = -\text{Tr}\ \rho_{GC} \ln \rho_{GC} = \frac{VP}{T} + \beta\langle H - \mu N \rangle$$

$$= \text{tr}\left[\pm \ln(1 \pm \exp[-\beta(h - \mu)]) + \beta \frac{h - \mu}{\exp[\beta(h - \mu)] \pm 1} \right]$$

$$= -\text{tr}[\rho_1 \ln \rho_1 \pm (1 \mp \rho_1) \ln(1 \mp \rho_1)].$$

Remarks (2.5.16)

1. The part in addition to the normal $-\text{tr}\ \rho \ln \rho$ in S reveals that the many-particle system has increased disorder. The addition shows up in the entropy of a spin-$\frac{1}{2}$ density matrix,

$$S\begin{pmatrix} \rho & 0 \\ 0 & 1 - \rho \end{pmatrix} = -\rho \ln \rho - (1 - \rho) \ln(1 - \rho),$$

where ρ is the probability for spin-up, and in the entropy of an oscillator,

$$S\left((1-x)\begin{pmatrix}1 & & & \\ & x & & \\ & & x^2 & \\ & & & \ddots\end{pmatrix}\right) = -\rho \ln \rho + (1+\rho)\ln(1+\rho),$$

where

$$\rho = \sum_{n=0}^{\infty} n\rho_{nn} = \frac{x}{1-x}$$

is the expectation value of the number of phonons.

2. In accordance with the maximum-entropy principle (2.5.3), the one-particle ρ_1 (2.5.13) is the $\rho \in \mathscr{C}_1(L^2(\mathbb{R}^3))$ that maximizes

$$\frac{PV}{T} = -\text{tr}[\rho \ln \rho \pm (1 \mp \rho)\ln(1 \mp \rho) + \rho\beta(h - \mu)]$$

(Problem 4). Also, on a formal level,

$$0 = -\frac{V}{T}\frac{\delta P}{\delta \rho}\bigg|_{\rho=\rho_1} = \beta(h - \mu) + \ln \rho_1 - \ln(1 \mp \rho_1)$$

$$\Rightarrow \rho_1 = [\exp[\beta(h-\mu)] \pm 1]^{-1}.$$

The density matrix ρ_1 describes the distribution of bosons or fermions. Its significance is brought out most clearly in the classical limit.

Classical Bounds for the Pressure of Particles in External Fields (2.5.17)

With notation like that of (2.2.7), *let*

$$h = |\mathbf{p}|^2 + v(\mathbf{x}) = \int d\Omega_z f(\mathbf{z})|\mathbf{z}\rangle\langle\mathbf{z}|, \qquad h(\mathbf{z}) = \langle\mathbf{z}|h|\mathbf{z}\rangle,$$

$$\rho(\mathbf{z}) = \text{Tr}\, a_z^* a_z \rho_{GC} = \langle\mathbf{z}|[\exp[\beta(h-\mu)] \mp 1]^{-1}|\mathbf{z}\rangle = \langle\mathbf{z}|\rho_1|\mathbf{z}\rangle,$$

where v is such that all expressions appearing are well defined. Then, with $\mathbf{z} = \mathbf{q} + i\mathbf{p}$, *for bosons,*

$$-\int d\Omega_z \ln(1 - \exp[-\beta(h(\mathbf{z}) - \mu)]) \le \beta P(\beta, \mu)V$$

$$\le -\int d\Omega_z \ln(1 - \exp[-\beta(|\mathbf{p}|^2 + v(\mathbf{q}) - \mu)]),$$

$$\beta P(\beta, \mu)V \le \int d\Omega_z \ln(1 + \rho(\mathbf{z})),$$

and for fermions,

$$\int d\Omega_z \ln(1 + \exp[-\beta(h(\mathbf{z}) - \mu)]) \leq \beta P(\beta, \mu)V$$

$$\leq \int d\Omega_z \ln(1 + \exp[-\beta(f(\mathbf{z}) - \mu)]),$$

$$-\int d\Omega_z \ln(1 - \rho(\mathbf{z})) \leq \beta P(\beta, \mu)V.$$

In analogy with (2.4.9), *one gathers that*

$$h(\mathbf{q} + i\mathbf{p}) = |\mathbf{p}|^2 + v_u(\mathbf{q}) + \int |\nabla u(\mathbf{x})|^2 \, d^3x$$

and

$$f(\mathbf{q} + i\mathbf{p}) = |\mathbf{p}|^2 + v^u(\mathbf{q}) - \int |\nabla u(\mathbf{x})|^2 \, d^3x,$$

where

$$v_u(\mathbf{q}) = \int v(\mathbf{x}) |u(\mathbf{x} - \mathbf{q})|^2 \, d^3x$$

and

$$v(\mathbf{x}) = \int v^u(\mathbf{q}) |u(\mathbf{x} - \mathbf{q})|^2 \, d^3q,$$

and u is an arbitrary vector of $L^2(\mathbb{R}^3)$ such that $\|u\|_2 = 1$ and $\|\nabla u\|_2 < \infty$.

Proof

Bosons. The first two inequalities are the analogues of (2.4.12), where the lower bound relies on (2.2.11) with the convex function $x \to -\ln(1 - \exp(-x))$. The upper follows from Corollary (2.1.8: 7) if it is borne in mind that $h - \mu$ must be positive, so $\|\exp[-(h - \mu)]\| < 1$, and the series

$$-\ln(1 - \exp[-\beta(h - \mu)]) = \sum_{n=1}^{\infty} \frac{\exp[-n\beta(h - \mu)]}{n}$$

converges in the norm $\|\cdot\|$. It must even converge in the norm $\|\cdot\|_1$, since it was assumed that $-\ln(1 - \exp[-\beta(h - \mu)]) \in \mathscr{C}_1$, and the series is monotonic. With recourse again to (2.4.9), each term is bounded by

$$-\int d\Omega_z (1/n) \exp[-n\beta(|\mathbf{p}|^2 + V(\mathbf{q}) - \mu)],$$

which also converges by assumption. Since all terms are positive, $\sum_n$ and

$\int d\Omega_z$ can be interchanged. The final inequality follows from the concavity of the function $x \to \ln(1 + x)$:

$$-\langle \mathbf{z}|\ln(1 - \exp[-\beta(h - \mu)])|\mathbf{z}\rangle = \langle \mathbf{z}|\ln(1 + \rho_1)|\mathbf{z}\rangle \leq \ln(1 + \langle \mathbf{z}|\rho_1|\mathbf{z}\rangle)$$

implies that

$$-\operatorname{tr}\ln(1 - \exp[-\beta(h - \mu)]) \leq \int d\Omega_z \ln(1 + \rho(\mathbf{z})).$$

Fermions. The first two inequalities again come from (2.2.11) with the convex function $x \to \ln(1 + \exp(-x))$, and the last one is a consequence of the convexity of $x \to -\ln(1 - x)$. □

Remarks (2.5.18)

1. If $x > 0$, then $(\exp(x) \pm 1)^{-1}$ is convex, and if $x < 0$, then it is concave. For bosons, $x > 0$, and so
$$\rho(\mathbf{z}) = \langle \mathbf{z}|(\exp[\beta(h - \mu)] - 1)^{-1}|\mathbf{z}\rangle \geq (\exp[\beta(h(\mathbf{z}) - \mu)] - 1)^{-1}.$$
The analogous inequality for fermions is true only if $h - \mu > 0$.
2. In Problem 3 it is shown that $\langle \mathbf{z}|(-\Delta)|\mathbf{z}\rangle = |\mathbf{p}|^2 + K, K = \int d^3x|\nabla u(\mathbf{x})|^2$, where $\mathbf{z} = \mathbf{q} + i\mathbf{p}$, and on the other hand, $-\Delta = \int d\Omega_z(|\mathbf{p}|^2 - K)|\mathbf{z}\rangle\langle\mathbf{z}|$. Similarly, $\langle \mathbf{z}|v|\mathbf{z}\rangle = \int d^3x|u(\mathbf{x} - \mathbf{q})|^2 v(\mathbf{x}) = v_u(\mathbf{q})$, and $v = \int d\Omega_z v^u(\mathbf{q}) \times |\mathbf{z}\rangle\langle\mathbf{z}|$, if $v(\mathbf{x}) = \int d^3q|u(\mathbf{x} - \mathbf{q})|^2 v^u(\mathbf{q})$. What goes on with the lower bound is thus that the classical Hamiltonian h is increased by the kinetic energy K of u, and the potential is smeared out by convolution with $|u|^2$. With the upper bound the classical Hamiltonian is reduced by K and the potential is unsmeared. If v is of slow enough variation that even for u with small K, $v^u(q)$ is approximately equal to $v_u(\mathbf{q}) = \langle \mathbf{z}|v|\mathbf{z}\rangle$, then the bounds draw close together.
3. In the very dilute limit of (2.5.11; 1) the bounds produce the classical result, if the indistinguishablity of the particles is accounted for by a $1/N!$ in the phase space:
$$\begin{aligned}\exp\left(\frac{PV}{T}\right) &= \sum_{N=0}^{\infty} \frac{1}{N!}\int d^3x_1 \cdots d^3p_N \\ &\quad \times \exp[-\beta(|\mathbf{p}_1|^2 + \cdots |\mathbf{p}_N|^2 + v(\mathbf{x}_1) + \cdots + v(\mathbf{x}_N) - N\mu)] \\ &= \exp[\exp(-\beta(F_{cl} - \mu))],\end{aligned}$$
so by (2.5.4),
$$\frac{PV}{T} = \int d^3x\, d^3p \exp[-\beta(|\mathbf{p}|^2 + v(\mathbf{x}) - \mu)] = N,$$
which is the ideal gas law. Unless $\exp(\beta(h - \mu)) \gg 1$, the statistics matter. They are built into the bounds, but the indeterminacy relation forces the bounds apart.

4. In the classical limit, in which Inequalities (2.5.17) become equalities, $\rho_1(\mathbf{x}, \mathbf{p}) = (\exp[-\beta(|\mathbf{p}|^2 + V(\mathbf{x}) - \mu)] \pm 1)^{-1}$ is the density on phase space that optimizes

$$\frac{PV}{T} = S(\rho_1) - \beta\langle h - \mu\rangle \equiv -\int d\Omega_z[\rho_1(\mathbf{z}) \ln \rho_1(\mathbf{z}) \pm (1 \mp \rho_1(\mathbf{z})) \ln(1 \mp \rho_1(\mathbf{z})) + \rho_1(\mathbf{z})\beta(h(\mathbf{z}) - \mu)]$$

(Problem 4).

5. If, more generally, ρ is a density matrix of the many particle system on Fock space, and

$$\rho_1 = \int d\Omega_z \, d\Omega_{z'} \, |\mathbf{z}\rangle\langle\mathbf{z}'| \operatorname{Tr}(\rho a^*_{z'} a_z)$$

and

$$\rho(\mathbf{z}) = \operatorname{Tr}(\rho a^*_z a_z) = \langle\mathbf{z}|\rho_1|\mathbf{z}\rangle$$

are the associated one-particle density matrix and density, then it follows from (2.5.3) and (2.5.15) that

$$S(\rho) = -\operatorname{Tr} \rho \ln \rho \le -\operatorname{tr}[\rho_1 \ln \rho_1 \pm (1 \mp \rho_1) \ln(1 \mp \rho_1)] \le -\int d\Omega_z[\rho(z) \ln \rho(z) \pm (1 \mp \rho(z)) \ln(1 \mp \rho(z))],$$

where the H in (2.5.3) is taken as the second quantization of $(1/\beta) \times [\ln(1 \mp \rho_1) - \ln \rho_1]$, and μ is set to 0. The first inequality becomes an equality with ρ_{GC}, which is the density matrix of greatest entropy for a given one-particle density matrix ρ_1. The second inequality follows from (2.2.11), since

$$x \to -[x \ln x \pm (1 \mp x) \ln(1 \mp x)]$$

is concave with the upper signs for $0 < x < 1$ and with the lower signs for $x < 0$.

The extent of the validity of the classical picture will be delineated through a series of examples.

Free Bosons and Fermions in a Box with Soft Walls (2.5.19)

With a harmonic potential $v(\mathbf{x}) = \omega^2|\mathbf{x}|^2$, the N-particle Hamiltonian is

$$H = \sum_{i=1}^{N} (|\mathbf{p}_i|^2 + \omega^2|\mathbf{x}_i|^2) = \sum_{i=1}^{N} |\mathbf{p}_i|^2 + \frac{\omega^2}{2N} \sum_{i,j=1}^{N} |\mathbf{x}_i - \mathbf{x}_j|^2 + \frac{\omega^2}{N}\left|\sum_{i=1}^{N} \mathbf{x}_i\right|^2,$$

containing harmonic forces between the particles and a harmonic force acting on the center of mass. As before (cf. (2.4.11) and (2.5.18; 1)), let $\mathbf{z} = \mathbf{q} + i\mathbf{k}$ and $u(\mathbf{x}) = \exp(-\omega|\mathbf{x}|^2/2)$: $h(\mathbf{z}) = |\mathbf{k}|^2 + \omega^2|\mathbf{q}|^2 + 3\omega$, $f(|\mathbf{z}|) = |\mathbf{k}|^2 + \omega^2|\mathbf{q}|^2 - 3\omega$. Because

$$\mp \int \frac{d^3k\, d^3q}{(2\pi)^3} \ln(1 \mp \exp[-\beta(|\mathbf{k}|^2 + \omega^2|\mathbf{q}|^2 - \mu)])$$
$$= \pm \frac{T^3}{(2\omega)^3} \sum_{\nu=1}^{\infty} (\pm 1)^\nu \frac{\exp(\nu\beta\mu)}{\nu^4},$$

(2.5.17) implies

$$\pm \frac{T^3}{(2\omega)^3} F_4(\pm(\exp[\beta(\mu - 3\omega)]) \leq \ln \operatorname{Tr}_{\substack{B\\F}} \exp[-\beta(H - \mu N)]$$
$$\leq \pm \frac{T^3}{(2\omega)^3} F_4(\pm\exp[\beta(\mu + 3\omega)]), \tag{2.5.20}$$

where

$$F_\sigma(x) \equiv \sum_{\nu=1}^{\infty} \frac{x^\nu}{\nu^\sigma}.$$

The result can be calculated exactly in this case, since the eigenvalues are $\varepsilon_{\mathbf{m}} = 3\omega + 2\omega(m_1 + m_2 + m_3)$, $\mathbf{m} \in (\mathbb{Z}^+)^3$, and so

$$\mp \sum_{\mathbf{m}} \ln(1 \mp \exp[-\beta(\varepsilon_{\mathbf{m}} - \mu)])$$
$$= \pm \sum_{\nu=1}^{\infty} (\pm 1)^\nu \frac{\exp[\nu\beta(\mu - 3\omega)]}{\nu} [(1 - \exp(-2\beta\omega\nu))^{-3} - 1].$$

The bounds draw together to this value in the limit $\omega \to 0$. This limit is related to the limit $V \to \infty$, since the average of, for instance, $|\mathbf{x}|^2$ is $\sim T/\omega^2$. Accordingly, we eliminate ω in favor of the effective volume $V = (\pi T)^{3/2}/\omega^3$ and take the limit $V \to \infty$. Then with $z = \exp(\beta\mu)$, (2.5.20) yields

$$P_{\substack{B\\F}}(T, z) = \pm \frac{T^{5/2}}{8\pi^{3/2}} F_4(\pm z). \tag{2.5.21}$$

Remarks (2.5.22)

1. As $\omega \to 0$, the potential v goes to zero pointwise, and the density (2.5.14; 3) on phase space turns into the well-known Bose or Fermi distribution,
$$\rho(\mathbf{x}, \mathbf{p}) = [\exp(\beta(|\mathbf{p}|^2 - \mu)) \mp 1]^{-1}.$$
2. The energy spectrum of this example resembles that of a massless particle in a box $\{\mathbf{x}: |x_i| < L/2\}$, $E = (p_1^2 + p_2^2 + p_3^2)^{1/2}$, $p_i = m_i\pi/L$, $\mathbf{m} \in (\mathbb{Z}^+)^3$.

In the limit $L \to \infty$, this E produces the same pressure up to a constant as $(m_1 + m_2 + m_3)\omega$, when ω is identified with π/L. Then

$$P_{\substack{B\\F}}(T, z) = \pm T^4 F_4(\pm z)\pi^{-3}.$$

A Box with Hard Walls (2.5.23)

Now suppose that the potential $v_L(\mathbf{x}) \geq 0$ is significantly smaller than $1/L^2$ for $|x_i| < L/2$ but increases exponentially as soon as $|x_i| > L/2$. Since what happens should not depend on the precise form of v_L, only certain bounds will be imposed on v_L. Because of the monotonic property, all the steps up to (2.5.17) and (2.5.18; 1) proceed as before.†

$$\gamma_-^3\, \varphi(x)\varphi(y)\varphi(z) \leq v_L(\mathbf{x}) \leq \gamma_+^3\, \varphi(x)\varphi(y)\varphi(z), \qquad 0 < \gamma_- < \gamma_+,$$

$$\varphi(x) = \exp\left(\frac{-cL}{2}\right)\cosh(cx),$$

$$\mathcal{N}^2 \int_{-\infty}^{\infty} dx' \exp(-bx'^2)\varphi(x + x') = \exp\left(\frac{c^2}{4b}\right)\varphi(x),$$

so for the other bound,

$$\varphi(x) = \int_{-\infty}^{\infty} dx \exp(-bx'^2) \exp\left(-\frac{c^2}{4b}\right)\varphi(x' + x)\mathcal{N}^2.$$

The x-space portion of the calculation of $\sum_{\nu=1}^{\infty}(-1)^{\nu+1}[\exp(\nu\beta\mu)/\nu] \times \int d\Omega_z \exp(-\beta\nu g(z))$, where $g(z) = f(z)$ or respectively $h(z)$ (cf. (2.5.17) and (2.2.11)), leads to

$$\int_{-\infty}^{\infty} dx \exp(-B_\pm \cosh cx) = \frac{2}{c}\int_1^{\infty} \frac{dv}{\sqrt{v^2 - 1}} \exp(-B_\pm v)$$

$$\overset{B_\pm \to 0}{=} \frac{2}{c}\left(\ln\frac{1}{B_\pm} + O(1)\right)$$

with $B_\pm = \gamma_\pm \beta \exp[\pm c^2/4b - cL/2])$, since it is being evaluated in the limit $V = L^3 \to \infty$. If a sequence $(v_L(\mathbf{x}))_{L\to\infty}$ of wall potentials has bounds of the above-mentioned form with $c(L) = o(L)$ and $\ln(\beta\gamma_\pm(L)) = o(c(L)\cdot L)$, then

$$\frac{2}{cL}\ln\frac{1}{B_\pm} = 1 \mp \frac{c}{2bL} - \frac{2}{cL}\ln\beta\gamma_\pm$$

converges to 1 for both bounds. The p-integral is the same as in (2.5.19), and so, finally,

$$P_{\substack{B\\F}}(T, z) = \pm\frac{T^{5/2}}{8\pi^{3/2}} F_{5/2}(\pm z). \qquad (2.5.24)$$

† From this point until right before (2.5.24), + and − will indicate upper and lower bounds for the potential due to the wall rather than Bose and Fermi statistics.

Remarks (2.5.25)

1. This is the same result as that of summing over all the eigenvalues of a free particle in a box with Dirichlet boundary conditions on the wall (Problem 5). The bounds (2.5.17) show that in very large part it is only the total volume of V that matters, rather than its detailed form.
2. The nature of the wall is expressed by $F_{5/2}$ in (2.5.23) and F_4 in (2.5.21). For lower densities, $z \ll 1$, they coincide, as $F_\sigma = z + O(z^2)$.

The Thermodynamic Functions of Free Particles (2.5.26)

All the thermodynamic functions can be obtained from $P(T, z)$, so (2.5.24) will allow the gaps left by (2.3.10) to be filled in, and the functions can be written down explicitly. We shall investigate the limiting cases where $z \to \infty$, $z \to 0$, and $z \to -1$, corresponding to the extremes of Fermi, Boltzmann, and Bose statistics. The limits $z \to \infty$, -1 are what is referred to as a **degenerate gas**. By Problem 1, F has the asymptotic forms

$$-F_{5/2}(-z) \begin{cases} \xrightarrow{z \to -1} & -\zeta(\tfrac{5}{2}) + (z+1)\zeta(\tfrac{3}{2}) \\ \xrightarrow{z \to 0} & z - z^2 \cdot 2^{-5/2} \\ \xrightarrow{z \to \infty} & \dfrac{4}{3\sqrt{\pi}}\left[\dfrac{2}{5}(\ln z)^{5/2} + \dfrac{\pi^2}{4}(\ln z)^{1/2}\right], \end{cases} \qquad (2.5.27)$$

where $\zeta(\sigma)$ is the Riemann zeta function,

$$\zeta(\sigma) = \sum_{\nu=1}^{\infty} \frac{1}{\nu^\sigma} = F_\sigma(1), \qquad \sigma \in \mathbb{C}, \quad \operatorname{Re} \sigma > 1.$$

The zeta function has an analytic continuation to the punctured complex plane $\{\sigma \in \mathbb{C} \mid \sigma \neq 1\}$. In the three limits,

$$\frac{P}{T} = \frac{2\varepsilon}{3T} = \begin{cases} \xrightarrow{\text{Bose}} & \dfrac{T^{3/2}}{8\pi^{3/2}}\left[\zeta(\tfrac{5}{2}) + (z-1)\zeta(\tfrac{3}{2})\right] \\ \xrightarrow{\text{Boltzmann}} & \dfrac{T^{3/2}}{8\pi^{3/2}} z(1 \pm z \cdot 2^{-5/2}) \\ \xrightarrow{\text{Fermi}} & \dfrac{T^{3/2}}{6\pi^2}\left[\dfrac{2}{5}(\ln z)^{5/2} + \dfrac{\pi^2}{4}(\ln z)^{1/2}\right], \end{cases} \qquad (2.5.28)$$

so, writing

$$\varepsilon_{\substack{B\\F}} = T^2 \frac{\partial}{\partial T} \frac{1}{T} P_{\substack{B\\F}}(T, z) = \tfrac{3}{2} P_{\substack{B\\F}}(T, z) = \pm \tfrac{3}{2} T^{5/2} F_{5/2}(\pm z) \frac{1}{8\pi^{3/2}},$$

$$\rho_{\substack{B\\F}} = z \frac{\partial}{\partial z} \frac{1}{T} P_{\substack{B\\F}}(T, z) = \pm T^{3/2} F_{3/2}(\pm z) \frac{1}{8\pi^{3/2}},$$

$$\sigma_{\substack{B\\F}} = \pm \frac{T^{3/2}}{8\pi^{3/2}} [\tfrac{5}{2} F_{5/2}(\pm z) - \ln z F_{3/2}(\pm z)], \tag{2.5.29}$$

to the lowest nonvanishing order,

$$\rho = \begin{cases} \xrightarrow{\text{Bose}} z\zeta(\tfrac{3}{2}) \dfrac{T^{3/2}}{8\pi^{3/2}} \\ \xrightarrow{\text{Boltzmann}} \dfrac{T^{3/2}}{8\pi^{3/2}} z(1 \pm z \cdot 2^{-3/2}) \\ \xrightarrow{\text{Fermi}} \dfrac{T^{3/2}}{6\pi^2} (\ln z)^{3/2} + \dfrac{T^{3/2}}{48} (\ln z)^{-1/2}, \end{cases} \tag{2.5.30}$$

$$\sigma = \begin{cases} \xrightarrow{\text{Bose}} \dfrac{T^{3/2}}{8\pi^{3/2}} \tfrac{5}{2}\zeta(\tfrac{5}{2}) \\ \xrightarrow{\text{Boltzmann}} \dfrac{T^{3/2}}{8\pi^{3/2}} (\tfrac{5}{2} z - z \ln z) \\ \xrightarrow{\text{Fermi}} \dfrac{T^{3/2}}{12} (\ln z)^{1/2}. \end{cases} \tag{2.5.31}$$

When expressed in terms of the more intuitively appealing variables ρ and T,

$$P = \frac{2}{3}\varepsilon = \begin{cases} \xrightarrow{\text{Bose}} \rho T + T^{5/2}(\zeta(\tfrac{5}{2}) - \zeta(\tfrac{3}{2}))/8\pi^{3/2} \\ \xrightarrow{\text{Boltzmann}} \rho T \quad (\text{"ideal gas"}) \\ \xrightarrow{\text{Fermi}} \dfrac{(6\pi^2 \rho)^{5/3}}{15\pi^2} + \dfrac{(6\pi^2 \rho)^{1/3}}{24} T^2, \end{cases}$$

$$\sigma = \begin{cases} \xrightarrow{\text{Bose}} \tfrac{5}{2}\zeta(\tfrac{5}{2}) \dfrac{T^{3/2}}{8\pi^{3/2}} \\ \xrightarrow{\text{Boltzmann}} \tfrac{5}{2}\rho - \rho \ln \dfrac{\rho 8\pi^{3/2}}{T^{3/2}} = \rho \ln \dfrac{T^{3/2} \exp(5/2)}{\rho 8\pi^{3/2}} \\ \xrightarrow{\text{Fermi}} \dfrac{T}{12} (6\pi^2 \rho)^{1/3}. \end{cases} \tag{2.5.32}$$

Remarks (2.5.33)

1. As $z \to 0$, (2.5.32) gives the classical result (2.3.9; 1) with an additional factor $1/N!$ in the volume of phase space. If V_P denotes the volume available in the one-particle phase space, and the $1/N!$ is incorporated into the general definition, then

$$S \sim \ln \frac{1}{N!} \left(\frac{V_P}{h^3}\right)^N$$

leads to

$$\frac{S}{N} \sim \ln \frac{V_P}{Nh^3}.$$

On the other hand, in configuration space and with units for which $\hbar = m = 1$, (2.5.32) informs us that $S \sim \ln VT^{3/2}/N$. Since $T^{-1/2}$ equals the thermal de Broglie wavelength λ with these units, the following rule of thumb applies to the entropy: Entropy per particle = ln{volume of phase space per particle, as measured in h^3} = ln{volume of configuration space per particle, as measured in λ^3}.

2. Fermions have a zero-point energy $E_0 = V\varepsilon_0$ left over when $T \to 0$, where $\varepsilon_0 \equiv (6\pi^2 \rho)^{5/3}/10\pi^2$, and a zero-point pressure $2\varepsilon_0/3$. Because

$$T = \frac{4(\varepsilon - \varepsilon_0)^{1/2}}{6(\pi^2 \rho)^{1/6}},$$

it is also possible to write

$$\sigma = \left(\frac{\varepsilon}{\varepsilon_0} - 1\right)^{1/2} \frac{\rho 2\pi}{\sqrt{10}},$$

showing that the number M of states in the interval $[E_0, E]$ is

$$M \cong \exp\left\{N\left(\frac{E}{E_0} - 1\right)^{1/2} \frac{2\pi}{\sqrt{10}}\right\}.$$

For example, in an atomic nucleus the kinetic energy is $E_0 \cong N \cdot 20$ MeV, so with a fixed kinetic excitation energy $\delta E = E - E_0$ the number of states in the interval is $\sim \exp 2\sqrt{N}\sqrt{\delta E/20 \text{ MeV}}$. If $\delta E \sim 1$ MeV, then for $N = 20$ there are about e^2, i.e., 7 or 8, states; whereas if $N \sim 200$, then the number increases to about $e^{6.5} \sim 0.5 \times 10^3$. This is in agreement with the experimental observation that the density of the energy states of heavy nuclei is on the order of $(\text{eV})^{-1}$.

3. If the energy of the ground state is redefined to zero, then z must be less than 1 for bosons—otherwise by (2.5.12) $n_0 \equiv \langle a_0^* a_0 \rangle = z/(1 - z)$ is either infinite or negative. Because $F_{3/2}(z) < \zeta(\frac{3}{2})$ when $0 < z < 1$, it follows from (2.5.29) that $T > T_c \equiv (8\pi^{3/2}\rho/\zeta(\frac{3}{2}))^{2/3}$. On the other hand, n_0 can be made arbitrarily big by taking z close enough to 1. The difficulty with this is that the two limits $z \to 1$ and $V \to \infty$ have to be taken jointly

if the density has been fixed. If $z(V) = 1 - 1/\rho_0 V$ and $T < T_c(\rho)$, then

$$\rho = \rho_0 + \zeta(\tfrac{3}{2})\frac{T^{3/2}}{8\pi^{3/2}},$$

$$P = \tfrac{2}{3}\varepsilon = \zeta(\tfrac{5}{2})\frac{T^{5/2}}{8\pi^{3/2}} = \lim_{V\to\infty}\frac{T}{V}\ln \operatorname{Tr}\exp\left[-\frac{1}{T}(H_V - \mu_V(T,\rho)N)\right],$$

with

$$\lim_{V\to\infty}\mu_V(T,\rho) = 0 \quad \text{for all } T \leq T_c(\rho),$$

$$\sigma = \tfrac{5}{2}\zeta(\tfrac{5}{2})\frac{T^{3/2}}{8\pi^{3/2}}.$$

This shows that a nonzero fraction $\rho_0/\rho = 1 - (T/T_c)^{3/2}$ of the particles reside in the ground state and contribute nothing to the energy, pressure, or entropy (provided H is replaced with $H - E_0$). The number of particles in the first excited state, $n_1 = 1/(z^{-1}\exp(\beta/L^2) - 1) \sim L^2$, is rather large, but $n_1/V \to 0$. For similar reasons, the relative mean-square deviation $(\Delta n_i)^2/\langle n_i\rangle^2$ remains positive for n_0 as $V \to \infty$, but goes to zero for the higher states. The specific heat

$$c_V = \left.\frac{\partial E}{\partial T}\right|_{V,N}$$

is continuous at T_c and $\partial c_V/\partial T$ is discontinuous (Problem 2). If $T = T_c$, then the choice of ρ_0 has to depend on V.

4. The values $\mu = 0$ and $z = 1$ apply to a situation where N is not conserved, such as a gas of photons or phonons (cf. (2.5.22; 2)). It is easy to calculate $\operatorname{Tr}\exp(-\beta H)$ with the H of (2.5.7). The pressure $P = -\varphi$, and

$$\left.\frac{\partial P}{\partial \rho}\right|_T = -\left.\frac{\partial \varphi}{\partial \rho}\right|_T = \mu = 0,$$

so the compressibility is infinite. The system behaves much like a gas at the condensation point, the vacuum state, i.e., no particles, being analogous to the condensed state. It therefore has $\varepsilon = \sigma = P = V = 0$, and the system can be compressed into the vacuum. The entropy density σ is then simply the quantity $\Delta s/\Delta v$ of the Clausius–Clapeyron equation which simply assumes the form

$$\left.\frac{\partial P}{\partial T}\right|_\rho = \sigma.$$

Since $P = -\varphi$, Theorem (2.4.14) implies that this equation holds identically. The quantities $\varepsilon/T \approx \rho \approx \sigma$ depend only on T and correspond to a particle of energy T in each wavelength cube. Consequently, entropy $\cong$ particle number $\cong$ energy/T.

Particles in a Magnetic Field (2.5.34)

The Hamiltonian was given in (III: 3.3.5; 3):

$$H = |\mathbf{p} - e\mathbf{A}|^2 = p_3^2 + 2eB(a^*a + \tfrac{1}{2}).$$

The boundary conditions are that the wave-function must vanish at $x_3 = 0$ and $x_3 = L$, the 3-axis pointing along B, so the eigenvalues of p_3 are $\pi m/L, m = 1, 2, 3, \ldots$. The center $\bar{\mathbf{x}}$ of the orbit is confined to $|\bar{\mathbf{x}}|^2 = (2/eB)(g + \frac{1}{2}) < R^2$ in the plane perpendicular to B, so the geometry is cylindrical. The "wall potential" $\infty \cdot \Theta(|\bar{\mathbf{x}}|^2 - R^2)$ confining the particle is not a multiplication operator by a real-valued function $V(x_1, x_2)$, but rather a function of the operator

$$|\bar{\mathbf{x}}|^2 = \tfrac{1}{4}(x_1^2 + x_2^2) + \frac{1}{e^2B^2}(p_1^2 + p_2^2) + \frac{1}{eB}(x_1 p_2 - x_2 p_1),$$

representing the sum of a two-dimensional harmonic oscillator in the $x_1 - x_2$-plane and the x_3-component of the angular momentum. The construction of such a momentum-dependent wall potential will be left to the ingenuity of the experimentalists. By (III: 3.5.3; 3), $|\bar{\mathbf{x}}|^2$ is quantized so that g is a whole number, and a^*a has the eigenvalues $n = 0, 1, 2, \ldots$. As $L \to \infty$, the sum $\sum_{g=0}^{R^2eB/2} \sum_{m=1}^{\infty} \sum_{n=0}^{\infty}$ turns into

$$\int_0^\infty dp_3 \frac{L}{\pi} \sum_{n=0}^{\infty} \frac{R^2 eB}{2} = \frac{VeB}{2\pi^2} \int_0^\infty dp_3 \sum_{n=0}^{\infty},$$

where V denotes the volume of the cylinder. The classical bounds amount to the replacement

$$\sum_{n=0}^{\infty} \to \int_0^\infty dn,$$

in which all magnetic effects are swept away. We have to resort to the exact expression (2.5.9), with which the grand canonical partition functions becomes

$$\begin{aligned}\beta P_{\substack{B\\F}}(z) &= \mp \frac{eB}{2\pi^2} \int_0^\infty dp_3 \sum_{n=0}^{\infty} \ln(1 \mp z \exp[-\beta(p_3^2 + eB(2n + 1))]) \\ &= \pm \frac{T^{3/2}}{8\pi^{3/2}} \sum_{\nu=1}^{\infty} \frac{(\pm z)^\nu}{\nu^{5/2}} \frac{\nu eB\beta}{\sinh \nu eB\beta},\end{aligned} \tag{2.5.35}$$

where the B in $P_{\substack{B\\F}}$ denotes Bose statistics as usual and has nothing to do with the magnetic field B. This reveals right away that, as in (2.3.33; 2), an arbitrarily weak magnetic field ruins the phase transition of the Bose gas, since for any T.

$$\rho_{\substack{B\\F}} = z\frac{\partial}{\partial z}\beta P_{\substack{B\\F}} = \pm\frac{T^{3/2}}{8\pi^{3/2}}\sum_{\nu=1}^{\infty}\frac{(\pm z)^{\nu}}{\nu^{1/2}}\frac{eB\beta}{\sinh \nu eB\beta}$$

can get arbitrarily big as $z \to \exp(\beta E_0) = \exp(\beta eB)$. This happens because the particles are free to move only parallel to $\mathbf{B}$ and are trapped in orbits in the direction perpendicular to $\mathbf{B}$ even though the radius of the cylinder goes to infinity. The system acts as though confined to a cylinder only the length of which tends to ∞, and in one dimension there is no Bose condensation. If the magnetic energy eB is much less than the thermal energy T, then the next correction to the foregoing result is $\sim B^2$:

$$\beta P_{\substack{B\\F}} \to \pm\frac{T^{3/2}}{8\pi^{3/2}}\left[F_{5/2}(\pm z) - \frac{1}{6}\left(\frac{eB}{T}\right)^2 F_{1/2}(\pm z)\right]. \tag{2.5.36}$$

If this is used to calculate the magnetization per volume in the limit $B \to 0$ with T fixed,

$$m = \frac{\partial P_{\substack{B\\F}}}{\partial eB} = \frac{1}{V}\left\langle \sum_{\substack{\text{all}\\\text{particles}}} (x_1 p_2 - x_2 p_1 - eB(x_1^2 + x_2^2))\right\rangle$$

$$= \mp\frac{T^{3/2}}{8\pi^{3/2}}\frac{eB}{3T}F_{1/2}(\pm z), \tag{2.5.37}$$

then with (2.5.26) and the formula $F_{\sigma-n}(z) = (z(d/dz))^n F_{\sigma}(z)$ (see (2.5.20)), its limits in the three extreme cases of the different statistics are

$$m \begin{cases} \xrightarrow{\text{Bose}} -eB\cdot\infty \\[2ex] \xrightarrow{\text{Boltzmann}} -\dfrac{eB}{3T}\rho \\[2ex] \xrightarrow{\text{Fermi}} -\dfrac{eB}{12\pi^2}(6\pi^2\rho)^{1/3}. \end{cases} \tag{2.5.38}$$

Remarks (2.5.39)

1. The negative sign indicates diamagnetism, which is to be expected quantum-mechanically: By Lenz's law the classical orbits rotate in the direction with negative L_z. However, a current appears in the other direction when particles bounce off the wall of the box (see Figure 20).

   With classical statistics the circulating currents cancel out at every point of the interior, leaving only a current circulating along the surface,

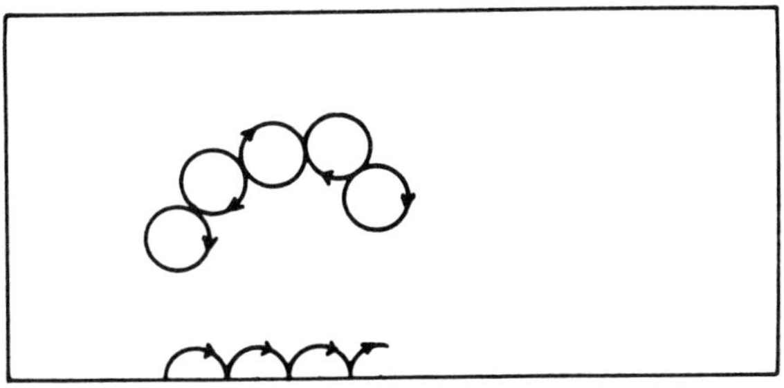

Figure 20 Classical trajectories of particles in a box with a magnetic field.

which is exactly compensated for by the "reflected" current, since the partition function

$$\int d^3x\, d^3p \exp[-\beta|\mathbf{p} - \mathbf{A}(\mathbf{x})|^2] = \int d^3x\, d^3p \exp(-\beta|\mathbf{p}|^2)$$

is completely independent of B. This means that if either ρ is fixed and $T \to \infty$ or T is fixed and $\rho \to 0$, then m tends to 0. Diamagnetism is therefore a characteristically quantum-mechanical effect; if the sum $\sum_{n=0}^{\infty}$ is replaced with an integral $\int_0^\infty dn$, and $2n + 1$ becomes $2n$, which is in essence the limit $\hbar \to 0$, then P becomes independent of B (a theorem of Bohr and van Leeuwen).

2. In quantum theory, states with negative L_z are energetically favored (III: 3.3.21; 4), so a quantum gas is diamagnetic. The reason that the magnetization m of a completely degenerate Bose gas tends to ∞ is that P fails to be analytic at $z = 1$, $B = 0$. This topic will shortly be discussed in more detail.
3. Since P depends only on R^2L,

$$R\frac{\partial}{\partial R}P = 2L\frac{\partial}{\partial L}P,$$

i.e., the pressure remains isotropic.

In order to make sense of the limit of degenerate Bose gas, let $\beta\mu = \ln z$, and write

$$\frac{\beta P_B}{V} = \frac{T^{3/2}}{8\pi^{3/2}}\sum_{\nu=1}^{\infty}\frac{\exp[-\beta\nu(eB - \mu)]}{\nu^{5/2}}\frac{2eB\beta\nu}{1 - \exp[-2eB\beta\nu]},$$

$$\rho = \frac{T^{3/2}}{8\pi^{3/2}}\sum_{\nu=1}^{\infty}\frac{\exp[-\beta\nu(eB - \mu)]}{\nu^{3/2}}\frac{2eB\beta\nu}{1 - \exp[-2eB\beta\nu]},$$

$$m = -\rho + \frac{T^{3/2}}{4\pi^{3/2}}\sum_{\nu=1}^{\infty}\frac{\exp[-\beta\nu(eB - \mu)]}{\nu^{3/2}(1 - \exp[-2eB\beta\nu])^2} \times [1 - \exp[-2eB\beta\nu](1 + 2eB\beta\nu)], \tag{2.5.40}$$

without expanding in B. The convergence of the series for m and ρ in (2.5.40) (by domination for $B \geq 0$ with μ fixed) implies that

$$\lim_{B \to +0} m(T, \mu, B) = 0$$

for all fixed $T > 0$ and $\mu < 0$. Yet if $B \to 0$ with T fixed and $\mu < 0$ then all the densities ρ are less than $\zeta(\frac{3}{2})T^{3/2}/8\pi^{3/2}$, as in (2.5.33; 3). If $T \leq T_c(\rho)$ (see (2.5.33; 3)), then the limits $B \to 0$ and $\mu \to 0$ must again be appropriately coordinated. Since for $B > 0$ and for all values $\rho > 0$ and $T > 0$ there exists a unique $\mu(T, \rho, B) < eB$ such that $\lim_{B \to 0} \mu(T, \rho, B) = 0$ for $T \leq T_c(\rho)$), and since the series for $m + \rho$ from (2.5.40) also converges uniformly in B on an interval containing $\mu = eB$, the limit $B \to 0$ can be taken term by term. This yields

$$\lim_{B \to 0} m(T, \rho, B) = -\rho_0 = -\rho\left[1 - \left(\frac{T}{T_c(\rho)}\right)^{3/2}\right],$$

provided that $T \leq T_c(\rho)$ (cf. (2.5.33; 3)). If $T \geq T_c(\rho)$ then the limit is zero as observed earlier.

Remarks (2.5.41)

1. The physical interpretation of this result is that in the limit $B \to 0$ only the particles in the ground state contribute to the magnetization. The ground state has $L_z = -1$, so for $B = 0$ the contribution to m is simply the sum of L_z over the particles in a unit volume in the ground state.
2. The notation B is perhaps misleading, since it stands only for the external field and not for that due to the system itself. Actually, the field due to the system has to be taken into account, as it screens B throughout the interior of the system.

Black-Body Radiation in Partial (i.e., Anisotropic) Equilibrium (2.5.42)

If the particles are massless, as in (2.5.22; 2) and (2.5.33; 4), and they have a density matrix like ρ_{GC} but containing only states in a certain dilatation-invariant part D of p-space, then we can still write

$$\varphi = T \int_D \frac{d^3p}{(2\pi)^3} \ln(1 - \exp[-\beta|\mathbf{p}|]) = -cT^4,$$

where the constant c depends on D (but not on T). It is then still true that

$$\varepsilon = 3P = -3\varphi = \tfrac{3}{4}T\sigma = 3cT^4.$$

A realistic example of this situation is sunlight falling on the earth, for which essentially all the p-vectors come from the direction of the sun. The constant c is reduced by a factor $\sim 10^{-5}$, the solid angle subtended by the sun, in comparison with the isotropic equilibrium value with $D = \mathbb{R}^3$. Once the radiation is made isotropic without changing ε significantly by the time it reaches the earth, T is lowered by a factor of about $10^{-5/4}$, from $\sim 6000°$ K to $\sim 300°$ K. At the same time, $\sigma = 4\varepsilon/3T$ is increased by this factor of 20. It is consistent with an increase in the total entropy that this physical process creates highly ordered structures with little entropy; their decrease of entropy is nothing compared with the gigantic increase of the radiation entropy. About 10^{20} photons per cm^2 arrive from the sun each minute, and this times 20 is the entropy increase/cm^2-min. In an hour this comes to roughly the total entropy of a cubic centimeter of matter for each square centimeter of ground, so, for example, a newly planted forest could grow to a height of 10 meters over a summer without violating the second law of thermodynamics. The sun thus expends entropy as well as energy. Although isotropic black-body radiation at 300° K would be just as energetic, the energy would be unusable for the creation of life (as would be the case as the universe subsided into heat death).

The grand canonical ensemble determines the expectation values of field operators as well as the thermodynamic functions. Equation (2.5.12) showed how to calculate quadratic expressions involving the field operators, and quartic expressions for particles in an external field can easily be calculated in the same way,

$$\begin{aligned}\langle a_m^* a_j^* a_{j'} a_{m'}\rangle &= (\delta_{mm'}\delta_{jj'} \pm \delta_{mj'}\delta_{jm'})(\exp[\beta(\varepsilon_m - \mu)] \mp 1)^{-1} \\ &\quad \times (\exp[\beta(\varepsilon_j - \mu)] \mp 1)^{-1} \\ &= \langle a_m^* a_{m'}\rangle\langle a_j^* a_{j'}\rangle \pm \langle a_m^* a_{j'}\rangle\langle a_j^* a_{m'}\rangle. \end{aligned} \tag{2.5.43}$$

Remark (2.5.44)

If the mean-square deviations of the occupation numbers are calculated in this way, then

$$\langle (a_m^* a_m)^2\rangle - \langle a_m^* a_m\rangle^2 = \langle a_m^* a_m\rangle(1 \pm \langle a_m^* a_m\rangle).$$

Independent particles would follow a Poisson distribution law $w(n) = \exp(-\bar{n})\bar{n}^n/n!$ for which the mean-square deviation would equal the expectation value of the occupation number. The deviation is greater with Bose statistics and less with Fermi statistics, which can be interpreted as meaning that bosons have a tendency to bunch up and fermions to keep at a distance.

In elementary quantum mechanics a state was characterized by the expectation values of the Weyl operators (cf. (III: 3.1.2; 1)), and likewise now the complete determination of the state requires the expectation value of, say, $\exp[i \int d^3x(a(\mathbf{x})f^*(\mathbf{x}) + a^*(\mathbf{x})f(\mathbf{x}))]$ for all $f \in C_0^\infty(\mathbb{R}^3)$. The best way for this to be calculated in the grand canonical ensemble for particles in an external field makes use of coherent states. In Problem 6 it is shown that

$$\frac{\operatorname{Tr}\exp[-\beta\omega a^* a]\exp[i(a^*\alpha + a\alpha^*)]}{\operatorname{Tr}\exp[-\beta\omega a^* a]}$$

$$= \exp\left[-|\alpha|^2\left(\frac{1}{2} + \frac{1}{\exp(\beta\omega) - 1}\right)\right] \quad \text{if } [a, a^*] = 1.$$

Therefore:

The Grand Canonical State for Bosons in an External Field (2.5.45)

is

$$\left\langle \exp\left[i\sum_m (a_m^*\alpha_m + a_m\alpha_m^*)\right]\right\rangle = \exp\left[-\sum_m |\alpha_m|^2\left(\frac{1}{2} + \frac{z}{\exp(\beta\varepsilon_m) - z}\right)\right].$$

Example (2.5.46)

Free bosons in a cube of volume $V = L^3$, with periodic boundary conditions. Let

$$a_V(\mathbf{k}) = \int_V \frac{d^3x}{L^{3/2}} \exp(-i\mathbf{k}\cdot\mathbf{x})a(\mathbf{x}),$$

and

$$a_f = \sum_{k\in((2\pi/L)\mathbb{Z})^3} L^{-3/2}\tilde{f}(\mathbf{k})a_V(\mathbf{k}), \qquad \tilde{f}(\mathbf{k}) = \int d^3x \exp(i\mathbf{k}\cdot\mathbf{x})f(\mathbf{x}),$$

for $f \in L^2(V)$. Then because $\omega = |\mathbf{k}|^2$,

$$\langle \exp[i(a_f^* + a_{f^*})]\rangle = \exp\left[-\sum_{k\in((2\pi/L)\mathbb{Z})^3} L^{-3}|\tilde{f}(\mathbf{k})|^2\left(\frac{1}{2} + \frac{z}{\exp(\beta|\mathbf{k}|^2) - z}\right)\right].$$

A more convenient expression in the calculation of ordered products is $\exp[i\sum_m a_m^*\alpha_m]\exp[i\sum_m a_m\alpha_m^*]$. Its expectation values can be read off from the formula $\exp(A + B) = \exp A \exp B \exp(\frac{1}{2}[B, A])$, which holds provided that $[A, [A, B]] = [B, [A, B]] = 0$, which in this case is in accordance with the Weyl relations (III: 3.1.2; 1):

The Generating Function for Ordered Products (2.5.47)

$$\left\langle \exp\left[i\sum_m a_m^*\alpha_m\right]\exp\left[i\sum_m a_m\alpha_m^*\right]\right\rangle = \exp\left[-\sum_m |\alpha_m|^2\frac{z}{\exp(\beta\varepsilon_m) - z}\right]$$

$$\equiv E(\alpha_i, \alpha_k^*),$$

which can be written

$$\langle \exp(ia_f^*)\exp(ia_f)\rangle = \exp(-\langle f|\rho_1 f\rangle)$$

with the use of ρ_1 from (2.5.13).

The expectation values of polynomials in the field operators can be obtained by differentiating the generating function by α or α^*. Note that all the factors within a given exponent of (2.5.47) commute, so nothing prevents the exponential functions from being differentiated:

$$\langle a_{m_1}^* \cdots a_{m_n}^* a_{j_1} \cdots a_{j_n}\rangle = (-i)^{n+n'} \frac{\partial}{\partial \alpha_{m_1}} \cdots \frac{\partial}{\partial \alpha_{m_n}} \frac{\partial}{\partial \alpha_{j_1}^*} \cdots \frac{\partial}{\partial \alpha_{j_n}^*} E \bigg|_{\substack{\alpha_i = 0 \\ \alpha_k^* = 0}}$$

$$= \delta_{nn'} \sum_P \prod_{i=1}^{n} \frac{\delta_{m_i j_{P_i}} z}{\exp(\beta \varepsilon_{m_i}) - z},$$

where P stands for any permutation of $(1, 2, \ldots, n)$.

We have been confronted again with a permanent, and it is easy to understand that the analogous expression for fermions contains $(-1)^P$ and thus involves a determinant. The $-z$ in the denominator is also turned into $+z$, but there are no other changes. Linear extension covers the cases of expectation values of products of arbitrary a_f, which are most conveniently written in terms of the one-particle density matrix ρ_1, as before:

The Grand Canonical Expectation Value of an Ordered Product (2.5.48)

$$\langle a_{f_1}^* \cdots a_{f_n}^* a_{g_1} \cdots a_{g_{n'}}\rangle = \delta_{nn'} \operatorname*{Per}_{\mathrm{Det}}(\langle f_i|\rho_1 g_j\rangle).$$

This section will conclude with a further investigation into the thermodynamic limit of the grand canonical state of a system of particles in an external field. Such a state will exist under the circumstances in which $\rho_{1,V}$ converges weakly, as for example with free particles, for which:

The Grand Canonical State of an Infinite System (2.5.49)

$$\langle a_{f_1}^* \cdots a_{f_n}^* a_{g_1} \cdots a_{g_{n'}}\rangle = \delta_{nn'} \operatorname*{Per}_{\mathrm{Det}}(\langle f_i|\rho_1 g_j\rangle),$$

$$\langle f|\rho_1 g\rangle = \int \frac{d^3k}{(2\pi)^3} \frac{\tilde{f}^*(\mathbf{k})\tilde{g}(\mathbf{k})z}{\exp(\beta|\mathbf{k}|^2) \mp z},$$

where $\beta > 0$, and for bosons, $0 \leq z < 1$, or for fermions, $z > 0$.

It was noticed in (2.5.33; 3) that with bosons at $T < T_c = (8\pi^{3/2}\rho/\zeta(\frac{3}{2}))^{2/3}$, the limits $V \to \infty$ and $z \to 1$ have to be taken jointly in order to have a given density ρ. This does not make the sum in (2.5.46) converge to the integral in

(2.5.49); rather, if $z = 1 - 1/\rho_0 V$, then the term with $k = 0$ survives separately:

$$\lim_{V\to\infty} \frac{1}{V} \sum_{k\in((2\pi/L)\mathbb{Z})^3} \frac{|\tilde{f}(\mathbf{k})|^2(1-(1/\rho_0 V))}{\exp(\beta|\mathbf{k}|^2) - 1 + (1/\rho_0 V)}$$

$$\to \rho_0 |\tilde{f}(\mathbf{0})|^2 + \int \frac{d^3k}{(2\pi)^3} \frac{|\tilde{f}(\mathbf{k})|^2}{\exp(\beta|\mathbf{k}|^2) - 1}.$$

This formula is justified if $f \in L^2(\mathbb{R}^3)$ with compact support, which makes $\tilde{f} \in L^2(\mathbb{R}^3) \cap C_0^\infty(\mathbb{R}^3)$, so the integrand remains integrable even at $\mathbf{k} = \mathbf{0}$. Therefore we have:

The Grand Canonical State in Bose Condensation (2.5.50)

$$\lim_{V\to\infty} \langle \exp(ia_f^*) \exp(ia_f) \rangle_{\beta, z=1-(1/\rho_0 V)}$$

$$= \exp\left[-\rho_0 |\tilde{f}(\mathbf{0})|^2 - \int \frac{d^3k}{(2\pi)^3} \frac{|\tilde{f}(\mathbf{k})|^2}{\exp(\beta|\mathbf{k}|^2) - 1}\right].$$

Remarks (2.5.51)

1. If $T < T_c$, then the grand canonical state of the Bose field algebra differs from the canonical state, which can be calculated as

$$\langle \exp(ia_f^*) \exp(ia_f) \rangle = \exp\left[-\int \frac{d^3k}{(2\pi)^3} \frac{|\tilde{f}(\mathbf{k})|^2}{\exp(\beta|\mathbf{k}|^2) - 1}\right]$$

$$\times \int_0^{2\pi} \frac{d\varphi}{2\pi} \exp[2i\sqrt{\rho_0}\, \mathrm{Re}(\tilde{f}(\mathbf{0}) \exp(i\varphi))]$$

   for $T < T_c$ [13].

2. Other than for bosons at $T < T_c$, the representations in the individual factors are thermal (1.4.7). According to Remark (1.4.17; 1) the factors are of type III in the infinite system. They form a reducible representation π, the tensor product $\pi_1 \otimes \pi_2$ of two Fock-like representations of the field algebra (cf. (1.4.7)):

$$\pi(a_f) = \pi_1\left(a \frac{\tilde{f}(\mathbf{p})}{\sqrt{\mp \exp[-\beta(|\mathbf{p}|^2 - \mu)] + 1}}\right) \otimes \mathbf{1}$$

$$+ (-\mathbf{1})^N \otimes \pi_2\left(a^* \frac{\tilde{f}^*(\mathbf{p})}{\sqrt{\exp[\beta(|\mathbf{p}|^2 - \mu)] \mp 1}}\right),$$

   where $a_f N = (N + 1)a_f$. It is straightforward to verify that

$$\langle a_{f_1}^* \cdots a_{f_n}^* a_{g_1} \cdots a_{g_{n'}} \rangle = \langle \Omega_1 \otimes \Omega_2 | \pi(a_{f_1}^*) \cdots \pi(a_{g_{n'}}) | \Omega_1 \otimes \Omega_2 \rangle.$$

3. For bosons at $T < T_c$ there is no factor representation; the analogue of the mean magnetization **s** (1.4.6: 2) is

$$a_0 \equiv \underset{V\to\infty}{w\text{-lim}}\, a_0^V, \quad \text{where } a_0^V \equiv \frac{1}{V}\int_V d^3x a(\mathbf{x}).$$

All bounded functions of a_0 lie in the center of the von Neumann algebra $\pi(\mathscr{A})''$. Now

$$\langle a_0^{*n} a_0^m \rangle = \left(\frac{\partial}{\partial \tilde{f}(\mathbf{0})}\right)^n \left(\frac{\partial}{\partial \tilde{f}^*(\mathbf{0})}\right)^m E|_{f\equiv 0},$$

so for instance $\langle a_0 \rangle = 0$, $\langle a_0^* a_0 \rangle = \rho_0$. Thus a_0 is not represented as a multiple of the identity.

4. The canonical state (2.5.51; 1) is an integral over states ω_φ for which the exponent in the generating function

$$\omega_\varphi(\exp(i\lambda a_0^*)\exp(i\lambda a_0)) = \exp(2i\lambda\sqrt{\rho_0}\cos\varphi)$$

is linear in $\lambda \in \mathbb{R}$. These states produce factor representations:

$$\pi_\varphi(a_0) = \sqrt{\rho_0}\exp(-i\varphi)\cdot \mathbf{1}.$$

5. If a term $V^\alpha(a_0^V - \sqrt{\rho_0}\exp(-i\varphi))^*(a_0^V - \sqrt{\rho_0}\exp(-i\varphi))$ with $0 < \alpha < 1$ is added to the local Hamiltonian H_V, then the $\mathbf{k} = \mathbf{0}$ component of βH_V becomes $\beta V^\alpha(a_0^V - \sqrt{\rho_0}\exp(-i\varphi))^*(a_0^V - \sqrt{\rho_0}\exp(-i\varphi))$. As will become more apparent below, the thermodynamic functions are unchanged for all $0 < T \le T_c(\rho)$ in the limit $V \to \infty$ if we set $z(V) \equiv 1$ and $\rho_0 = \rho(1 - (T/T_c(\rho))^{3/2})$ (cf. (2.5.33; 3)). Because

$$\begin{aligned}
&\mathrm{Tr}\{\exp[-\beta V^\alpha(a_0^V - \sqrt{\rho_0}\exp(-i\varphi))^*(a_0^V - \sqrt{\rho_0}\exp(-i\varphi))] \\
&\quad\times \exp(i\tilde{f}(\mathbf{0})a_0^{V*})\cdot\exp(i\tilde{f}^*(\mathbf{0})a_0^V)\} \\
&= \mathrm{Tr}[\exp(-\beta V_\alpha a_0^{V*}a_0^V)\exp(i\tilde{f}(\mathbf{0})a_0^{V*})\exp(i\tilde{f}^*(0)a_0^V)] \\
&\quad\times \exp(2i\sqrt{\rho_0}\,\mathrm{Re}(\tilde{f}(\mathbf{0})\exp(i\varphi))
\end{aligned}$$

and

$$\frac{\mathrm{Tr}[\exp(-\beta V^\alpha a_0^{V*}a_0^V)\cdot\exp(i\tilde{f}(\mathbf{0})a_0^{V*})\cdot\exp(i\tilde{f}^*(\mathbf{0})a_0^V)]}{\mathrm{Tr}\exp(-\beta V^\alpha a_0^{V*}a_0^V)} = \exp\left[-|\tilde{f}(\mathbf{0})|^2/\beta V^\alpha + o\left(\frac{1}{V^\alpha}\right)\right]$$

(see Problem 6), in the limit $V \to \infty$ the perturbed grand canonical state reduces to ω_φ, the integrand of the canonical state in the decomposition (2.5.51; 1), since the contribution to the generating function from the components of H_V with $\mathbf{k} \neq \mathbf{0}$ is not affected by the extra term. Since the exponent in this generating function is linear in $\tilde{f}(\mathbf{0})$ and $\tilde{f}^*(\mathbf{0})$,

$$\pi_{\omega_\varphi}(a_0) = \sqrt{\rho_0}\exp(-i\varphi)\cdot\mathbf{1}.$$

This shows that ω_φ is a factor state, and the density of the particles in the ground state is represented by the (dispersionless) multiplication operator $\rho_0 \cdot \mathbf{1}$. Although the assumption that $\alpha > 0$ is essential (the limit state is not changed by perturbations bounded uniformly in V), the bound $\alpha < 1$ only serves to illustrate that a surface effect is enough to single out any given pure phase from a mixture as the limit $V \to \infty$ is taken.

This example appears at first only academic from the physical point of view. Since constant phases of the wave-functions are not observable properties, at least for free particles, the Bose algebra should be replaced with the gauge-invariant subalgebra $\mathscr{E}$, i.e., the subalgebra invariant under the automorphism induced by $f \to \exp(i\varphi)f$. All the states ω_φ are the same on the subalgebra, and the phase mixture of the ground state is not observable. However, these phases do have experimental consequences in superconductors, in the Josephson effect.

Problems (2.5.52)

1. Calculate the asymptotic forms of $F_{5/2}(z)$ (for $z \to 1$ use $zF'_\sigma(z) = F_{\sigma-1}(z)$, $F_\sigma(1) = \zeta(\sigma)$).
2. Calculate the heat capacity per particle of an ideal Bose gas at constant density, as well as its derivative by the temperature.
3. Verify (2.5.18; 2).
4. Show the maximum properties of (2.5.16; 2) and (2.5.18; 4).
5. Calculate P_B and P_F for particles in a box. Show that the result agrees with (2.5.24) in the limit $V \to \infty$.
6. Calculate $\operatorname{Tr} \exp[i(a^*\alpha + a\alpha^*)] \exp[-\beta a^* a]/\operatorname{Tr} \exp[-\beta a^* a]$, assuming that $[a, a^*] = 1$.

Solutions (2.5.53)

1. $z \to 0$: $F_{5/2}(z) = \sum_{\nu=1}^{\infty} \frac{z^p}{\nu^{5/2}} \sim z + \frac{z^\nu}{2^{5/2}} + \cdots$

$z \to 1$: $F_{5/2}(z) \sim F_{5/2}(1) + (z-1)F'_{5/2}(1) + \cdots = \zeta(\frac{5}{2}) + (z-1)\zeta(\frac{3}{2}) + \cdots$

$z \to \infty$: Let $\alpha = \ln z > 0$

$$\int_0^\infty dt\sqrt{t}\ln(1+\exp(-t+\alpha)) = \int_{-\alpha}^\infty dt\sqrt{t+\alpha}\ln(1+\exp(-t))$$

$$= \frac{2}{3}\int_{-\alpha}^\infty dt(t+\alpha)^{3/2}(1+\exp(t))^{-1}$$

$$= \frac{2}{3}\left[\int_0^\alpha dt(\alpha-t)^{3/2} - \int_0^\alpha \frac{dt(\alpha-t)^{3/2}}{1+e^t} + \int_0^\infty \frac{dt(t+\alpha)^{3/2}}{1+e^t}\right]$$

$$= \frac{2}{3}\left[\int_0^\alpha dt(\alpha-t)^{3/2} + \int_0^\infty \frac{dt((t+\alpha)^{3/2} - |t-\alpha|^{3/2})}{1+\exp(t)}\right] + O(\exp(-\alpha));$$

because

$$|(\alpha + t)^{3/2} - |\alpha - t|^{3/2} - 3t\alpha^{1/2}| \leq 2t^2\alpha^{-1/2}$$

and

$$\int_0^\infty \frac{dt\, t^{\sigma-1}}{1 + \exp(t)} = (1 - 2^{1-\sigma})\Gamma(\sigma)\zeta(\sigma),$$

with $\zeta(2) = \pi^2/6$, $\Gamma(2) = 1$, it follows that

$$\int_0^\infty dt\sqrt{t}\ln(1 + \exp(-t + \alpha)) = \tfrac{2}{3}\left[\frac{2}{5}\alpha^{5/2} + \alpha^{1/2}\frac{\pi^2}{4}\right] + O(\alpha^{-1/2}).$$

2.

$$\varepsilon = \begin{cases} \frac{3}{2}T^{5/2}\dfrac{1}{8\pi^{3/2}}F_{5/2}(z), & T > T_c, \text{ i.e., } 0 < z < 1, \\ \frac{3}{2}T^{5/2}\dfrac{1}{8\pi^{3/2}}\zeta(\frac{5}{2}), & T \leq T_c, \text{ i.e., } z = 1, \end{cases}$$

which implies

$$\gamma \equiv \lim_{N\to\infty}\frac{C_V}{N} = \begin{cases} \dfrac{15}{4}\dfrac{1}{8\pi^{3/2}\rho}T^{3/2}F_{5/2}(z) - \dfrac{9}{4}\dfrac{F_{3/2}(z)}{F_{1/2}(z)}, & T > T_c, \text{ i.e., } 0 < z < 1, \\ \dfrac{15}{4}\dfrac{T^{3/2}}{8\pi^{3/2}\rho}\zeta(\frac{5}{2}), & T \leq T_c, \text{ i.e., } z = 1, \end{cases}$$

because of the formula $F_{3/2}(z) = 8\pi^{3/2}\rho T^{-3/2}$ for $T > T_c$. The function γ is continuous at $T = T_c$ and equals $(15/4)\zeta(\frac{5}{2})/\zeta(\frac{3}{2}) = 1.93$, and as $T \to \infty$, $F_\sigma(z) \sim z \sim 8\pi^{3/2}\rho T^{-3/2}$, and

$$\gamma \sim \frac{15}{4}\frac{1}{8\pi^{3/2}\rho}T^{3/2}z - \tfrac{9}{4} \sim \tfrac{15}{4} - \tfrac{9}{4} = \tfrac{3}{2}.$$

With the expansion $F_{5/2}(z) = 2.363t^{3/2} + 1.342 - 2.612t - 0.730t^2\ldots$, where $t \equiv -\ln z$, valid for $z \lesssim 1$, and the recursion formula

$$F_{\sigma-1}(\exp(-t)) = -(d/dt)F_\sigma(\exp(-t)),$$

there results

$$\left(\frac{\partial\gamma}{\partial T}\right)_{T=T_c-0} - \left(\frac{\partial\gamma}{\partial T}\right)_{T=T_c+0} = \frac{3.66}{T_c}.$$

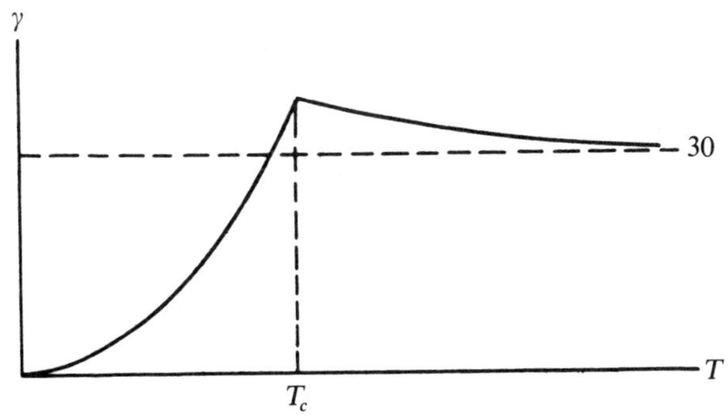

Figure 21 Specific heat of an ideal Bose gas.

3. If the wave-function of $|\mathbf{z}\rangle$ is $\exp(i\mathbf{k}\cdot\mathbf{x})u(\mathbf{x}-\mathbf{q})$ with u real-valued, then $\langle\mathbf{z}||\mathbf{p}|^2|\mathbf{z}\rangle = \int d^3x\,|i\mathbf{k}u(\mathbf{x}-\mathbf{q}) - \nabla u(\mathbf{x}-\mathbf{q})|^2 = |\mathbf{k}|^2 + \int d^3x\,|\nabla u|^2$. At the same time, the expectation value of $\int d\Omega_z|\mathbf{z}\rangle\langle\mathbf{z}||\mathbf{k}|^2$ in a normalized ψ equals

$$\int\frac{d^3q\,d^3k\,|\mathbf{k}|^2}{(2\pi)^3}\int d^3x\,d^3x'\psi^*(\mathbf{x})\exp(i\mathbf{k}\cdot\mathbf{x})u(\mathbf{x}-\mathbf{q})\exp(-i\mathbf{k}\cdot\mathbf{x}')u(\mathbf{x}'-\mathbf{q})\psi(\mathbf{x}')$$

$$= \int d^3q\,d^3x\nabla(\psi^*(\mathbf{x})u(\mathbf{x}-\mathbf{q}))\cdot\nabla(u(\mathbf{x}-\mathbf{q})\psi(\mathbf{x}))$$

$$= \int d^3x\,|\nabla\psi(\mathbf{x})|^2 + \int d^3q\,|\nabla u(\mathbf{q})|^2,$$

because the mixed terms drop out in the q integration. Therefore,

$$\int d\Omega_z|\mathbf{z}\rangle\langle\mathbf{z}||\mathbf{k}|^2 = |\mathbf{p}|^2 + \int d\Omega_z|\mathbf{z}\rangle\langle\mathbf{z}|\int d^3q\,|\nabla u|^2.$$

4. Klein's inequality (2.1.8; 5) with $K(\rho) = \rho\ln\rho \pm (1\mp\rho)\ln(1\mp\rho)$, $K'(\rho) = -\ln(1/\rho\mp 1)$ and $\tilde{\rho} = [\exp(\beta(h-\mu))\pm 1]^{-1}$ leads to

$$\mathrm{Tr}[K(\rho) - K(\bar{\rho}) + (\rho-\bar{\rho})\beta(h-\mu)] \geq 0,$$

proving (2.5.16; 2). In the classical case, i.e., $\rho = \rho(z)$, $h = h(z)$, $\bar{\rho} = \bar{\rho}(z) =$ all being real,

$$K(\rho(z)) - K(\bar{\rho}(z)) + (\rho(z) - \bar{\rho}(z))\beta(h(z)-\mu) \geq 0$$

for all z, and consequently (2.5.18; 4).

5. Particles in a box. If the shape of the box is a parallelepiped with sides L_1, L_2, and L_3, and the wave-functions satisfy Dirichlet boundary conditions, then the eigenvalues are

$$\varepsilon_m = \pi^2\left(\frac{m_1^2}{L_1^2} + \frac{m_2^2}{L_2^2} + \frac{m_3^2}{L_3^2}\right), \qquad m_i \in \mathbb{Z}^+.$$

Consequently

$$\beta V P_{\substack{B\\F}}(z) = \mp\sum_{m_i=1}^{\infty}\ln(1\mp z\exp(-\beta\varepsilon_m)),$$

and in the thermodynamic limit $L_i \to \infty$ the sum over m_i becomes $L_1\cdot L_2\cdot L_3(2\pi)^{-2}\times\int_0^\infty d\varepsilon\sqrt{\varepsilon}\ldots$, so

$$P_{\substack{B\\F}}(T, z) = \mp T^{5/2}(2\pi)^{-2}\int_0^\infty dt\sqrt{t}\,\ln(1\mp z\exp(-t)) = \pm T^{5/2}\frac{1}{8\pi^{3/2}}F_{5/2}(\pm z).$$

6. Because $\exp A\exp B = \exp(A+B)\exp(\frac{1}{2}[A, B]) = \exp B\exp A\exp[A, B]$ for $[A, B] = c\cdot\mathbf{1}$, the coherent states (2.2.6) with $|u\rangle = |0\rangle$, $a|0\rangle = 0$, can be written

$$|z\rangle = \exp\left(\frac{a^*z}{\sqrt{2}}\right)|0\rangle\exp\left(\frac{-|z|^2}{4}\right).$$

As in Remark (III: 3.1.14; 1), with $\exp(-\beta a^* a) f(a^*)|0\rangle = f(a^* \exp(-\beta))|0\rangle$ it follows that

$$\begin{aligned}
&\operatorname{Tr} \exp(\alpha a^*) \exp(-\alpha^* a) \exp(-\beta a^* a) \\
&= \int \frac{dz}{2\pi} \langle 0| \exp\left(\frac{a z^*}{\sqrt{2}}\right) \exp(-\alpha^* a) \exp(-\beta a^* a) \exp(\alpha a^*) \exp\left(\frac{a^* z}{\sqrt{2}}\right)|0\rangle \exp\left(\frac{-|z|^2}{2}\right) \\
&= \int \frac{dz}{2\pi} \langle 0| \exp\left[a\left(\frac{z^*}{\sqrt{2}} - \alpha^*\right)\right] \exp\left[\exp(-\beta) a^*\left(\frac{z}{\sqrt{2}} + \alpha\right)\right]|0\rangle \exp\left(\frac{-|z|^2}{2}\right) \\
&= \int \frac{dz}{2\pi} \exp\left[-\frac{|z|^2}{2}(1 - \exp(-\beta)) + \exp(-\beta)\left(\frac{1}{\sqrt{2}}(z^*\alpha - z\alpha^*) - |\alpha|^2\right)\right] \\
&= \exp\left[-|\alpha|^2 \frac{1}{\exp(\beta) - 1}\right] \Big/ (1 - \exp(-\beta)),
\end{aligned}$$

so by changing α to $i\alpha$,

$$\begin{aligned}
\langle \exp[i(a^*\alpha + a\alpha^*)]\rangle = \langle \exp[\alpha a^* - \alpha^* a]\rangle &= \langle \exp(\alpha a^*) \exp(-\alpha^* a)\rangle \exp(-\tfrac{1}{2}|\alpha|^2) \\
&= \exp\left[-|\alpha|^2\left(\frac{1}{2} + \frac{1}{\exp(\beta) - 1}\right)\right].
\end{aligned}$$

3 Thermodynamics

3.1 Time-Evolution

Whereas small systems evolve almost periodically in time, large systems appear chaotic and their time-evolution mixes the observables thoroughly.

The framework for this discussion will be an algebra $\mathscr{A}$ of observables with a strongly continuous time-automorphism and a time-invariant state ρ. In the GNS representation the invariant state is made into a vector $|\Omega\rangle$, and the time-automorphism is represented as a unitary group of operators $U = \{\exp(iHt)\}$, $U|\Omega\rangle = |\Omega\rangle$. The time-evolution then extends to the weak closure $\mathscr{A}''$. If the representation is reducible, then it may occur that $U \not\subset \mathscr{A}''$, even if $U_t^{-1}\mathscr{A}U_t \subset \mathscr{A}$. The von Neumann algebra

$$\mathscr{R} \equiv \{\mathscr{A} \cup U\}'', \qquad \mathscr{R}' = \mathscr{A}' \cap U',$$

generated by $\mathscr{A}$ and U is known as the **covariance algebra** and will figure prominently in what follows. If the only invariant elements of $\mathscr{A}'$ are of the form $\alpha \cdot \mathbf{1}$, then it is all of $\mathscr{B}(\mathscr{H})$, as $\mathscr{R}' = \alpha \cdot \mathbf{1} \Rightarrow \mathscr{R}'' = \mathscr{R} = \mathscr{B}(\mathscr{H})$.

An initial orientation to the various possibilities can be obtained by looking at some

Examples (3.1.1)

1. Classical dynamical systems. The Abelian algebra $\mathscr{A}$ of C^∞ functions $a(p, q)$ on the phase space $T^*(M)$ is a special case of the general schema. If $d\mu$ is a probability measure on $T^*(M)$, then the elements $a \in \mathscr{A}$

are represented as multiplication operators on the Hilbert-space. $L^2(T^*(M), d\mu)$. The advantage of the Hilbert-space approach to classical mechanics is that it ignores exceptional trajectories making up null sets. If a time-invariant measure $d\mu$, such as the Liouville measure $dq_1 \cdots dp_{3N}$ is restricted to a time-invariant region Ω of finite volume and normalized, then the time-evolution $a(p, q) \to a(p(t), q(t))$ is represented unitarily on $L^2(\Omega, d\mu)$. It can be written formally as $U_t = \exp(-iht)$, where $h = iL_{X_H}$ is the Liouville operator (I:2.2.25; 1), and this unitary group of transformations extends to the von Neumann algebra $\mathscr{A}'' = L^\infty(\Omega, d\mu)$. Of course U_t does not belong to $\mathscr{A}''$, which is maximally Abelian, $\mathscr{A}'' = \mathscr{A}' = \mathscr{Z}$. The algebra $\mathscr{R}$ is all of $\mathscr{B}(\mathscr{H})$ if and only if the system is ergodic, for then the only time-invariant functions are constant almost everywhere, and are thus the constant functions of $L^\infty(\Omega, d\mu)$.

2. A single spin in a magnetic field, cf. (1.1.1):

$$\mathscr{A} = \mathscr{B}(\mathbb{C}^2) = \{\mathbf{1}, \sigma, \sigma^\pm\}'', \qquad \rho(\cdot) = \left\langle \begin{pmatrix}1\\0\end{pmatrix} \middle| \cdot \middle| \begin{pmatrix}1\\0\end{pmatrix} \right\rangle,$$

$$U_t = \exp(iB(\mathbf{1} - \sigma)t) \qquad \mathscr{A}' = \mathscr{Z} = \mathscr{R}' = \{\alpha \cdot \mathbf{1}\}, \qquad \mathscr{A}'' = \mathscr{A} = \mathscr{R}.$$

Observe that while there is only one invariant vector, there is a second pure invariant state, $\langle \binom{0}{1} | \cdot | \binom{0}{1} \rangle$.

3. A single spin in a magnetic field, in a thermal representation (1.4.7):

$$\mathscr{A} = \{\mathbf{1}, \sigma, \sigma^\pm\}'' \otimes \mathbf{1}, \qquad \rho(\cdot) = \langle \Omega | \cdot | \Omega \rangle,$$

$$\Omega = \sqrt{\frac{1+s}{2}} \begin{pmatrix}1\\0\end{pmatrix} \otimes \begin{pmatrix}1\\0\end{pmatrix} + \sqrt{\frac{1-s}{2}} \begin{pmatrix}0\\1\end{pmatrix} \otimes \begin{pmatrix}0\\1\end{pmatrix},$$

$$\mathscr{A}' = \mathbf{1} \otimes \{\mathbf{1}, \tau, \tau^\pm\}'', \qquad U_t = \exp(iB(\tau - \sigma)t),$$

$$\mathscr{A}'' = \mathscr{A},\ \mathscr{Z} = \{\alpha \cdot \mathbf{1}\},\ \mathscr{R}' = \mathbf{1} \otimes \{\mathbf{1}, \tau\}'',\ \mathscr{R} = \{\mathbf{1}, \sigma, \sigma^\pm\}'' \otimes \{\mathbf{1}, \tau\}''.$$

This factor representation on $\mathbb{C}^4$ has a two-dimensional invariant subspace and a five-dimensional manifold of invariant states. Two of these are pure states corresponding to noninvariant vectors. Notice that the formal equation $h = -B\sigma$ has to be normalized not only with a constant but also by $B\tau \in \mathscr{A}'$, to ensure that $U|\Omega\rangle = |\Omega\rangle$. With a different choice of the basis for $\mathbb{C}^4$, Ω can also be written as $\binom{1}{0} \otimes \binom{1}{0}$, which makes the representation π of $\mathscr{A}$ somewhat more complicated (cf. (2.5.51; 2)):

$$\pi(\sigma^\pm) = \sqrt{\frac{1+s}{2}}\, \sigma^\pm \otimes \mathbf{1} - (-1)^\sigma \otimes \sqrt{\frac{1-s}{2}}\, \tau^\mp,$$

$$\pi(\sigma) = \frac{1+s}{2} \sigma \otimes \mathbf{1} - \mathbf{1} \otimes \tau \frac{1-s}{2} + \sqrt{1-s^2}\{\sigma^- \otimes \tau^- + \sigma^+ \otimes \tau^+\}.$$

It is easy to verify the algebraic relationships

$$\pi(\sigma^+)\pi(\sigma^-) \pm \pi(\sigma^-)\pi(\sigma^+) = \begin{cases} \mathbf{1}, \\ \pi(\sigma), \end{cases} \qquad \pi(\sigma^+)^2 = 0,$$

4. An infinite, interacting spin system. Consider the model of a ferromagnet (2.3.32; 2) in the limit $N \to \infty$. It is not hard to discover that the thermal expectation values converge to those with the vector

$$\otimes\left(\begin{pmatrix}1\\0\end{pmatrix} \otimes \begin{pmatrix}1\\0\end{pmatrix}\sqrt{\frac{1+s}{2}} + \begin{pmatrix}0\\1\end{pmatrix} \otimes \begin{pmatrix}0\\1\end{pmatrix}\sqrt{\frac{1-s}{2}}\right),$$

as with a type-III representation (1.4.7). The quantities

$$s = \langle \sigma \rangle = -\tanh B_{\text{eff}}\beta, \qquad B_{\text{eff}} = B - 2s,$$

are to be determined self-consistently, for the interaction can be written as

$$\frac{1}{N}\sum_{i,j} \boldsymbol{\sigma}_i \cdot \boldsymbol{\sigma}_j = \frac{1}{N}\sum_i (\boldsymbol{\sigma}_i - \langle \boldsymbol{\sigma}_i \rangle) \cdot \sum_j (\boldsymbol{\sigma}_j - \langle \boldsymbol{\sigma}_j \rangle) - 2\langle \sigma \rangle \sum_i \sigma_i + \text{const.}$$

If now $N \to \infty$, the first term on the right describes the fluctuations and becomes negligible compared with $-2\langle \sigma \rangle \sum_i \sigma_i$, and the commutators of H approach those of $B_{\text{eff}} \sum_i \sigma_i$, $B_{\text{eff}} = -Z\langle \sigma \rangle$ (cf. (1.1.11)). The time-evolution is accordingly given by

$$U_t = \bigotimes_j \exp(iB_{\text{eff}}(\tau_j - \sigma_j)t).$$

The Hilbert space $\mathscr{H}$ contains infinitely many invariant vectors, viz., all the ones that differ from Ω in the replacement of finitely many factors with an invariant vector from Example 3. Since B_{eff} depends on β, the time-automorphisms on representations with different β are different. Therefore there is not any automorphism of the algebra $\mathscr{A}$ generated by the σ's on the sum of two representations with different β. Although an isomorphism of $\pi(\mathscr{A})$, as a subalgebra of $\mathscr{B}(\mathscr{H}_\pi)$, is given by

$$\alpha_{-t}(\pi(\mathscr{A})) = U_t^{\beta_1, \beta_2}\pi(\mathscr{A})(U_t^{\beta_1, \beta_2})^{-1}$$

with

$$U_t^{\beta_1, \beta_2} = U_t^{\beta_1} \oplus U_t^{\beta_2}, \qquad \pi = \pi_{\beta_1} \oplus \pi_{\beta_2},$$

it is not an automorphism, since there are times t at which $\alpha_t(\pi(\mathscr{A})) \neq \pi(\mathscr{A})$. The smallest subalgebra of $\mathscr{B}(\mathscr{H}_\pi)$ for which $(\alpha_t)_{t \in \mathbb{R}}$ becomes a group of automorphisms is clearly $\bigcup_t \alpha_t(\pi(\mathscr{A}))$. If $B = 0$ and $T < 2$, then there is such a sum, or even an integral. There are nonzero solutions to the equation $B_{\text{eff}} = 2\tanh \beta B_{\text{eff}}$, but nothing favors any direction. Expectation values are averages over the unit sphere of expectation values with $\mathbf{B}_{\text{eff}} = \mathbf{n}B_{\text{eff}}$, by means of which the representation takes on the form

$$\pi(\mathscr{A}) = \int_{S_2} d\mathbf{n}\,\pi_{\mathbf{n}}(\mathscr{A}),$$

where $\pi_{\mathbf{n}}$ is specified by (1.4.7) with $\sigma \equiv (\boldsymbol{\sigma}\cdot\mathbf{n})$. The time-evolution on $\pi_{\mathbf{n}}(\mathscr{A})$ is the rotation $\sigma_j^\alpha(t) = (\exp(tR))^{\alpha\beta}\sigma_j^\beta$ having the matrix

$$R = B_{\text{eff}}\begin{pmatrix} 0 & n_3 & -n_2 \\ -n_3 & 0 & n_1 \\ n_2 & -n_1 & 0 \end{pmatrix}.$$

However, as the strong limit of $(1/N)\sum_{j=1}^N \boldsymbol{\sigma}_j$ as $N \to \infty$, $\mathbf{n}$ is contained in $\pi(\mathscr{A})''$ and lies in the center of this algebra but is not a multiple of $\mathbf{1}$. It is constant in time, and the $\mathbf{n}$-dependent time-evolution of the σ's can be viewed as an automorphism of $\pi(\mathscr{A})''$.

5. Free fermions. The algebra $\mathscr{A}$ is generated by the field operators a_f (1.3.2), and as in (1.3.3; 5) the free time-evolution

$$f(\mathbf{p}) \to \exp(-i|\mathbf{p}|^2 t)f(\mathbf{p}) \equiv f_t(\mathbf{p}),$$

provides a group of automorphisms on $\mathscr{A}: a_f \to a_{f_t}$. The thermal state (2.5.49) is clearly invariant in time and leads to a unitary time-evolution $U_t = \exp(-iHt)$. In order to tell the type of the representation, we can write it in a form like the one in Example 3. Let $|\Omega_{1,2}\rangle$ be two Fock vacua and $\pi_{1,2}(a_f)$ be the representations formed with $|\Omega_{1,2}\rangle$. Then with the tensor product

$$|\Omega\rangle = |\Omega_1\rangle \otimes |\Omega_2\rangle$$

we get

$$\pi(a(f)) = \pi_1\left(a\left(\frac{\tilde{f}(\mathbf{p})}{\sqrt{1+\exp(-\beta(|\mathbf{p}|^2-\mu))}}\right)\right)\otimes \mathbf{1}$$
$$+ (-\mathbf{1})^N \otimes \pi_2\left(a^*\left(\frac{\tilde{f}^*(\mathbf{p})}{\sqrt{1+\exp(\beta(|\mathbf{p}|^2-\mu))}}\right)\right),$$

where $aN = (N+1)a$ (cf. (1.3.13)). It can be verified that

$$\langle a_{f_1}^* \cdots a_{f_n}^* a_{g_1} \cdots a_{g_n}\rangle = \langle \Omega|\pi(a_{f_1}^*)\cdots \pi(a_{f_n}^*)\pi(a_{g_1})\cdots \pi(a_{g_n})|\Omega\rangle,$$

so this representation is equivalent to the thermal representation with infinitely many spins. Consequently, if $T > 0$, then it is a factor of type III. The local field operators in momentum space can be used to write H_π as

$$H_\pi = \int \frac{d^3p}{(2\pi)^3}|\mathbf{p}|^2\{\pi_1(a^*(\mathbf{p})a(\mathbf{p}))\otimes \mathbf{1} - \mathbf{1}\otimes \pi_2(a^*(\mathbf{p})a(\mathbf{p}))\}.$$

The operator a^*a differs from the usual one not only in that the infinite zero-point energy of field theory has been subtracted off, but also in the removal of an operator of $\mathscr{A}'$.

The Time-Evolution of Open Systems (3.1.2)

It seems illusory to consider every single local property of a large system as belonging to the algebra of observables. It is certainly true that practically anything can be measured, but not all at once, and putting the system into a state that is dispersionless with respect to a maximally Abelian subalgebra is actually impossible. In reality only fairly small subsystems get measured, so it is of practical interest to divide the total system into the subsystem that is observed, called an "open" system, and all the rest, acting as a reservoir. Accordingly, let $\mathscr{H} = \mathscr{H}_S \otimes \mathscr{H}_R$ and let Tr^{S+R}, Tr^S, and Tr^R be the traces on $\mathscr{H}$, $\mathscr{H}_S$, and $\mathscr{H}_R$. The time-evolution U_t will mix $\mathscr{H}_S$ and $\mathscr{H}_R$, so it does not create an automorphism of $\mathscr{B}(\mathscr{H}_S)$. However, if the initial state postulated can be factorized and written in terms of a density matrix $\rho \otimes \omega$, then a time-evolution τ_t: $\mathscr{B}(\mathscr{H}_S) \to \mathscr{B}(\mathscr{H}_S)$ can be defined for the open system in the Heisenberg picture, or the dual time-evolution for the density matrices τ_t^*: $\mathscr{C}_1(\mathscr{H}_S) \to \mathscr{C}_1(\mathscr{H}_S)$ can be defined in the Schrödinger picture. If $a \in \mathscr{B}(\mathscr{H}_S) \otimes \mathbf{1}$, then the time-dependence of the expectation values can be written as

$$\langle a(t) \rangle \equiv \mathrm{Tr}^{S+R}(\rho \otimes \omega)U_{-t}(a \otimes \mathbf{1})U_t = \mathrm{Tr}^S \rho\tau_t(a) = \mathrm{Tr}^S \tau_t^*(\rho)a,$$

where by definition

$$\tau_t(a) \equiv \mathrm{Tr}^R(\mathbf{1} \otimes \omega)U_{-t}(a \otimes \mathbf{1})U_t,$$
$$\tau_t^*(\rho) = \mathrm{Tr}^R U_t(\rho \otimes \omega)U_{-t}. \tag{3.1.3}$$

Note that the states transform with $U_t^* = U_{-t}$ rather than U_t.

Properties of the Time-Evolution of the Subsystem (3.1.4)

The operators τ_t and τ_t^* are

(i) one-parameter, strongly continuous families of completely positive linear mappings;
(ii) *not* groups: $\tau_{t_1} \circ \tau_{t_2} \neq \tau_{t_1+t_2}$;
(iii) *not* isomorphisms of the algebra: $\tau_t(a \cdot b) \neq \tau_t(a) \cdot \tau_t(b)$.

Equality holds in (ii) and (iii) only if U_t factorizes.

Gloss (3.1.5)

A linear mapping Φ: $\mathscr{B}(\mathscr{H}) \to \mathscr{B}(\mathscr{H})$ is said to be ***n*-positive** iff $\Phi \otimes \mathbf{1}$ acting on $\mathscr{B}(\mathscr{H}) \otimes \mathscr{B}(\mathbb{C}^n)$: $a \otimes M \to \Phi(a) \otimes M$ is positive for all $M \in \mathscr{B}(\mathbb{C}^n)$, i.e., it maps the cone of positive elements of $\mathscr{B}(\mathscr{H}) \otimes \mathscr{B}(\mathbb{C}^n)$ into itself. The mapping Φ is **completely positive** iff it is positive for all $n = 1, 2, \ldots$. It can be shown [14] that all completely positive mappings are obtained by taking tensor products of positive operators, composing with unitary operators, and then taking partial traces, just as in the construction of τ_t and τ_t^*. The completely positive mappings form a semigroup with respect to composition.

Examples (3.1.6)

1. The classical harmonic oscillator
The observables are chosen as the position coordinates q, so

$$\mathrm{Tr}^{S+R} \to \int dp\, dq, \qquad \mathrm{Tr}^{R} \to \int dp, \qquad \mathrm{Tr}^{S} \to \int dq.$$

Let $\rho(q) = \pi^{-1/2}\exp(-(q-q_0)^2)$ be the probability distribution function of the coordinates and $\omega(p) = \pi^{-1/2}\exp(-(p-p_0)^2)$ be that of the momenta. The time-evolution of the total system, $q(t) = q\cos t + p\sin t$, $p(t) = p\cos t - q\sin t$, induces

$$\tau_t(q) = q\cos t + p_0 \sin t,$$

$$\tau_t^*(\rho) = \pi^{-1/2}\exp[-(q - q_0\cos t - p_0 \sin t)^2]$$

on the subsystem. However, τ_t is not an isomorphism,

$$\tau_t(q^2) = (q\cos t + p_0\sin t)^2 + \tfrac{1}{2}\sin^2 t \neq \tau_t(q)^2,$$

since ω is not free of fluctuations. The choice of equal widths for ρ and ω, as with quantum-mechanical coherent states, causes a rigid oscillation of ρ. If, instead, $\omega(p) = \delta(p - p_0)$, then there would be a periodic focusing and defocusing of ρ,

$$\tau_t^*(\rho) = \frac{\exp[-(q - q_0\cos t - p_0\sin t)^2\cos^{-2} t]}{\sqrt{\pi}\cos t}.$$

2. Quantum-mechanical coupled oscillators.
Let us return to the chain of oscillators (1.1.13) and take ξ_0 and ξ_1 as the open system. Instead of the pure state (1.1.21), suppose the system is in a thermal state

$$\left\langle \exp\left[i \sum_{n=-\infty}^{\infty} (\xi_{2n} r_n + \xi_{2n+1} s_n)\right]\right\rangle$$
$$= \exp\left[-\frac{1}{4}\tanh\frac{\eta}{2}\sum_{n=-\infty}^{\infty}(r_n^2 + s_n^2) + i\sum_{n=\infty}^{\infty}(r_n s_n' - r_n' s_n)\right].$$

As in (2.5.53.6),

$$\frac{\mathrm{Tr}\exp[-\eta((p-\bar{p})^2 + (q-\bar{q})^2)]\exp[i(pr+qs)]}{\mathrm{Tr}\exp[-\eta((p-\bar{p})^2 + (q-\bar{q})^2)]}$$
$$= \exp\left[-\frac{r^2+s^2}{4}\tanh\frac{\eta}{2} + i(\bar{p}r + \bar{q}s)\right], \quad (3.1.7)$$

so this state is a Gibbs state with harmonic forces centered at s', $-r'$. Under the time-evolution (1.1.18), the expectation values of the Weyl operators of the open system are

$$\langle \exp i(r\xi_0(t) + s\xi_1(t))\rangle = \exp\left\{\sum_n \left\{-\tfrac{1}{4}\tanh\frac{\eta}{2}[(rJ_{2n} + sJ_{2n+1})^2 + (rJ_{2n+1} + sJ_{2n})^2] + is'_n(rJ_{2n} + sJ_{2n-1}) - ir'_n(rJ_{2n+1} + sJ_{2n})\right\}\right\}.$$

At time t the subsystem is in a state of the form (3.1.7) with

$$s'_0(t) = \sum_n (s'_n(0)J_{2n}(t) - r'_n(0)J_{2n+1}(t)),$$

$$r'_0(t) = \sum_n (r'_n(0)J_{2n}(t) - s'_n(0)J_{2n-1}(t)).$$

The average values $s'_0(t)$, $r'_0(t)$ move classically as in Example 1. They converge to zero, but not monotonically.

3. Coupled spins
Consider spin 1 of the chain (1.1.1) as the open system and the infinitely many others as the thermal reservoir. The coupling constants $\varepsilon(n)$ are chosen as in (1.1.9). The initial state

$$\rho_1 = \tfrac{1}{2}(1 + \sigma_1^+ \exp(-i\alpha) + \sigma_1^- \exp(i\alpha)),$$
$$\omega = \prod_{k\neq 1} \tfrac{1}{2}(1 + \sigma_k^+ \exp(-i\alpha) + \sigma_k^- \exp(i\alpha)),$$

((1.17) with $s = 0$) evolves as

$$\tau_t^*(\rho) = \tfrac{1}{2}\left(1 + \frac{\sin^2 t}{t^2}[\sigma^+ \exp(-i(\alpha + 2Bt) + \sigma^- \exp(i(\alpha + 2Bt)]\right)$$

if $N \to \infty$. The state ρ oscillates as it approaches the equilibrium state $\frac{1}{2}\cdot\mathbf{1}$ as $T \to \infty$.

Remarks (3.1.8)

1. The failure of the time-evolution τ or τ^* to be a group is due to the effect of the system on the reservoir and the reaction of the reservoir on the system. The reaction influences the system at later times, so $(\partial/\partial t)\tau_t^*(\rho)$ depends on $\tau_s^*(\rho)$ not only for $s = t$ but for all $s \leq t$, i.e., on its whole history. The time-evolution of the density matrix of the reservoir can be written down formally and substituted into the equation for $(\partial/\partial t)\tau_t^*(\rho)$. The resulting **master equation** is an integrodifferential equation for ρ including the memory effects just mentioned.
2. The requirement of complete positivity of the time-evolution is not a mere technicality but a genuine restriction, and it even has some experi-

mentally verifiable consequences. For instance, its implications for the motion of a spin in a thermal reservoir have been confirmed experimentally [15].

The retrospective effects of (3.1.8; 1) disappear in certain limiting cases, so the time-evolution τ becomes a semigroup. The limits involve the time-scale or the coupling constants. The most understandable case is that of a simplified version of electrodynamic radiative reaction of volume II, §2.4.

Example (3.1.9)

Model of Brownian motion
We modify Example (1.1.13) to take a single harmonic oscillator in three dimensions as the system and represent the rest of the system, functioning as a reservoir, as a continuous scalar field $\Phi(\mathbf{x})$. Suppose initially that the oscillator is coupled to an averaged field $\int d^3x\Phi(\mathbf{x})c(\mathbf{x})$, $c \in C_0^\infty(\mathbb{R}^3)$, and later take the limit $c(\mathbf{x}) \to \gamma\delta(\mathbf{x})$, $\gamma \in \mathbb{R}$. We shall study the quantum-theoretical time-evolution from the outset; since the equations of motion are linear it agrees with the classical time-evolution. If Q, P and $\Phi(\mathbf{x})$, $\Pi(\mathbf{x})$ are the canonically conjugate coordinate and field variables, then the Hamiltonian is

$$H_S = \tfrac{1}{2}(P^2 + \omega_0^2 Q^2),$$

$$H_R = \tfrac{1}{2}\int d^3x\{\Pi(\mathbf{x})^2 + |\nabla\Phi(\mathbf{x})|^2\},$$

$$H' = \int d^3x c(\mathbf{x})\Phi(\mathbf{x})Q.$$

The resulting equations of motion,

$$\left(\frac{\partial^2}{\partial t^2} - \Delta\right)\Phi(\mathbf{x}, t) = c(\mathbf{x})Q(t),$$

$$\left(\frac{\partial^2}{\partial t^2} + \omega_0^2\right)Q(t) = \int d^3x\Phi(\mathbf{x}, t)c(\mathbf{x}),$$

can be integrated immediately with Green's formula (II: 1.2.36). This is the trivial case of a scalar field on $\mathbb{R}^4$, so with the Green function

$$D(\mathbf{x}, t) = \frac{\delta(r - t)}{4\pi r}$$

(II: 2.2.7), the solution of the initial-value problem is

$$\Phi(\mathbf{x}, t) = \int d^3x'(\Phi(\mathbf{x}', 0)\dot{D}(\mathbf{x} - \mathbf{x}', t) + \dot{\Phi}(\mathbf{x}', 0)D(\mathbf{x} - \mathbf{x}', t))$$

$$+ \int d^3x' \int_0^t dt' D(\mathbf{x} - \mathbf{x}', t - t')c(\mathbf{x}')Q(t')$$

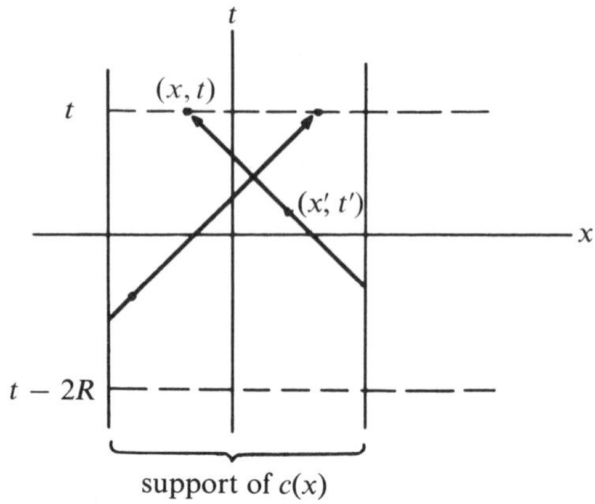

Figure 22 The domain of influence of F_{reaction}.

for all $t > 0$, where $\dot{\Phi} = \partial\Phi/\partial t$, etc. Hence the force exerted by the field on the oscillator is

$$\int d^3x\Phi(\mathbf{x}, t)c(\mathbf{x}) = F_{\text{field}}(t) + F_{\text{reaction}}(t),$$

$$F_{\text{field}}(t) = \int d^3x\, d^3x' c(\mathbf{x})(\Phi(\mathbf{x}', 0)\dot{D}(\mathbf{x} - \mathbf{x}', t) + \dot{\Phi}(\mathbf{x}', 0)D(\mathbf{x} - \mathbf{x}', t))$$

$$F_{\text{reaction}}(t) = \int \frac{d^3x d^3x'}{4\pi|\mathbf{x} - \mathbf{x}'|} c(\mathbf{x})c(\mathbf{x}')Q(t - |\mathbf{x} - \mathbf{x}'|)\Theta(t - |\mathbf{x} - \mathbf{x}'|).$$

In the reaction force $F_{\text{reaction}}(t)$, $Q(t')$ contributes only for $t - 2R \leq t' \leq t$ if $c(\mathbf{x}) = 0$ for all $\mathbf{x}$ such that $|\mathbf{x}| > R$ (see Figure 22).

Now if $c(\mathbf{x}) \to 2\sqrt{\pi}\gamma\delta(\mathbf{x})$, so $R \to 0$, then the retrospective effects disappear, and when the expansion

$$Q(t - |\mathbf{x} - \mathbf{x}'|) = Q(t) - |\mathbf{x} - \mathbf{x}'|\dot{Q}(t) + \tfrac{1}{2}|\mathbf{x} - \mathbf{x}'|^2\ddot{Q}(t) - \cdots$$

is substituted into F_{reaction},

$$F_{\text{reaction}}(t) \to \delta\omega^2 Q(t) - \gamma^2\dot{Q}(t).$$

The quantity $\delta\omega^2$ is the formally infinite integral $\gamma^2 \int (d^3x\, d^3x'/|\mathbf{x} - \mathbf{x}'|) \times \delta(\mathbf{x})\delta(\mathbf{x}')$, so the limit $c(\mathbf{x}) \to \gamma\delta(\mathbf{x})$ must be taken jointly with a change in ω_0^2. If $\bar{\omega}^2 \equiv \omega_0^2 - \delta\omega^2$, then the equation of motion becomes

$$\left(\frac{\partial^2}{\partial t^2} + \bar{\omega}^2 + 2\Gamma\frac{\partial}{\partial t}\right)Q(t) = F_{\text{field}}(t), \qquad \Gamma = \frac{\gamma^2}{2}, \quad t \geq 0.$$

For a thermal state with $\langle\Phi(\mathbf{x}, 0)\rangle = \langle\dot{\Phi}(\mathbf{x}, 0)\rangle = 0$, $\langle F_{\text{field}}(t)\rangle = 0$, and the time-evolution of the expectation value of Q for $t \geq 0$ is

$$\langle Q(t)\rangle = \exp(-\Gamma t)\left(\langle Q(0)\rangle\left(\cos\omega t + \frac{\Gamma}{\omega}\sin\omega t\right) + \langle\dot{Q}(0)\rangle\frac{\sin\omega t}{\omega}\right),$$

provided that $\omega^2 \equiv \bar{\omega}^2 - \Gamma^2 > 0$. The expectation values of the canonical variables $\langle Q(t)\rangle$ and $\langle \dot{Q}(t)\rangle$ then evolve according to a symplectic semigroup,

$$\begin{pmatrix} \langle Q(t)\rangle \\ \langle \dot{Q}(t)\rangle \end{pmatrix} = \exp(-\Gamma t)\begin{pmatrix} \cos\omega t + \frac{\Gamma}{\omega}\sin\omega t & \frac{\sin\omega t}{\omega} \\ -\left(\omega + \frac{\Gamma^2}{\omega}\right)\sin\omega t & \cos\omega t - \frac{\Gamma}{\omega}\sin\omega t \end{pmatrix}\begin{pmatrix} \langle Q(0)\rangle \\ \langle \dot{Q}(0)\rangle \end{pmatrix}.$$

The time-evolution of an open system is not generally a unitary transformation of the density matrix, and so the entropy of a subsystem is not necessarily constant. Nothing can be said *a priori* about the sign of the change in entropy; the system might start off hotter than the reservoir and lose entropy as the temperature equalizes. However, the relative entropy introduced in (2.2.22) turns out to be a Liapunov function [16] for the time-evolution (3.1.3).

The Decrease of the Relative Entropy (3.1.10)

For the time-evolution τ^* *of* (3.13),

$$S(\tau_t^*(\sigma)\,|\,\tau_t^*(\rho)) \le S(\sigma\,|\,\rho).$$

Proof

With Definition (2.2.22) and the unitary invariance,

$$S(\mathrm{Tr}^R U_{-t}\sigma\otimes\omega U_t\,|\,\mathrm{Tr}^R U_{-t}\rho\otimes\omega U_t) \overset{\text{(iv)}}{\le} S(U_{-t}\sigma\otimes\omega U_t\,|\,U_{-t}\rho\otimes\omega U_t)$$
$$= S(\sigma\otimes\omega\,|\,\rho\otimes\omega) \overset{\text{(iii)}}{=} S(\sigma\,|\,\rho). \qquad \square$$

Remarks (3.1.11)

1. The relative entropy is always positive, and in the special case of (2.2.23; 1), it is β times the difference between the free energy of the state ρ and the free energy at equilibrium. Its decrease reflects the tendency of the system to equilibrium.
2. Monotony in time cannot be claimed if $\tau_{t_1+t_2} \ne \tau_{t_2}\circ\tau_{t_1}$. In Example (3.1.9) friction returned the oscillator monotonically to its rest-point, owing to the semigroup property, which was in turn a consequence of the absence of retrospective effects. The general fact is

Monotony of the Relative Entropy with a Dynamic Semigroup (3.1.12)

If $\tau_{t_1+t_2} = \tau_{t_2}\circ\tau_{t_1}$ *for all* t_1 *and* $t_2 \ge 0$, *then* τ_t *is said to be a dynamical semigroup. The function* $S(\tau_t^*(\sigma)\,|\,\tau_t^*(\rho))$ *is then a monotonically decreasing function of* t.

Proof

This is a direct consequence of (3.1.10). □

Remarks (3.1.13)

1. Because $S(\sigma|\rho) \geq 0$, the limit of $S(\tau_t^*(\sigma)|\tau_t^*(\rho))$ as $t \to \infty$ exists.
2. It cannot yet be claimed that the free energy approaches its equilibrium value as $t \to \infty$; $S(\sigma/\rho)$ might stop at some positive value and never fall to zero.
3. The apparent asymmetry in the direction of time comes from the requirement of (3.1.3) that the initial state factorizes. Starting at $t < 0$, the later state at $t = 0$ is factorized, so the relative entropy increases.
4. If the dynamical semigroup is governed by a master equation of the type of (2.1.11; 3), then $S(\rho)$ increases monotonically.

That finishes the orientation toward various phenomena connected with the time evolution. Let us now return to more global questions of time-dependence. The problem, put concisely, is that a finite system the Hamiltonian of which has pure point spectrum $\{\varepsilon_i\}$ has observables whose expectation values $\langle a(t)\rangle = \sum_{j,k} a_{jk} \exp(i(\varepsilon_j - \varepsilon_k)t)$ are almost-periodic functions, as superpositions of periodic functions. Only the average over time makes sense; the time-limit exists only for infinite systems the Hamiltonians of which have absolutely continuous spectra. Although in actuality only finite systems come under observation, the recurrence times are so long that they are indistinguishable from infinite systems within the times of relevance to human beings. In any event, the first issue to settle is how to define the time-average of a function $f(t) \in C(\mathbb{R})$, the set of bounded, continuous functions on $\mathbb{R}$. The obvious guesses would be

$$\lim_{T\to\infty} \frac{1}{2T}\int_{-T}^{T} dt f(t) \quad \text{or} \quad \lim_{\varepsilon\to 0} \frac{\varepsilon}{2}\int_{-\infty}^{\infty} dt \exp(-\varepsilon|t|) f(t),$$

but these limits do not converge for such functions as $\sin(\ln(|t| + 1)) \in C(\mathbb{R})$. Any suitable average would have to be linear, positive, and invariant under displacements in time. Every invariant state on the C^* algebra $C(\mathbb{R})$ has the right qualifications, and the existence of many invariant states on $C(\mathbb{R})$ means that there are many possible time-averages. There is thus no question whether a time-average exists, but it is not unique.

The Time-Average of an Observable (3.1.14)

Let η be an average over $C(\mathbb{R})$ and $t \to a_t$ be a weakly continuous mapping $\mathbb{R} \to \mathscr{B}(\mathscr{H})$ such that $\|a_t\| \leq \|a_0\|$ for all t. Then the average $\eta(a)$ is defined by

$$\langle x|\eta(a)|y\rangle = \eta(\langle x|a_t|y\rangle) \quad \text{for all } x, y \in \mathscr{H}.$$

Remarks (3.1.15)

1. Since $|\eta(\langle x|a_t|y\rangle)| \leq \|x\| \cdot \|y\| \cdot \|a_0\|$, this sesquilinear form defines a bounded operator $\eta(a)$.
2. In the Schrödinger picture, the average $\eta(\sigma)$ of a state σ on the algebra generated by the operators a_t is defined by $\eta(\sigma)(a) = \eta(\sigma(a_t))$.

Examples (3.1.16)

1. If $a_t = \exp(-iHt) \equiv U(t)$, then $\eta(U) = E_0 \equiv$ the projection onto the eigenvectors of H with eigenvector 0.

Proof

(i) $\langle x|E_0\eta(U)y\rangle = \eta\langle E_0 x|U_t y\rangle = \langle x|E_0 y\rangle \Rightarrow E_0\eta(U) = E_0$.
(ii) $\langle x|U(t_0)\eta(U)y\rangle = \eta\langle x|U(t + t_0)y\rangle = \langle x|\eta(U)y\rangle \Rightarrow$
$U(t_0)\eta(U) = \eta(U) \Rightarrow E_0\eta(U) = \eta(U) = E_0$ by part (i). □

2. $a_t = U(t)aU^{-1}(t)$, where $U(t)$ has pure point spectrum. If the projections onto the eigenspaces are E_i, then $\eta(a) = \sum_i E_i a E_i$.

Proof

Take matrix elements with the eigenvectors of H and note that $\eta(\exp(i\alpha t)) = 0$ for all η and all $\alpha \in \mathbb{R}$ different from 0. □

3. $\eta(a_t E_0) = E_0 a E_0$, since $\eta(a_t E_0) = \eta(U(t)aE_0) = E_0 a E_0$, as in Example 1.

Remarks (3.1.17)

1. In these examples the concrete averages $(1/2T)\int_{-T}^{T} dt \exp(iHt)$ and $(\varepsilon/2)\int_{-\infty}^{\infty} dt \exp(-\varepsilon|t|)\exp(iHt)$ converge strongly (Problem 1). Hence E_0 belongs to U'' as well as U'.
2. In the Schrödinger picture the time-average of a vector $|x\rangle$ is defined by $|\eta(x)\rangle \equiv \eta(U(t)|x\rangle) = E_0|x\rangle$. It can be characterized as the vector with the least norm in the convex hull of its trajectory $\{U(t)|x\rangle, t \in \mathbb{R}\}$ (Problem 2). It is not, however, true in general for the state $\sigma(a) = \langle x|a|x\rangle$ formed with $|x\rangle$ that $\eta(\sigma)(a) = \langle\eta(x)|a|\eta(x)\rangle$.
3. There is no definition of $\eta(a)$ independent of the representation; since $\lim_{T\to\infty}(1/T)\int_0^T dt\, a_t$ belongs only to the weak closure of the algebra, η may send operators out of their C^* algebra. Our representations will usually be such that the time-automorphism α_t can be implemented

unitarily, and the image of E_0 will contain a cyclic vector for $\mathscr{A}$. If the averages $\eta(a)$ belong to $\mathscr{A}'$, then they are determined uniquely by

$$\eta(a)E_0 = \lim_{T\to\infty} \frac{1}{T}\int_0^T dta_t E_0 = \lim_{T\to\infty}\frac{1}{T}\int_0^T U_t a E_0 = E_0 a E_0,$$

since $E_0\mathscr{H}$ separates $\mathscr{A}'$ (Problem 5). However, as will be seen in (3.1.22; 4), $\eta(a)$ in general depends on the representation.

4. The time-average may be nonunique if $f(t)$ converges, as $t \to +\infty$ and $t \to -\infty$, but to different values. This situation is familiar to us from scattering theory. Whenever the time-average of a function f is unique, it agrees with the "concrete average"

$$\lim_{T\to\infty}\frac{1}{2T}\int_{-T}^{T} dtf(t), \qquad \lim_{\varepsilon\to 0}\frac{\varepsilon}{2}\int_{-\infty}^{\infty} dt \exp(-\varepsilon|t|)f(t),$$

or even

$$\lim_{T\to\infty}\frac{1}{T}\int_0^T dtf(t).$$

These averages exist in classical ergodic theory, in which the Liouville measure on phase space provides the invariant cyclic vector. Some ergodic systems will be defined later, and for them E_0 is one-dimensional, projecting onto the cyclic vector. This projection is then constant on the energy shell, so the time-average $E_0 a E_0$ equals the average over the energy shell.

5. The point spectrum of H can be turned into a continuum by an arbitrarily small perturbation, so averaging over time focuses unduly on the exact form of H, since η is quite different depending on whether the spectrum is pointlike or continuous: If in the spectral representation of H the operator a on the subspace belonging to σ_{ac} has a continuous integral kernel, then η projects this part of a to 0, and by Remark 2 only its point-spectrum part remains (cf. (I: 3.3.4; 6)).

6. Pure states of classical systems are points in phase space, and averages over pure states are averages over classical trajectories.

7. If the spectrum of H is pure point and nondegenerate, then every normal, invariant state can be written as the time-average of a pure state. Normal, invariant states are of the form

$$\sigma(a) = \sum_i c_i\langle x_i|a|x_i\rangle, \quad 0 \le c_i \le 1, \qquad \sum_i c_i = 1, \quad H|x_i\rangle = \varepsilon_i|x_i\rangle;$$

so

$$\sigma(a) = \langle x|\eta(a)|x\rangle, \qquad x = \sum_i \sqrt{c_i}|x_i\rangle.$$

Although the canonical state $\rho = \exp(-\beta(H - F))$ is an average over the trajectory of a pure state, it is certainly not true that every averaged pure state is the canonical state.

Our reasoning until this point has applied indifferently to all sorts of quantum systems, but not all quantum systems exhibit thermodynamic behavior. An isolated atom is rather like a frictionless perpetual-motion machine; only large systems are dissipative. The concept introduced in (1.3.10) of asymptotic commutativity turns out to be a useful characteristic of dissipative systems. If the local observables are asymptotically Abelian with respect to the time-automorphism α_t, that means that local perturbations dissipate through the system as time passes. Of course, this is possible only if H has continuous spectrum, and hence only if the system is infinite. We shall remain with Definition (1.3.10), although many of its consequences can be derived with weaker assumptions. Definition (1.3.10) applies to a system of free fermions, but it has not been possible to prove that even weakened versions of it apply to more realistic, interacting systems. It is trivial that classical systems are asymptotically Abelian, and (1.3.10) means roughly that asymptotically Abelian systems behave classically on a macroscopic time scale.

Properties of Asymptotically Abelian Systems (3.1.18)

Let $\mathscr{A}$ be an asymptotically Abelian C^* algebra with respect to a group of automorphisms $a \to a_t$, and let ω be an invariant state having a representation on a Hilbert space $\mathscr{H}$ with a cyclic vector $|\Omega\rangle$. Then, abbreviating $\mathscr{A}' = \pi_w(\mathscr{A})'$, etc.,

1. *the invariant elements of $\mathscr{A}$ belong to $\mathscr{A}'$;*
2. *the invariant elements of $\mathscr{A}'$ lie in the center (i.e., $\mathscr{R}' = \mathscr{A}' \cap U' = \eta(\mathscr{A}')$ is a subalgebra of the center $\mathscr{A}' \cap \mathscr{A}''$), and so $\mathscr{R}' = \eta(\mathscr{A}'')$;*
3. *$E_0 \mathscr{A}'' E_0$ is maximally Abelian in $E_0 \mathscr{H}$, where E_0 is the projection onto the invariant vectors of $\mathscr{H}$; and*
4. *if σ produces a factor (i.e., the* GNS *representation $\pi_\sigma(\mathscr{A})$ and $\pi_\sigma(\mathscr{A})'$ constructed with the cyclic vector Ω_σ generate all of $\mathscr{B}(\mathscr{H})$), then*

$$\lim_{t \to \pm\infty} (\sigma(a_t b) - \sigma(a_t)\sigma(b)) \to 0,$$

even if $\sigma(a_t) \neq \sigma(a)$.

Remarks (3.1.19)

1. Neither E_0 nor $E_0 \mathscr{A}'' E_0$ necessarily belongs to $\mathscr{A}''$. Moreover, $E_0 \mathscr{A}'' E_0$ may fail to be an algebra, and the somewhat loose phrasing of Property 3 is intended to mean that the algebra generated by $E_0 \mathscr{A}'' E_0$ is the same as its commutant.
2. The point of (3.1.18) is that invariant elements such as time-averages and time-limits form an Abelian algebra, and thus equal its center. Factor states are pure when restricted to the center, and are therefore

characters (see Definition (III: 2.2.25)), which explains why they factorize in time-limits and time-averages.

Proof

1. $[a, b] = \lim_{t\to\infty} [a_t, b] = 0$ for all invariant $a \in \mathscr{A}$ and all $b \in \mathscr{A}$.
2. By Property 3, $E_0 \mathscr{R} E_0 = E_0 \mathscr{A}'' E_0$ is maximally Abelian and so equal to $(E_0 \mathscr{R} E_0)' E_0$. Since $E_0 \in \mathscr{R}$, $(E_0 \mathscr{R} E_0)' E_0 = E_0 \mathscr{R}' E_0$ [17], and therefore $E_0 \mathscr{R}' E_0 = E_0(\mathscr{R}' \cap \mathscr{R}) E_0$. Since $|\Omega\rangle$ separates $\mathscr{A}'$, the equation $E_0 a' E_0 = a' E_0$ determines every $a' \in \mathscr{R}'$ uniquely, so $a' \in \mathscr{R}$. However, $\mathscr{R} \cap \mathscr{R}'$ is $\mathscr{A}'' \cap \mathscr{A}' \cap U'$, because $U \cap \mathscr{A}' = \{\mathbf{1}\}$.
3. The set $E_0 \mathscr{A} E_0$ must be Abelian, as otherwise some commutator would fail to vanish as $t \to \pm\infty$:

$$\eta_t[a_t, b] = 0 \Rightarrow \eta_t E_0(aU_t b - bU_{-t} a)E_0 = 0$$
$$\Rightarrow [E_0 a E_0, E_0 b E_0] = 0 \qquad \text{for all } a, b \in \mathscr{A}.$$

   Hence $E_0 \mathscr{A}'' E_0 = (E_0 \mathscr{A} E_0)''$ is also Abelian, and in fact maximally Abelian, as otherwise $E_0 a E_0$ would be $\sim \mathbf{1}$ on a subspace of dimension greater than one for all $a \in \mathscr{A}$, and $|\Omega\rangle = E_0 |\Omega\rangle$ would not be cyclic.
4. For every $b \in \pi_\sigma(\mathscr{A})$ there exist two operators b_1 and b_2 such that $b_2 |\Omega_\sigma\rangle = b_1^* |\Omega_\sigma\rangle = 0$ and $b = \mathbf{1}\langle\Omega_\sigma | b | \Omega_\sigma\rangle + b_1 + b_2$. This is obvious for finite matrices:

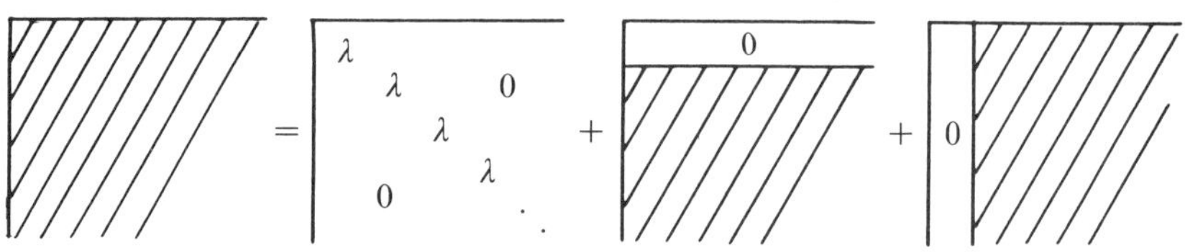

$b \quad = \quad \lambda \cdot \mathbf{1} \quad + \quad b_1 \quad + \quad b_2$

   and it carries over to $\mathscr{B}(\mathscr{H})$. Then $\sigma(a_t b) - \sigma(a_t)\sigma(b) = \sigma([a_t, b_1])$. If σ produces a factor, then b_1 can be approximated with a finite sum

$$\sum_{i=1}^{n} d_i d_i', \qquad d_i \in \pi_\sigma(\mathscr{A}), \qquad d_i' \in \pi_\sigma(\mathscr{A})',$$

   and $\sum_i \sigma([a_t, d_i] d_i')$ tends to 0 as $t \to \pm\infty$ by Definition (1.3.10). Although the subalgebra of $\mathscr{B}(\mathscr{H})$ generated by $\pi_\sigma(\mathscr{A}) \cup \pi_\sigma(\mathscr{A})'$ is only strongly dense, operators with these properties can be approximated even in the norm sense ([18], V.1.4), which justifies these conclusions. □

The set of invariant states is convex, so any invariant state is a convex combination of the extremal points of the set or a limit of such combinations. As the purest among the time-invariant states, the extremal elements deserve a special term:

Definition (3.1.20)

An invariant state is **ergodic**, or **extremal invariant**, if it can not be written as a convex combination of other invariant states.

Remarks (3.1.21)

1. In classical dynamics an invariant submanifold $\mathcal{N}$ of phase space corresponds to an invariant state (= measure) $\mu_{\mathcal{N}} = \prod_i dq^i \wedge dp^i_{|\mathcal{N}}$, which is ergodic if $\mathcal{N}$ cannot be decomposed into invariant pieces with strictly positive measures $\mu_{\mathcal{N}}$.
2. A classical system is said to be ergodic if the surface of the energy shell $\rho(p, q) = \delta(E - H(p, q)) \exp(-S(E))$ corresponds to an ergodic state.
3. Every time-invariant state is a sum or integral of ergodic states, so it is tempting to interpret the ergodic states as the pure phases of the system. Mixtures would then be incoherent superpositions in the sense of quantum theory rather than coexisting, spatially separated phases. With any reasonable definition of pure phases, the decomposition into ergodic states should be unique, and the set of time-invariant states must be a simplex. This is indeed the case for asymptotically Abelian systems, which follows from the observation that $\mathscr{R}' = \mathscr{A}' \cap \{U_t\}'$ is Abelian: As was seen in (1.4.9) and (III: 2.3.24; 2), every Abelian subalgebra of $\mathscr{A}'$ corresponds to a unique decomposition of a state ω; if $\{P_i\}$, $\sum_i P_i = \mathbf{1}$, are the orthogonal projections of this algebra, and
$$\omega_i(a) = \frac{\omega(P_i a)}{\omega(P_i)} \quad \text{for all } a \in \mathscr{A},$$
provided that $\omega(P_i) > 0$, and is otherwise arbitrary, then $\omega = \sum_i \lambda_i \omega_i$, $\lambda_i = \omega(P_i)$ and $\pi_\omega = \bigoplus_i \pi_{\omega_i}$, where π_{ω_i} acts on $P_i \mathscr{H}_\omega$. Now if ω is invariant and is to have a decomposition into other invariant states, then the projections P_i must belong to $\mathscr{A}' \cap \{U_t\}'$, and in fact the extremal states correspond to the minimal projections. Since $\mathscr{A}' \cap \{U_t\}' \subset \mathscr{Z}$, the decomposition into ergodic states is never as fine as the factor decomposition. Hence if a factor representation is given by the invariant state ω, it is necessarily ergodic.

Ergodicity in fact singles out the desired properties. This is shown by the

Characterization of the Ergodic States (3.1.22)

Let $\mathscr{A}$ be an algebra that is asymptotically Abelian in time, ρ an invariant state on $\mathscr{A}$, and $|\Omega\rangle$ the vector of the state ρ in the GNS *representation. Then the following conditions are equivalent:*

1. *ρ is ergodic;*
2. $\mathscr{R}' = \{\alpha \cdot \mathbf{1}\}$;

3. *given any decomposition* $\rho = \int \sigma \, d\mu(\sigma)$ *and a* μ*-measurable mean* η, $\eta(\sigma) = \rho$ *almost everywhere for* μ;
4. $\eta(a) = \mathbf{1} \cdot \rho(a)$ *for all* $a \in \mathscr{A}$ *and all invariant means* η;
5. $(\mathscr{A} \cup \mathscr{A}') \cap U' = \{\alpha \cdot \mathbf{1}\}$;
6. $E_0 = |\Omega\rangle\langle\Omega|$;
7. ρ *is a unique, invariant, normal state on* $\pi_\rho(\mathscr{A})''$;
8. $\eta(\rho(ab_t)) = \rho(a)\rho(b)$ *for all* a *and* $b \in \mathscr{A}$ *and all invariant means* η.

Remarks (3.1.23)

1. If the quantum system is finite, H has pure point spectrum, with eigenvectors $\{|x_i\rangle\}$. As we have learned, the invariant states are of the form $a \to \sum_i c_i \langle x_i | a x_i \rangle$, so the extremal invariant states are of the form $a \to \langle x_i | a x_i \rangle$ and therefore pure. If the system is either infinite or classical, then ergodic does not imply pure. For example, the state of free fermions (2.5.49) produces a factor and is therefore ergodic, but $\mathscr{A}'$ is isomorphic to $\mathscr{A}$ and thus different from $\{\alpha \cdot \mathbf{1}\}$. It will be discovered later that this is the normal situation for equilibrium states.
2. According to (III: 2.3.10; 5), Condition 2 means that ρ is a pure state on $\mathscr{R}$, and can also be written as $\mathscr{R} \cap \mathscr{R}' = \{\alpha \cdot \mathbf{1}\}$; in particular, every factor state is ergodic.
3. Condition 3 can be sharpened for classical systems with Birkhoff's ergodic theorem, according to which almost every trajectory fills the energy shell densely. In this case, with the decomposition into pure states, the Cesàro mean exists; $\eta(\sigma)$ is μ-measurable, and the order of η and $\int d\mu$ can be reversed.
4. By Condition 4 the time-average of operators in this situation is unique and a multiple of the identity. More particularly, the classical time-average of any set of positive ρ-measure is spread out over the whole support of ρ. Hence the time-average of states with a density function equals the equilibrium state. Since averaged observables are multiples of the identity, they exhibit no deviation.
5. The implication of Condition 5 for classical dynamics is that if the system is ergodic, then every measurable, time-independent function is constant on the energy shell. Note that $(\mathscr{A} \cup \mathscr{A}')''$ might contain additional time-invariant operators; for instance, for a factor this set is $\mathscr{B}(\mathscr{H})$ and therefore also contains U.
6. Condition 6 implies that 1 is a simple eigenvalue of U.
7. By Condition 7, all the other eigenvectors of U lead to the same state as ρ. Classically, the eigenfunctions $\varphi(p, q)$ must always have $|\varphi|^2$ constant independently of p and q. Thus ergodicity does not make it impossible for the spectrum to be purely pointlike, but only prevents 0 from being a degenerate eigenvalue of H. The extra word "normal" of Condition 7 is important. In Example (3.1.1; 5) of free fermions, equilibrium states at

different temperatures from that of the specified representation are invariant in time, but not normal. This means classically that different energy shells have disjoint support.

8. Condition 8 means that the autocorrelation function $\rho(ab_t) - \rho(a)\rho(b)$ has time-average 0. Also, according to Condition 4 the expectation values of operators in states of the form $a|\Omega\rangle$ have the same time-averages as those with the state ρ. Since the states $a|\Omega\rangle$ are dense, the time-average of every normal state is ρ. This is a sort of converse to Condition 3, in so far as $\eta(\sigma) = \rho$ for all σ's that are pure and normal (as states on $\pi_\rho(\mathscr{A})''$). It may happen that the set of such σ's is empty (cf. (1.4.17; 3)), and some non-normal, pure states converging to something other than the equilibrium state will make their appearance later.

Proof

$1 \Rightarrow 2$: Let $t \in \mathscr{R}'$, $0 < t < 1$; then the vector $|\Omega_\rho\rangle$ associated with ρ in the GNS representation is cyclic for $\mathscr{R}$ and therefore separates $\mathscr{R}'$. With $|\Omega_\rho\rangle$,

$$0 < ||t^{1/2}\Omega_\rho\rangle|^2 = \langle\Omega_\rho|t\Omega_\rho\rangle \equiv \lambda < 1,$$

so if

$$\rho_1(a) = \frac{1}{\lambda}\langle\Omega_\rho|at\Omega_\rho\rangle,$$

and

$$\rho_2(a) = \frac{1}{1-\lambda}\langle\Omega_\rho|a(1-t)\Omega_\rho\rangle \quad \text{for all } a \in \mathscr{A},$$

then $\rho = \lambda\rho_1 + (1-\lambda)\rho_2$ has a genuine decomposition into invariant states.

$2 \Rightarrow 1$: Let $\rho = \lambda\rho_1 + (1-\lambda)\rho_2$, where $0 < \lambda < 1$. Then according to (III: 2.3.24; 2) there exists a $t \in \pi_\rho(\mathscr{A})'$ such that $0 \le t \le 1$ and $\rho_1(a) = \langle\Omega_\rho|t\Omega_\rho\rangle^{-1}\langle\Omega_\rho|at\Omega_\rho\rangle$ for all $a \in \mathscr{A}$. If ρ_1 is invariant, then t is in $\mathscr{R}'$, and it follows from Condition 2 that $\rho = \rho_1 = \rho_2$.

$2 \Leftrightarrow 4$: $\mathscr{R}' \supset \{\eta(a)\colon a \in \mathscr{A}\}$. (Cf. (3.1.18; 2).)

$1 \Rightarrow 3$: The state $\rho = \int \sigma\, d\mu(\sigma)$ is invariant in time, so $\rho(a) = \int d\mu(\sigma)\eta(\sigma(a))$. Therefore $\rho = \int d\mu(\sigma)\eta(\sigma)$, and, since ρ is an extremal invariant, it equals the invariant state $\eta(\sigma)$ almost everywhere in μ.

$3 \Rightarrow 1$: Suppose that ρ is not ergodic. Then there exist invariant states $\rho_1 \ne \rho_2$ such that $\rho = \lambda\rho_1 + (1-\lambda)\rho_2$. This is a special case of a decomposition with $\rho_i = \eta(\rho_i) \ne \rho$, so Condition 3 would be violated.

$2 \Leftrightarrow 5$: The invariant elements of $\mathscr{A}$ and $\mathscr{A}'$ compose $\mathscr{R}'$.

$6 \Rightarrow 1$: Suppose that $\rho = \lambda\rho_1 + (1-\lambda)\rho_2$; then by (III: 2.3.24; 2), ρ_1 is of the form $\rho_1(a) = \langle\Omega_\rho|t\Omega_\rho\rangle^{-1}\langle t^{1/2}\Omega_\rho|at^{1/2}\Omega_\rho\rangle$ for $a \in \mathscr{A}$, and t is in $\pi_\rho(\mathscr{A})' \cap U'_\rho$ if ρ_1 is invariant. Condition 6 implies that $|t^{1/2}\Omega_\rho\rangle \propto |\Omega_\rho\rangle$, because $|t^{1/2}\Omega_\rho\rangle \in E_0\mathscr{H}$, so $\rho = \rho_1 = \rho_2$.

$6 \Rightarrow 8$: $\eta(\rho(ab_t)) = \eta(\langle\Omega|aU_t b|\Omega\rangle) = \langle\Omega|aE_0 b|\Omega\rangle = \rho(a)\rho(b)$.

$7 \Rightarrow 6$: If there existed a second invariant vector $|\Omega'\rangle$, then all vectors $\sqrt{\alpha}|\Omega\rangle + \sqrt{1-\alpha}|\Omega'\rangle$ for $0 \le \alpha \le 1$ would give rise to the same state, but by Property (3.1.18; 3), since the algebra is maximally Abelian on the subspace, this would mean that $|\Omega\rangle = |\Omega'\rangle$.

$4 \Rightarrow 7$ and 8: ω invariant $\Rightarrow \omega = \eta(\omega) \Rightarrow \eta(\omega)(a) = \rho(a)$.

$8 \Rightarrow 4$: From $\eta([b_t, c]) = 0$ it follows that $\rho(ac)\rho(b) = \eta(\rho(acb_t)) = \eta(\rho(ab_t c))$, so the matrix elements of $\rho(b) \cdot \mathbf{1}$ and $\eta(b)$ are equal on a dense set.

□

Examples (3.1.24)

1. The only possible ergodic states on classical systems are those concentrated on $\delta(E - H(p, q))$; otherwise $\mathscr{A}$ would contain the additional invariant $F(H)$, contradicting Condition 4. Let us examine a chain of N coupled oscillators (1.1.14). The Hamiltonian can be written in terms of action and angle variables K_i (see (I: 3.3.3) and (I: 3.3.14)) and $\varphi_i \in T^1$ as

$$H = \sum_{i=1}^{N} \omega_i K_i,$$

and the time-evolution is $\varphi_i \to \varphi_i + \omega_i t$. If $N > 1$, the state $\sim \delta(E - H)$ is not ergodic, although the state $\sim \prod_i \delta(K_i - c_i)$ concentrated on T^N is, provided that the angular velocities ω_i are rationally independent (cf. (I: 3.3.3)). To understand why, observe that the operator h on $L^2(T^N)$ introduced in (3.1.1; 1) arises when K_i is interpreted as the displacement operator, the eigenvalues of which are $2\pi n$, $n \in \mathbb{Z}$. The spectrum of h is therefore purely pointlike, with eigenvalues $2\pi \sum_i \omega_i n_i$. If the ω_i are rationally independent, then the eigenvalue 0 (all $n_i = 0$) is nondegenerate and otherwise it is degenerate. According to (3.1.22; 6) this is a criterion for ergodicity. This example is also useful for illustrating the other criteria. For instance, Condition 4 states that every invariant L^∞ function is constant almost everywhere on T^N. Roughly speaking, a function assuming one value on half the trajectories and a different value on the other half is not measurable.
2. Of the quantum-mechanical examples of (3.1.1), only the free fermions (3.1.1; 5) fall within the category covered by (3.1.22), as the others are not asymptotically Abelian. Since (3.1.1; 5) has a factor state, it is ergodic according to Condition 5. If we go through the other criteria, we notice

that Condition 8 holds in the sharpened form $\lim_{t \to \pm\infty} \rho(ab_t) = \rho(a)\rho(b)$ for all a and $b \in \mathscr{A}$. This means that normal states approach ρ not only in the mean, but also actually in the limit $t \to \pm\infty$. The situation is as described intuitively in §1.1, where the states converge to the equilibrium state.

Even though Example 1 is ergodic, it does not exhibit the sort of behavior appropriate for a thermodynamic system. The time-evolution is a rigid displacement in T^N, and this submanifold does not get thoroughly mixed. States like those given by pieces of T^N do not converge as $t \to \infty$; only their means converge. Example 2 conforms better to the notion of a thermodynamic system, which suggests sharpening some of Criteria (3.1.22) as much as possible, by replacing the time-average with the time-limit.

Definition (3.1.25)

An invariant state on an asymptotically Abelian system is called **mixing** iff one of the following equivalent conditions is satisfied:

4′. $w\text{-}\lim_{t \to \pm\infty} \pi_\rho(a_t) = \mathbf{1} \cdot \rho(a)$ for all $a \in \mathscr{A}$ (The weak limit is that of the GNS representation);
6′. $U_t \xrightarrow{t \to \pm\infty} |\Omega\rangle\langle\Omega|$;
8′. $\lim_{t \to \pm\infty} \rho(ab_t) = \rho(a)\rho(b)$.

Remarks (3.1.26)

1. By Condition 4′, every operator converges to its equilibrium value and its deviation goes to zero. Hence, in the Schrödinger picture every normal state approaches the equilibrium state ρ. In classical dynamics probability distributions of normal states are described by functions—i.e., not by δ distributions—and so they spread out through all of ρ.
2. Criterion 6′ is satisfied if the spectrum of U is absolutely continuous other than the eigenvalue associated with $|\Omega\rangle$. In any case, $|\Omega\rangle$ must be the only eigenvector.
3. Concerning Condition 8′, we have learned that for a factor the correlation functions vanish automatically as $t \to \pm\infty$. Therefore, for factors ergodic is equivalent to mixing. In general it is only true that mixing implies ergodic. It is also not true to say that mixing implies a factor, since there are classical mixing systems. However, it will be shown in the next section that in quantum theory equilibrium states are mixing iff the algebra is a factor. In the case of free particles with the spatial translations, as the group of automorphisms with respect to which their algebra of observables is asymptotically Abelian, this reasoning implies that the spatial correlation function goes to zero for factors.

4. If a state is a limit of pure states, then it is mixing: If σ is pure and $\sigma_t \to \rho$ then $\rho(ab_t c) - \lim_{s\to\infty} \sigma(a_s b_{s+t} c_s) + \lim_{s\to\infty} \sigma(a_s c_s)\sigma(b_{t+s}) - \rho(ac)\rho(b) = 0$. A pure state is a factor state, so (3.1.18; 4) applies, showing that $\rho(ab_t c) \to \rho(ac)\rho(b)$. The converse is not true in general, since the pure states into which ρ is decomposed need not converge as $t \to \pm\infty$. For example, the pure states for classical systems are points in phase space, which will keep moving forever.

Proof of the Equivalence in (3.1.25)

$8' \Leftrightarrow \rho(ab_t c) = \rho(a[b_t, c]) + \rho(acb_t) \to \rho(ac)\rho(b) \Leftrightarrow 4'$, and $\rho(a_t b) = \rho(aU_t b)$, hence $6' \Leftrightarrow 8'$. □

Classical systems that mix are of necessity complicated, and it requires a rather demanding example to show that the concept of (3.1.25) is not empty:

Motion on a Surface of Constant, Negative Curvature (3.1.27)

The ergodic system (3.1.24; 1) is not mixing; the spectrum of U_t is purely discrete. This agrees with the perception that displacements in T^2 do not mix its parts together:

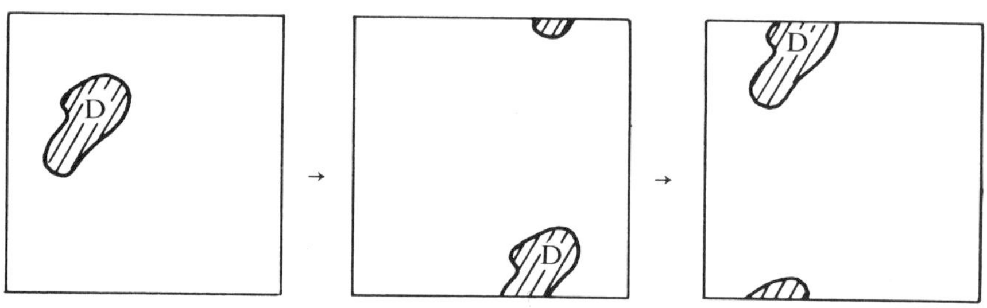

t_0 $t_1 > t_0$ $t_2 > t_1$

To produce mixing we need a somewhat geometrically irregular configuration; fortunately, as will now be demonstrated, it suffices to have a surface of constant negative curvature. The construction of the example makes use of the following more abstract reformation of (3.1.24; 1). Treat $\mathbb{R}^2$ as a two-dimensional group and the trajectory as a one-dimensional subgroup, and consider its image in the quotient space $T^2 = \mathbb{R}^2/\mathbb{Z}^2$. Conservation of angular momentum gets lost, and the trajectory can be dense in T^2. The present example will have an energy shell that is diffeomorphic to the Lorentz group SO(2, 1), and the trajectory will be a one-parameter subgroup. In order to destroy the other constants of the motion and have an energy shell of finite volume, map the space to SO(2, 1)/$\mathscr{Z}$, where $\mathscr{Z}$ is a discrete subgroup of SO(2, 1). The dynamics furnishes a unitary representation

$U_t = \exp(mt)$ of a one-parameter subgroup of SO(2, 1), but, unlike with $\mathbb{R}^2$, U has only absolutely continuous spectrum other than the point 1, and so the system is mixing by (3.1.26; 2).

We realize these ideas in a classical system the Lagrangian of which is quadratic in the velocities. The motion thus proceeds in the absence of forces, but the invariance under SO(2, 1) brings about some unusual signs. The extended configuration space is the submanifold of $\mathbb{R}^3$ for which

$$(x|x) \equiv x_1^2 + x_2^2 - x_0^2 = -1. \tag{3.1.28}$$

If $\dot{x}$ denotes the derivative of x by the proper time t, then the Lagrangian is

$$L = \tfrac{1}{2}(\dot{x}|\dot{x}).$$

The constraint (3.1.28) enters into the Euler–Lagrange equations through a Lagrange multiplier,

$$\ddot{x}_i = \lambda x_i, \tag{3.1.29}$$

and there are the following constants:

$$(x|x) = -1, \qquad (\dot{x}|x) = 0, \qquad (\dot{x}|\dot{x}) = 1 \tag{3.1.30}$$

(which normalize t). The three-dimensional manifold defined by the constants corresponds to the energy shell (recall that the configuration space is two-dimensional and the phase space is four-dimensional), and on it is the SO(2, 1)-invariant Liouville measure

$$d\Omega = d^3x\, d^3\dot{x}\,\delta((\dot{x}|x))\delta((x|x) + 1)\delta((\dot{x}|\dot{x}) - 1)\Theta(x_0). \tag{3.1.31}$$

There are also three constants associated with the angular momentum,

$$l_i = \varepsilon_{ikm} x_k \dot{x}_m, \tag{3.1.32}$$

which are connected by an algebraic relationship,

$$(l|l) = -(x|x)(\dot{x}|\dot{x}) = 1.$$

One dimension is left for the trajectory. Because $(l_i|x) = 0$, the projection of the trajectory onto configuration space is the intersection of the hyperboloid (3.1.28) with a plane passing through the origin and making an angle less than 45° with the x_0-axis (see Figure 23).

The energy is only apparently indefinite; x_0 can be eliminated, and then

$$L = \frac{\dot{x}_1^2 + \dot{x}_2^2 + (\dot{x}_1 x_2 - \dot{x}_2 x_1)^2}{x_1^2 + x_2^2 + 1}$$

describes motion in the $x_1 - x_2$-plane without forces, but with a positive effective mass that depends on the position.

The indefinite scalar product $(\cdot|\cdot)$ and consequently also the formalism that has been developed are invariant under SO(2, 1). The group SO(2, 1)

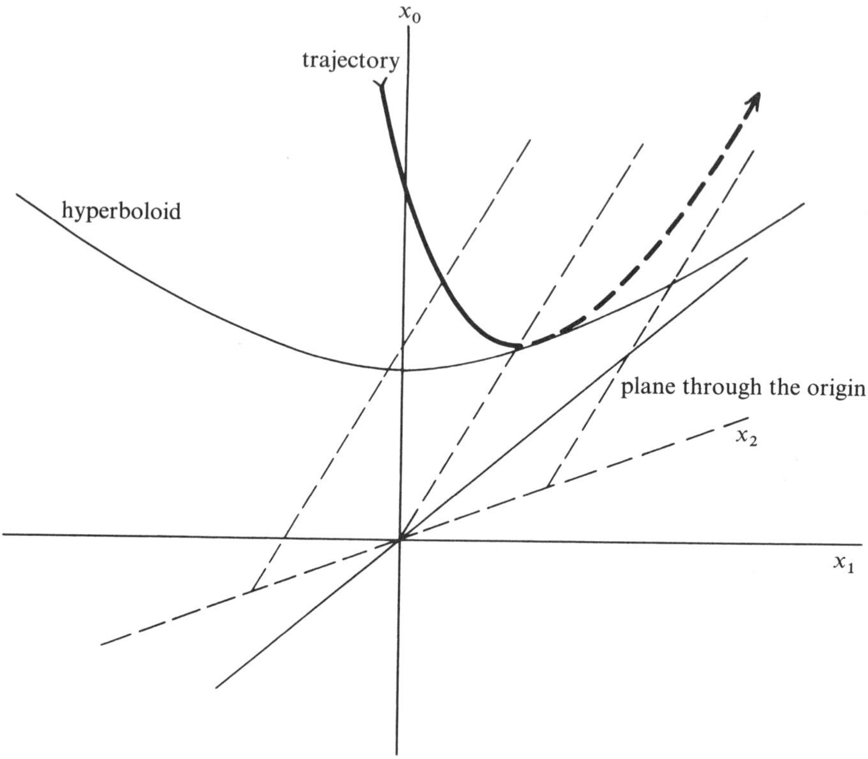

Figure 23 The trajectory in configuration space.

acts transitively on the energy shell (3.1.30), and every point can be written

$$\{x, \dot{x}\} = \{M(1, 0, 0), M(0, 1, 0)\} \tag{3.1.33}$$

for some $M \in SO(2, 1)$. It is easy to see that M is determined uniquely, and this creates the diffeomorphism between the energy shell and SO(2, 1) that was mentioned above. Accordingly, every trajectory can be obtained by making Lorentz transformations of the group generated by

$$M(t) = \begin{bmatrix} \cosh t & \sinh t & 0 \\ \sinh t & \cosh t & 0 \\ 0 & 0 & 1 \end{bmatrix}.$$

The most convenient construction of the discrete subgroup makes use of the isomorphism between SO(2, 1) and $SL(2, \mathbb{R})/\{1, -1\}$, since 2×2 matrices are easier to handle than 3×3 matrices. The source of this isomorphism, like that of $SO(3) = SU(2, \mathbb{C})/\{1, -1\}$, lies in the observation that

$$\begin{pmatrix} \alpha & \beta \\ \gamma & \delta \end{pmatrix} \in SL(2, \mathbb{R}), \quad \text{i.e., } (\alpha, \beta, \gamma, \delta) \in \mathbb{R}^4 : \alpha\delta - \beta\gamma = 1, \tag{3.1.34}$$

produces the Lorentz transformation $x \to x'$ by

$$\begin{pmatrix} x_0' + x_2' & x_1' \\ x_1' & x_0' - x_2' \end{pmatrix} = \begin{pmatrix} \alpha & \beta \\ \gamma & \delta \end{pmatrix} \begin{pmatrix} x_0 + x_2 & x_1 \\ x_1 & x_0 - x_2 \end{pmatrix} \begin{pmatrix} \alpha & \gamma \\ \beta & \delta \end{pmatrix}. \tag{3.1.35}$$

It is necessary to take the quotient by the center $\{1, -1\}$, since the Lorentz transformations corresponding to the matrix $m \in \mathrm{SL}(2, \mathbb{R})$ and $-m$ are the same. It is not hard to come up with discrete subgroups of $\mathrm{SL}(2, \mathbb{R})$, such as

$$\mathscr{Z} = \left\{ \begin{pmatrix} \alpha & \beta \\ \gamma & \delta \end{pmatrix} \in \mathrm{SL}(2, \mathbb{R}) \colon \alpha, \beta, \gamma, \delta \text{ integers} \right\}.$$

Now let us investigate the motion on the quotient space $\Omega_0 = \mathrm{SO}(2, 1)/\mathscr{Z} \cong \mathrm{SL}(2, \mathbb{R})/\{1, -1\}/\mathscr{Z}$. Unlike the case of T^2, the quotient space is not a group, since $\mathscr{Z}$ is not a normal divisor, though for our purposes this does not matter. Thus Ω_0 is the energy shell (3.1.30), on which points are identified if they are transformed into each other by $\mathscr{Z}$. For the trajectory this means that if it goes out one end of the domain of periodicity it reappears at the other. Conservation of angular momentum breaks down, leaving the possibility that the trajectory fills Ω_0 densely.

To get a clearer picture of Ω_0 we have to find out what corresponds to the square $0 \le \varphi_1, \varphi_2 \le 1$ of the earlier example, that is, a region containing no points equivalent under $\mathscr{Z}$, but for each boundary point of which there is a $z \ne 1$ of $\mathscr{Z}$ mapping it to another boundary point. The subgroup $\mathscr{Z}$ is generated by the matrices

$$\begin{pmatrix} 1 & 1 \\ 0 & 1 \end{pmatrix}, \quad \begin{pmatrix} 0 & 1 \\ -1 & 0 \end{pmatrix},$$

the latter of which is the reflection $(x_1, x_2) \to (-x_1, -x_2)$. It is therefore possible to restrict attention to the upper half plane $\{x_2 > 0\}$ in configuration space and choose a region symmetric about the x_2-axis. The boundary curves can be obtained by transforming the x_2-axis with the matrices

$$\begin{pmatrix} 1 & \pm\frac{1}{2} \\ 0 & 1 \end{pmatrix}$$

of $\mathrm{SL}(2, \mathbb{R})$. They have the parametric representation

$$\begin{aligned} h_\pm &= \left\{ x_1' \colon \begin{pmatrix} x_0' + x_2' & 0 \\ x_1' & x_0' - x_2' \end{pmatrix} \right. \\ &= \left. \begin{pmatrix} 1 & \pm\frac{1}{2} \\ 0 & 1 \end{pmatrix} \begin{pmatrix} \sqrt{1 + x_2^2} + x_2 & 0 \\ 0 & \sqrt{1 + x_2^2} - x_2 \end{pmatrix} \begin{pmatrix} 1 & 0 \\ \pm\frac{1}{2} & 1 \end{pmatrix}, x_2 > 0 \right\}; \end{aligned} \tag{3.1.36}$$

note that

$$\begin{pmatrix} 1 & 1 \\ 0 & 1 \end{pmatrix} \cdot h_- = h_+,$$

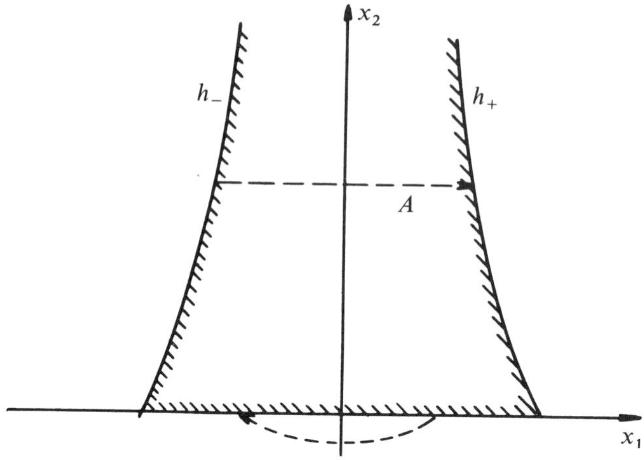

Figure 24

so $x_2' = \pm\frac{1}{4}(1/x_1' - 3x_1')$. The projection of Ω_0 onto configuration space looks as depicted in Figure 24, where the lines A indicate the identifications.

The identification of the boundary points by $\begin{pmatrix} 1 & 1 \\ 0 & 1 \end{pmatrix}$ means that if the trajectory leaves through one side, it reappears at the corresponding point of the other side (see Figure 25).

Now we are in a position to verify that the measure of Ω_0 with $d\Omega$ (3.1.31) is actually finite. This follows from

$$\int d^3\dot{x}\delta[(\dot{x}|x)]\delta[(\dot{x}|\dot{x}) - 1] \equiv F(x, x) < \infty$$

and

$$F(-1)\int d^3x\delta[(x|x) + 1] < \infty,$$

where the integral runs over the region bounded by (3.1.36).

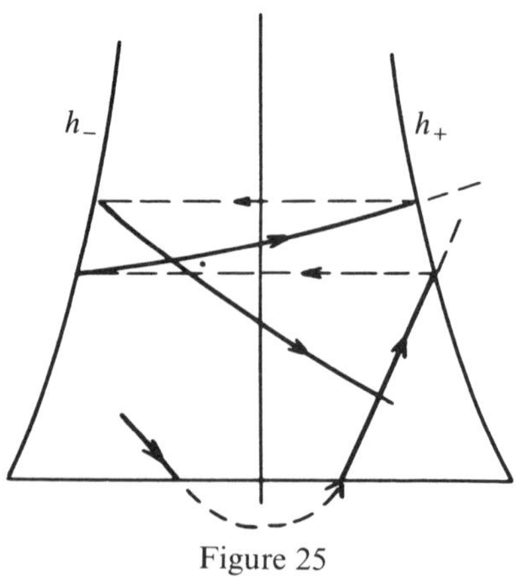

Figure 25

The time-evolution is controlled by the unitary group

$$U_t = \exp(mt), \qquad m = \left.\frac{\partial M}{\partial t}\right|_{t=0},$$

where the anti-Hermitian operator m is one of the generators of SO(2, 1). If the other two generators are combined into $m_\pm \equiv m_1 \pm m_2$, then $m_\pm$ satisfy the commutation relations

$$[m, m_\pm] = \pm m_\pm \quad \text{and} \quad [m_+, m_-] = 2m.$$

Note that in contradistinction to SO(3), this time $(m_\pm)^* = -m_\pm$. This fact will be crucial, since the generators of SO(3) have purely discrete spectra. Instead of SO(2, 1), let us now examine the simpler two-parameter subgroups

$$U_\pm(a, t) = \exp(am_\pm)\exp(tm)$$

with the multiplication law

$$U_\pm(a, t)U_\pm(a', t') = U_\pm(a + \exp(\pm t)a', t + t').$$

Because $[m_+, m_-] = 2m$, the operators $U_+(a, 0)$ and $U_-(a, 0)$ generate the whole group, and $U(t) = U_+(0, t) = U_-(0, t)$.

Next consider the representation (3.1.1; 1) of classical dynamics on $\mathscr{H} = L^2(\Omega_0, d\Omega)$. Not just U_t, but in fact all of SO(2, 1) is represented unitarily on $\mathscr{H}$ by $f(x) \to f(Mx)$, and we shall now reduce this representation according to the irreducible representations of the subgroups $U_\pm$. We start by observing that $U_\pm(a, 0)$ is a normal divisor, and the factor groups $U_\pm(a, t)/U_\pm(a, 0)$ are isomorphic to $\mathbb{R}$. Hence there are irreducible, one-dimensional representations of the type

$$I: \quad U_\pm(a, t) = \exp(i\lambda t), \qquad \lambda \in \mathbb{R}.$$

In addition it is readily seen that $U_\pm$ can also be represented on $L^2(\mathbb{R}, dx)$ by

$$II: \quad [U_+(a, t)\psi](x) = \exp(iae^x)\psi(x + t), \qquad \psi \in L^2(\mathbb{R}, dx),$$

and similarly for U_-. It can be shown [19] that these possibilities exhaust the irreducible representations of SO(2, 1), so, decomposing into the irreducible representations of $U_\pm$,

$$L^2(\Omega_0, d\Omega) = \mathscr{H}_I^+ \oplus \mathscr{H}_{II}^+ = \mathscr{H}_I^- \oplus \mathscr{H}_{II}^-.$$

On the subspaces $\mathscr{H}_{II}^+$ and $\mathscr{H}_{II}^-$ the operator $U(t)$ acts as a translation on $L^2(\mathbb{R}, dx)$, and thus its spectrum is continuous. A discrete spectrum could only be found on $\mathscr{H}_I^+ \cap \mathscr{H}_I^-$, but every vector ψ of $\mathscr{H}_I^+ \cap \mathscr{H}_I^-$ satisfies the equation

$$U_+(a, 0)\psi = \psi = U_-(a, 0)\psi.$$

Since $U_+(a, 0)$ and $U_-(a, 0)$ together suffice to generate all of SO(2, 1), ψ is invariant under the action of every group element. Since the group acts transitively on Ω_0, ψ must be a constant. Because Ω_0 has finite measure, any constant function belongs to $L^2(\Omega_0, d\Omega)$, so the situation is like that of (3.1.26; 2). Unless the quotient by $\mathscr{Z}$ is taken, U has no point spectrum, as constant functions would not be integrable. In sum the argument is that the system is mixing because the spectrum of U consists of a single nondegenerate eigenvalue 1 and an absolutely continuous portion. This is in contrast to the motion on the torus, for which the spectrum of U_t was purely discrete, and the system was only ergodic, not mixing.

Example (3.1.37)

The quantum-mechanical example of an infinite system of free fermions was seen to be mixing. Despite the absence of interaction, a local perturbation spreads out to infinity through the diffusion of free wave-packets. From among the characterizations of ergodic states (3.1.22), let us look in particular at the third. It holds in the sharper form of (3.1.26; 4); the grand canonical state (2.5.49) is the time-limit of a pure state. The proof of this fact uses the transformations

$$a_\uparrow(f) = b_\uparrow(\beta f) + b_\downarrow^*(\sqrt{1 - |\beta|^2} f^*)$$

and

$$a_\downarrow(f) = b_\downarrow(\beta f) - b_\uparrow^*(\sqrt{1 - |\beta|^2} f^*). \tag{3.1.38}$$

We have directly taken up the realistic case of spin-$\frac{1}{2}$ fermions, where $\uparrow$ and $\downarrow$ indicate the direction of the spin that the field operator describes. In Fourier-transformed space β is a function $\mathbf{k} \to \beta(\mathbf{k})\colon \mathbb{R}^3 \to \{z \in \mathbb{C}\colon |z|^2 \le 1\}$, and βf is the function $\beta(\mathbf{k}) f(\mathbf{k})$. In x-space β is a convolution. It is straightforward to verify that the a's satisfy the usual commutation relations (1.3.3; 2),

$$[a_\uparrow(f), a_\uparrow^*(g)]_+ = [a_\downarrow(f), a_\downarrow^*(g)]_+ = (f|g),$$
$$[a_\uparrow(f), a_\uparrow(g)]_+ = [a_\uparrow(f), a_\downarrow(g)]_+ = [a_\uparrow(f), a_\downarrow^*(g)]_+ = [a_\downarrow(f), a_\downarrow(g)]_+ = 0, \tag{3.1.39}$$

supposing that the b's satisfy the commutation relations. Clearly the a's and the b's generate the same C^* algebra. The expectation values of the a's in the Fock state $|0\rangle$ (1.3.2) for the b's: $b_\uparrow(f)|0\rangle = b_\downarrow(f)|0\rangle = 0$, are

$$\langle 0|a_\uparrow(f)a_\uparrow^*(g)|0\rangle = \langle 0|a_\downarrow(f)a_\downarrow^*(g)|0\rangle = \int \frac{d^3\mathbf{k}}{(2\pi)^3} |\beta(\mathbf{k})|^2 f^*(\mathbf{k}) g(\mathbf{k}),$$
$$-\langle 0|a_\uparrow(f)a_\downarrow(g)|0\rangle = \langle 0|a_\downarrow(f)a_\uparrow(g)|0\rangle$$
$$= \int \frac{d^3\mathbf{k}}{(2\pi)^3} f^*(\mathbf{k}) g^*(\mathbf{k}) \beta(\mathbf{k}) \sqrt{1 - |\beta(\mathbf{k})|^2};$$
$$\langle 0|a_\uparrow(f)a_\uparrow(g)|0\rangle = \langle 0|a_\uparrow(f)a_\downarrow^*(g)|0\rangle = 0. \tag{3.1.40}$$

The state $|0\rangle$ was seen to be pure in (1.3.16; 1). Under the time-evolution $f(\mathbf{k}) \to \exp(-it|\mathbf{k}|^2) f(\mathbf{k})$, the quantity $-\langle 0|a_\uparrow a_\downarrow|0\rangle = \langle 0|a_\downarrow a_\uparrow|0\rangle$ goes to 0 as $t \to \pm\infty$ by the Riemann–Lebesque lemma. If

$$\beta(\mathbf{k}) = (1 + \exp(-\beta(|\mathbf{k}|^2 - \mu)))^{-1/2},$$

then in the limit $t \to \pm\infty$ the generalization of the state (2.5.49) for spin $\frac{1}{2}$ is all that is left over.

Remarks (3.1.41)

1. The limit of a pure state is clearly not always an equilibrium state; other functions could be chosen for $\beta(\mathbf{k})$.
2. Since the thermal representation of free fermions (3.1.1; 5) is a factor of type III, the pure state $|0\rangle$ associated with the thermal representation cannot be normal (cf. (1.4.17; 3)). Likewise, any other states of the latter formed with different $\beta(\mathbf{k})$ are not normal because of (3.1.22; 7), even though they are invariant.
3. The state given by $|0\rangle$ is not invariant in time, and in this representation the time-evolution is certainly not a unitary group (cf. (1.3.16; 7)). If it were, then the time displacement $\tau_t: a \to a_t$ would be weakly continuous and hence extensible to $\pi(\mathscr{A})''$, which would lead to a contradiction: $\mathscr{A}_e$ is asymptotically Abelian with respect to the spatial translation T_x, so in the representation with the translation-invariant state $|0\rangle$, $\lim_{|x|\to\infty} T_x a = \mathbf{1}\cdot\langle 0|a|0\rangle$ for all $a \in \mathscr{A}_e$. Since T_x commutes with τ_t, it would follow that $\lim_{x\to\infty} T_x \tau_t(a) = \mathbf{1}\cdot\langle 0|a_t|0\rangle = \lim_{x\to\infty} \tau_t T_x(a) = \mathbf{1}\cdot\langle 0|a|0\rangle$, which would then imply that the state $\langle 0|\cdot|0\rangle$ would be invariant in time.

Problems (3.1.42)

1. (i) Prove von Neumann's statistical ergodic theorem, $(1/2T)\int_{-T}^{T} \exp(iHt)\,dt \to E_0$. (Show that on all vectors of the form $x = \exp(iHs)y - y$, $y \in \mathscr{H}$, $s \in \mathbb{R}$, we have $(1/2T)\int_{-T}^{T} \exp(iHt)x\,dt \to 0$. Let $\mathscr{H}_1$ be the closed linear hull of these vectors, and note that the same fact applies to all $x \in \mathscr{H}_1$. Finally, show that $\mathscr{H}_1^{\perp} = \{x: \exp(iHs)x = x \text{ for all } s\} = E_0\mathscr{H}$.)
   (ii) Show similarly that $(\varepsilon/2)\int_{-\infty}^{\infty} \exp(-\varepsilon|t|)\exp(iHt)\,dt \to E_0$.

2. Show that in the Schrödinger picture the time-average of a vector x has the following characterization: $\eta(x)$ is the vector of least norm of the norm-closed, convex hull of $\{U(t)x\}$, denoted $\mathscr{K}$. (Hint: see the example given earlier for $\eta(x) \in \mathscr{K}$. Show (i) that $\mathscr{K}$ contains a unique vector ξ of least norm; (ii) that ξ is invariant under all $U(t)$; and (iii) that $\mathscr{K}$ contains no other fixed point.)

3. Show that $\mathscr{Z} = \{\alpha\cdot\mathbf{1}\}$ iff $w(ab) = w(a)w(b)$ for all $w \in \mathscr{A}^*$, $a \in \mathscr{A}$, and $b \in \mathscr{Z}$.

4. Show that for a classical system, if there exists a constant $f(p, q)$ not of the form $\alpha\cdot\mathbf{1}$, then ρ is not ergodic.

5. Show that a set $E \subset \mathscr{H}$ is a totalizer for $\mathscr{A}$ iff E separates $\mathscr{A}'$. (Cf. (III: 2.3.4); a totalizer is a set E such that $\mathscr{A}E$ is dense in $\mathscr{H}$, and separating means that $a'E = 0 \Rightarrow a' = 0$.)

6. Boson states of the form (2.5.49) with $\langle f|\rho g\rangle = \int d^3k\rho(\mathbf{k})\tilde{f}^*(\mathbf{k})\tilde{g}(\mathbf{k})$, $0 \le \rho(\mathbf{k})$, are factor states and consequently mixing. Express such a state as a time-limit of a pure state (cf. (3.1.37)).

Solutions (3.1.43)

1. (i) If $x = \exp(iHs)y - y$, then

$$\left\|\frac{1}{2T}\int_{-T}^{T}\exp(iHt)x\,dt\right\| = \left\|\frac{1}{2T}\left\{\int_{T}^{T+s}\exp(iHt)y\,dt - \int_{-T}^{-T+s}\exp(iHt)y\,dt\right\}\right\|$$
$$\le \frac{|s|\,\|y\|}{T} \to 0.$$

Because $\|(1/2T)\int_{-T}^{T}\exp(iHt)\,dt\| \le 1$, this holds for all $x \in \mathscr{H}_1$.

$$x \in \mathscr{H}_1^{\perp} \Leftrightarrow (x|\exp(iHs)y - y) = (\exp(-iHs)x - x|y) = 0 \qquad \text{for all } y \in \mathscr{H}$$
$$\Leftrightarrow \exp(iHs)x = x \qquad \text{for all } s \Leftrightarrow E_0 x = x$$

by the spectral theorem.

(ii) It suffices to show that $\varepsilon\int_0^\infty \exp(-\varepsilon t)\exp(iHt)\,dt \to E_0$, which will follow if $\varepsilon\int_0^\infty \exp(-\varepsilon t)\exp(iHt)x\,dt \to 0$ for vectors $x = \exp(iHs)y - y$. This integral equals

$$\varepsilon\exp(\varepsilon s)\int_s^\infty \exp(-\varepsilon t)\exp(iHt)y\,dt - \varepsilon\int_0^\infty \exp(-\varepsilon t)\exp(iHt)y\,dt$$
$$= (\exp(\varepsilon s) - 1)\varepsilon\int_s^\infty \exp(-\varepsilon t)\exp(iHt)y\,dt - \varepsilon\int_0^s \exp(-\varepsilon t)\exp(iHt)y\,dt \to 0,$$

since $\|\varepsilon\int_0^\infty \exp(-\varepsilon t)\exp(iHt)y\,dt\| \le \|y\|$.

2. (i) Let $\lambda = \inf\{\|x\| : x \in \mathscr{K}\}$. There exists a sequence $\{x_n\}$ in $\mathscr{K}$ such that $\|x_n\| \to \lambda$. By the parallelogram law,

$$\left\|\frac{x_n - x_m}{2}\right\|^2 + \left\|\frac{x_n + x_m}{2}\right\|^2 = \tfrac{1}{2}(\|x_n\|^2 + \|x_m\|^2),$$

x_n is a Cauchy sequence, so it has a limit ξ. If $\|x\| = \|\xi\|$, then

$$\left\|\frac{x - \xi}{2}\right\|^2 = \tfrac{1}{2}(\|x\|^2 + \|\xi\|^2) - \left\|\frac{x + \xi}{2}\right\|^2 \le 0, \text{ which implies that } x = \xi.$$

(ii) $\|U(t)\xi\| = \|\xi\| \Rightarrow U(t)\xi = \xi$.

(iii) Suppose that η is a second fixed point. For all $\varepsilon > 0$, there exist $\lambda_1, \ldots, \lambda_n$ and $\lambda'_1, \ldots, \lambda'_m$ such that $\sum_i \lambda_i = \sum_i \lambda'_i = 1$, with $\lambda_i, \lambda'_i \ge 0$, and there exist $t_1, \ldots, t_n$ and $t'_1, \ldots, t'_m$ such that if $V \equiv \lambda_1 U(t_1) + \cdots + \lambda_n U(t_n)$, and $W \equiv \lambda'_1 U(t'_1) + \cdots + \lambda'_m U(t'_m)$, then $\|Vx - \xi\| < \varepsilon$, and $\|Wx - \eta\| < \varepsilon$. However, then

$$\|\xi - \eta\| \le \|\xi - VWx\| + \|VWx - \eta\| = \|W\xi - VWx\| + \|VWx - V\eta\|$$
$$\le \|W\|\,\|Vx - \xi\| + \|V\|\,\|Wx - \eta\| < 2\varepsilon,$$

so $\xi = \eta$.

Remark: The strong and weak closures of a convex set are identical.

3. $\Rightarrow$: This part is trivial.
   $\Leftarrow$: Let P_1 and P_2 be projections in $\mathscr{Z}$, such that $P_1 \perp P_2$, and let $w_i(\cdot) = w(P_i \cdot)$, $a_i = P_i a$, $b_i = P_i b$ for $i = 1, 2$. If $w = \alpha w_1 + (1 - \alpha)w_2$, then

$$\begin{aligned} w(ab) &= \alpha w(a_1)w(b_1) + (1 - \alpha)w(a_2)w(b_2) \\ &\neq (\alpha w(\alpha_1) + (1 - \alpha)w(a_2))(\alpha w(b_1) + (1 - \alpha)w(b_2)). \end{aligned}$$

4. Let $\bar{f}(p, q) = \inf(1, |f(p, q)|)$ (if necessary multiply f by a suitable constant to ensure that $\bar{f}$ is not identically 1). Then $d\rho$ is the sum of two invariant states,

$$d\rho = \tfrac{1}{2}(1 + \bar{f})\, d\rho + \tfrac{1}{2}(1 - \bar{f})\, d\rho.$$

5. $\Rightarrow$: Let $a' \in \mathscr{A}'$. $a'E = 0 \Rightarrow a'\mathscr{A}E = 0 \Rightarrow a' = 0$ on a dense set, which implies that $a' = 0$.
   $\Leftarrow$: Let $E_\perp$ be the orthogonal complement of $\mathscr{A}E$. Then $\mathscr{A}E_\perp = E_\perp$, so the projection $P_\perp$ onto $E_\perp$ belongs to $\mathscr{A}'$, but $P_\perp E = 0$, so E does not separate $\mathscr{A}'$.

6. In a Fock representation of the free fields b, $b(k)|0\rangle = 0$, write

$$a(\mathbf{k}) = \sqrt{\rho(\mathbf{k})}\, b^*(\mathbf{k}) + \sqrt{1 + \rho(\mathbf{k})}\, b(\mathbf{k}),$$

and

$$a^*(\mathbf{k}) = \sqrt{\rho(\mathbf{k})}\, b^*(\mathbf{k}) + \sqrt{1 + \rho(\mathbf{k})}\, b(\mathbf{k}).$$

These operators a likewise satisfy the commutation relations

$$a(\mathbf{k})a^*(\mathbf{k}') - a^*(\mathbf{k}')a(\mathbf{k}) = \delta(\mathbf{k} - \mathbf{k}'),$$

and

$$\begin{aligned} \langle 0|a(\mathbf{k})a^*(\mathbf{k}')|0\rangle &= \delta(\mathbf{k} - \mathbf{k}')\rho(\mathbf{k}), \\ \langle 0|a(\mathbf{k})a(\mathbf{k}')|0\rangle &= \delta(\mathbf{k} - \mathbf{k}')\sqrt{\rho(\mathbf{k})}\sqrt{1 + \rho(\mathbf{k})}. \end{aligned}$$

Hence

$$\langle|a_{f_t} a^*_{g_t}|0\rangle = \int dk \rho(\mathbf{k}) \tilde{f}^*(\mathbf{k})\tilde{g}(\mathbf{k}),$$

$$\langle 0|a_{f_t} a_{g_t}|0\rangle = \int dk \exp(2i|\mathbf{k}|^2 t)\sqrt{\rho(\mathbf{k})}\sqrt{1 + \rho(\mathbf{k})}\, \tilde{f}^*(\mathbf{k})\tilde{g}^*(\mathbf{k});$$

this last integral goes to zero as $t \to \pm\infty$ by the Riemann–Lebesgue lemma, and therefore its time-average is zero. The analogous fact holds for the higher correlation functions, so the time-average of the pure Fock state $|0\rangle$ is of the form (2.5.49).

3.2 The Equilibrium State

In the course of time the Maxwell–Boltzmann distribution has proved more and more fundamental, and has become deeply rooted in the mathematical description of infinite quantum systems.

With a certain normalization of H the canonical state has the form $w(a) = \operatorname{Tr} \exp(-\beta H)a$, as we have seen. The appearance of the Hamiltonian H in both the time-evolution and the state creates all sorts of important connections between them. To avoid technical complications at first we shall concentrate only on the finite-dimensional case. The commutativity of the trace gives rise to a symmetry between the representation of the algebra and its commutant.

The GNS Representation of $\mathscr{B}(\mathbb{C}^n)$ with a Faithful State (3.2.1)

Let $\mathscr{A} = \mathscr{B}(\mathbb{C}^n)$ be given the inner product $\langle a|b\rangle = \operatorname{Tr} a^*b$ so that it becomes a Hilbert space isomorphic to $\mathbb{C}^{n^2}$, and define

$$\pi: \mathscr{A} \to \mathscr{B}(\mathbb{C}^{n^2}): \pi(a)|b\rangle = |ab\rangle,$$

$$\pi': \mathscr{A} \to \mathscr{B}(\mathbb{C}^{n^2}): \pi'(a)|b\rangle = |ba^*\rangle,$$

$$J: \mathbb{C}^{n^2} \to \mathbb{C}^{n^2}: J|b\rangle = |b^*\rangle.$$

Then

(i) *π is a factor representation (*-isomorphism);*
(ii) *π' is a *-antiisomorphism, i.e.,*

$$\pi'(ab) = \pi'(a)\pi'(b), \qquad \pi'(\lambda a) = \bar{\lambda}\pi'(a), \qquad \pi'(a^*) = (\pi'(a))^*,$$
$$\pi'(a+b) = \pi'(a) + \pi'(b)) \quad \text{with } \pi'(\mathscr{A}) = \pi(\mathscr{A})';$$

(iii) *the conjugate-linear operator J preserves norms and $J^2 = \mathbf{1}$;*
(iv) *$J\pi(\mathscr{A})J = \pi'(\mathscr{A})$, $J\pi'(\mathscr{A})J = \pi(\mathscr{A})$;*
(v) *let w be a faithful state, that is, if $a > 0$, then $w(a) > 0$, so by (2.1.5(ii)), $w(a) = \operatorname{Tr} \rho a = \langle\sqrt{\rho}|a|\sqrt{\rho}\rangle$, $\rho > 0$, $\operatorname{Tr}\ \rho = 1$. The vector $|\sqrt{\rho}\rangle$ is cyclic and separating for π and π', i.e., $\pi(a)|\sqrt{\rho}\rangle = 0 \Rightarrow a = 0$. Hence the* GNS *representation using w is unitarily equivalent to π.*

Proof

The isomorphism and antiisomorphism properties are obvious.

(ii) $\pi'(a)\pi(b)|c\rangle = \pi'(a)|bc\rangle = |bca^*\rangle = \pi(b)\pi'(a)|c\rangle$, and therefore $\pi'(\mathscr{A}) \subset \pi(\mathscr{A})'$. On the other hand, if $B \in \pi(\mathscr{A})'$ then $B|1\rangle$ is $|b^*\rangle$ for some $b \in \mathscr{A}$. Hence

$$B|a\rangle = B\pi(a)|1\rangle = \pi(a)B|1\rangle = \pi(a)|b^*\rangle = \pi(a)\pi'(b)|1\rangle = \pi'(b)|a\rangle$$

for all $a \in \mathscr{A}$, so $B = \pi'(b)$ and $\pi'(\mathscr{A}) = \pi(\mathscr{A})'$.

(i) Let $\pi(a) \in \pi(\mathscr{A})'$. Then by part (ii) it equals $\pi'(b^*)$ for some b. Hence $\pi(a)|c\rangle = |ac\rangle = \pi'(b^*)|c\rangle = |cb\rangle$, so $ac = cb$ for all $c \in \mathscr{A}$, and therefore $a = b = \alpha \cdot \mathbf{1}$. Thus $\pi(\mathscr{A})$ is a factor.

(iii) $\|J|a\rangle\|^2 = \text{Tr}\, aa^* = \text{Tr}\, a^*a = \||a\rangle\|^2$, and $J^2 = \mathbf{1}$ since $b^{**} = b$.

(iv) $J\pi(a)J|b\rangle = J\pi(a)|b^*\rangle = J|ab^*\rangle = |ba^*\rangle = \pi'(a)|b\rangle \Rightarrow J\pi(a)J = \pi'(a) \Rightarrow \pi(a) = J\pi'(a)J$, because $J^2 = \mathbf{1}$.

(v) Since ρ^{-1} exists, $|a\rangle$ may be written as $|b\sqrt{\rho}\rangle = \pi(b)|\sqrt{\rho}\rangle, b = a\rho^{-1/2}$, which shows that $\sqrt{\rho}$ is cyclic for π. If $\rho_i > 0$ are the eigenvalues of ρ, then in the diagonal representation of ρ,

$$\|\pi(a)|\sqrt{\rho}\rangle\|^2 = \text{Tr}\, \rho a^* a = \sum_{i,k} \rho_i |a_{ik}|^2 = 0,$$

which implies that $a_{ik} = 0$, and similarly for π'. By (III: 2.3.10; 6) π_ρ is equivalent to π. □

Remarks (3.2.2)

1. An anti-isomorphism came up once before, in the reversal of the motion (III: 3.3.18), and J is like the conjugate-linear operator Θ' (3.3.19; 2).
2. The representation π, being a finite-dimensional factor of type I, is of the form $\pi(a) = a \otimes \mathbf{1}_{|\mathbb{C}^n}$, so $\pi'(a)$ is $\mathbf{1}_{|\mathbb{C}^n} \otimes a^*$.

Consider next how to represent the time-evolution $a \to a_t = \exp(iht)\, a \exp(-iht)$. At first thought it might be represented by $\exp(i\pi(h)t)$, but this would not leave the cyclic vector $|\sqrt{\rho}\rangle$ invariant. The correct way to proceed is as in Example (3.1.1; 3).

The Time-Evolution on $\mathscr{B}(\mathbb{C}^n)$ (3.2.3)

The unitary representation (1.3.5) *of the time-evolution $a \to a_t$ on the invariant state $a \to \text{Tr}\, \rho a$, $\rho = \exp(-\beta h)$, is given by $U_t = \exp(-iHt)$, $H = \pi(h) - \pi'(h)$. It satisfies the following*:

(i) $JHJ = -H, JU_tJ = U_t$;

(ii) $U_{-i\beta/2}\pi(a)|\sqrt{\rho}\rangle = J\pi(a^*)|\sqrt{\rho}\rangle$;

(iii) $\langle\sqrt{\rho}|\pi(a)\pi(b)|\sqrt{\rho}\rangle = \langle\sqrt{\rho}|\pi(b)\pi(a_{i\beta})|\sqrt{\rho}\rangle$.

Proof

It is immediately clear that $\exp(iHt)\pi(a)\exp(-iHt) = \pi(a_t)$. Moreover, $\exp(iHt)|\sqrt{\rho}\rangle = |\exp(iht)\exp(-\beta h)\exp(-iht)\rangle = |\sqrt{\rho}\rangle$.

(i) This follows from (3.2.1(iv)).

(ii) $U_{-i\beta/2}\pi(a)|\sqrt{\rho}\rangle = U_{-i\beta/2}|a \exp(-\beta h/2)\rangle = |\exp(-\beta h/2)a\rangle = J|a^*\exp(-\beta h/2)\rangle = J\pi(a^*)|\sqrt{\rho}\rangle$.

(iii) $\text{Tr}\exp(-\beta h)ab = \text{Tr}\exp(-\beta h)a\exp(\beta h)\exp(-\beta h)b = \text{Tr}\exp(-\beta h)ba_{i\beta}$. □

Remarks (3.2.4)

1. The density matrix ρ was written simply as $\exp(-\beta H)$ under the assumption that h had been redefined by the addition of a multiple of the identity so that $\operatorname{Tr}\exp(-\beta h) = 1$. This affects neither the time-evolution nor H.
2. Note that J does not reverse the direction of time.
3. The operator $\rho = \exp(-h)$ is always positive. Conversely, if $\rho > 0$ (i.e., all eigenvalues $\rho_i > 0$), then $\ln \rho = -h$ is well defined. This shows that groups of automorphisms and faithful states are bijectively related. There is a special term for their relationship.

The Modular Automorphism (3.2.5)

For each faithful state w on $\mathscr{B}(\mathbb{C}^n)$ there is a unique one-parameter group of automorphisms $\tau_t: a \to a_t$ such that

(i) w is invariant in the sense that $w(a_t) = w(a)$.
(ii) w satisfies the **Kubo–Martin–Schwinger** (KMS) condition, $w(ab) = w(ba_i)$.
(iii) there exists an anti-isomorphism $\pi_w(\mathscr{A}) \to J\pi_w(\mathscr{A})J$ onto $\pi_w(\mathscr{A})'$ such that

$$U_{-i/2}\,\pi(a)|\Omega\rangle = J\pi(a^*)|\Omega\rangle,$$

where $|\Omega\rangle$ is the cyclic vector and U_t is the unitary operator representing τ_t in the GNS representation with w.

If the dimension of the Hilbert space is now infinite, but the state is still given by a density matrix $\rho = \exp(-\beta h)$, then there are a few technical difficulties to clear up.

The Temporal Correlation Functions of Finite Quantum Systems (3.2.6)

If the time is made complex, then in general

$$a_{x+iy} \equiv \exp((ix - y)h)a\exp(-(ix - y)h)$$

is unbounded, and hence does not belong to the algebra. However, we shall continue to use this notation, as this operator will never act on anything outside its domain of definition.

(i) Continuity in the strip $-\beta \le \operatorname{Im} t \le 0$. $w(a_t b) = \langle\Omega|a\exp(-iHt)b|\Omega\rangle$, and if t is complex, then by (3.2.3(ii)), $b|\Omega\rangle$ is in the form domain of $\exp(yH)$ for $y \ge -\beta$. In a spectral representation it is apparent that the vector $\exp(yH/2)b|\Omega\rangle$ is norm-continuous in y, so $\rho(a_t b)$ is norm-continuous in t.

(ii) Boundedness in the strip $-\beta \le \operatorname{Im} t \le 0$. Let $H = \pi(h) - \pi'(h)$ as in (3.2.3), so $H|\Omega\rangle = 0$. Because

$$\begin{aligned} a_{x+iy} &= \exp((ix - y)H)a\exp(-(ix - y)H), \\ |w(a_{x+iy}b)|^2 &= |\langle\Omega|a_x \exp(yH)b|\Omega\rangle|^2 \\ &\le \langle\Omega|a_x \exp(yH)a_x^*|\Omega\rangle\langle\Omega|b^* \exp(yH)b|\Omega\rangle. \end{aligned}$$

The function $\langle\Omega|a\exp(yH)a^*|\Omega\rangle$ is positive and, because

$$\frac{\partial^2}{\partial y^2}\langle\Omega|a\exp(yH)a^*|\Omega\rangle = \|H\exp(yH/2)a^*|\Omega\rangle\|^2 \ge 0,$$

convex, achieving its maximum at $y = 0$ or $y = -\beta$. It is clear that $w(aa^*) \le \|a\|^2$, but even at the lower edge it is bounded, as shown by

$$\begin{aligned} w(a_{i\beta/2}a^*_{-i\beta/2}) &= \operatorname{Tr}\exp(-\beta h)\exp(\beta h/2)a^*\exp(-\beta h)a\exp(\beta h/2) \\ &= \operatorname{Tr}\exp(-\beta h)aa^* \le \|a\|^2, \end{aligned}$$

since $\operatorname{Tr}\exp(-\beta h) = 1$. Therefore

$$|w(a_t b)| \le \|a\|\,\|b\| \quad \text{for } -\beta \le \operatorname{Im} t \le 0.$$

(iii) Analyticity in the strip $-\beta < \operatorname{Im} t < 0$. The function $w(a_t b)$ is not differentiable on the real axis for generic a's, but only for complex times within the strip. The proof is similar to that of (2.4.7) and will not be repeated here. The relationship $w(ab) = w(ba_{i\beta})$, named for Kubo, Martin, and Schwinger, which follows from the invariance of the trace, can be continued analytically to the strip: The functions $w(a_t b)$ and $w(ba_t)$ are analytic respectively in $-\beta < \operatorname{Im} t < 0$ and $0 < \operatorname{Im} t < \beta$, where they satisfy the KMS condition $w(a_t b) = w(ba_{t+i\beta})$, which determines the value of $w(a_t b)$ at $y = -\beta$ as $w(ba)$ (see Figure 26).

(iv) The physical significance of the KMS condition. For a finite system the canonical state with $\rho = \exp(-\beta H)$ is not an eigenstate of the energy. The modular Hamiltonian (also denoted H) has $|\Omega\rangle$ as an eigenvector, $H|\Omega\rangle = 0$. This operator H is not generally bounded below; however, the KMS condition distinguishes positive energies because of the positive sign of β. The energy spectrum of $\pi(a)|\Omega\rangle$ for $a = a^* \in \mathscr{A}$ consists predominately of positive energies,

$$\begin{aligned} f(E) \equiv \langle\Omega|\pi(a)\delta(H - E)\pi(a)|\Omega\rangle &= \int_{-\infty}^{\infty} \frac{dt}{2\pi}\exp(iEt)\rho(a_t a) \\ &= \int_{-\infty}^{\infty} \frac{dt}{2\pi}\exp(iEt)\rho(aa_{t+i\beta}) \\ &= \exp(\beta E)\langle\Omega|\pi(a)\delta(H + E)\pi(a)|\Omega\rangle, \end{aligned}$$

and therefore

$$\frac{f(E)}{f(-E)} = \exp(\beta E).$$

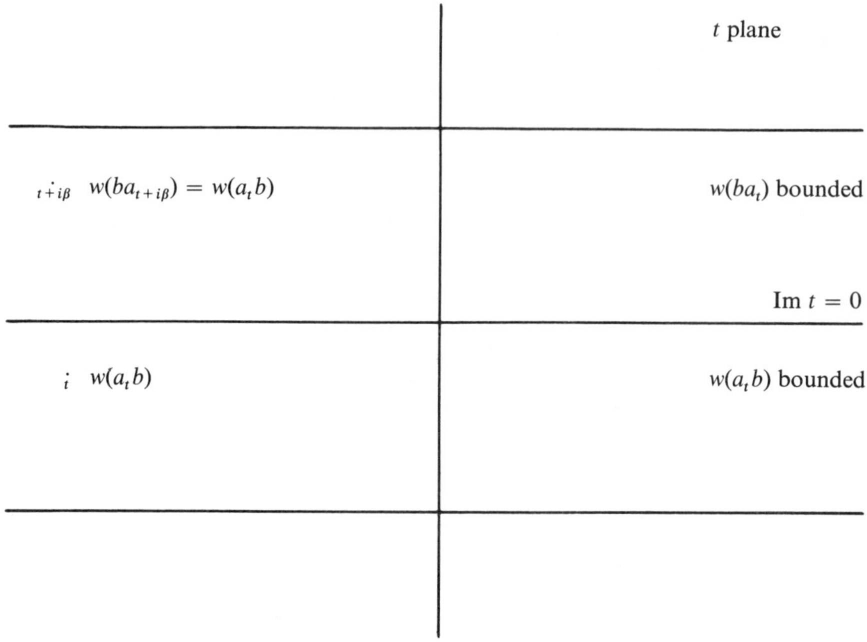

Figure 26 The connection between $w(ba_t)$ and $w(a_t b)$ on their domain of analyticity.

It is thus not possible to remove arbitrary amounts of energy from a system in equilibrium, even though $|\Omega\rangle$ is not its ground state.

(v) Analytic operators. If the dimension of the space is finite, the mapping $t \to a_t$ is analytic, and thus so is $t \to w(a_t b)$. If it is only known that h is semibounded, this is not necessarily the case, and the question arises of which a's are analytic in t. One way to construct such elements of $\mathscr{A}$ is to average over time,

$$a(f) \equiv \int_{-\infty}^{\infty} dt' a(t') f(t').$$

If the Fourier transform $\tilde{f} \in \mathscr{C}^2$, and $\operatorname{supp} \tilde{f} \subset [-\alpha, \alpha]$, then $f(t)$ is analytic and satisfies the estimate

$$|f(x + iy)| \leq \frac{\exp(\alpha |y|)}{(1 + x^2)} \gamma, \quad \text{where } \gamma = (2\pi)^{-1/2}(\|\tilde{f}\|_1 + \|\tilde{f}''\|_1).$$

The time-translate of $a(f)$,

$$\tau_t(a(f)) = \int_{-\infty}^{\infty} dt' a(t') f(t' - t),$$

is then an entire function in t such that $\|\tau_{x+iy}(a(f))\| \leq \pi\gamma \|a\| \exp(\alpha |y|)$. It is easy to see from the continuity of τ_t that the set $\tilde{\mathscr{A}}$ of such regularized

a's (for variable f and α) is dense in $\mathscr{A}$ in norm. Within the set $\tilde{\mathscr{A}}$ it is always possible to continue analytically with controlled growth.

If we now think about an infinite system, the density matrix

$$\exp(-\beta H)/\text{Tr}\exp(-\beta H)$$

no longer makes sense. However, the characterization of certain states made in (3.2.5(ii)) may continue to work in the infinite limit.

Definition (3.2.7)

Given a C^* algebra $\mathscr{A}$ with a continuous time-automorphism $a \to a_t$, a state w on the algebra is called a **KMS state** with respect to temperature $1/\beta$ whenever the functions $t \to w(a_t b)$ and $t \to w(ba_t)$ can be continued analytically to the strips $-\beta < \text{Im}\, t < 0$ and, respectively, $0 < \text{Im}\, t < \beta$, and are continuous on the closures of the strips, where they satisfy the condition

$$w(a_t b) = w(ba_{t+i\beta}).$$

Examples (3.2.8)

1. Free fermions. The grand canonical state (2.5.49) is KMS with respect to the combination of free time-evolution and gauge transformations,

$$\tau_t\colon a_f \to a_{f_t}, \qquad \tilde{f}_t(\mathbf{k}) = \exp[it(|\mathbf{k}|^2 - \mu)]\tilde{f}(\mathbf{k}).$$

First, note that clearly

$$\begin{aligned}\rho(a_f a^*_{g_{i\beta}}) &= \int \frac{d^3k}{(2\pi)^3} \tilde{f}^*(\mathbf{k})\tilde{g}(\mathbf{k}) \exp[-\beta(|\mathbf{k}|^2 - \mu)] \\ &\quad \times \left(1 - \frac{1}{\exp[\beta(|\mathbf{k}|^2 - \mu)] + 1}\right) \\ &= \rho(a^*_g a_f),\end{aligned}$$

and likewise

$$\begin{aligned}\rho(a^*_g a_{f_{i\beta}}) &= \int \frac{d^3k}{(2\pi)^3} \tilde{f}^*(\mathbf{k})\tilde{g}(\mathbf{k}) \exp[\beta(|\mathbf{k}|^2 - \mu)]\left(\frac{1}{\exp[\beta(|\mathbf{k}|^2 - \mu)] + 1}\right) \\ &= \rho(a_f a^*_g).\end{aligned}$$

(If f and g are arbitrary functions in L^2, then in general f_t and $\rho(a_f a^*_{g_t})$ have maximal analytic continuations only into the upper half-plane $\{z = t + iy \,|\, y > 0\}$, and $\rho(a^*_g a_{f_t})$ only into the region $\{z = t + iy \,|\, y < \beta\}$. However, if either $\tilde{f}$ or $\tilde{g}$ has compact support, for example, then the maximal analytic continuation of any of the expressions above is in fact an entire function.) The proof of the KMS property of ρ for arbitrary elements of the algebra will not be given here, because of the amount of combinatorics it requires. The gauge transformation makes an appearance because of the extension of the state to the whole field algebra. If one deals only with the gauge-invariant algebra of observables $\mathscr{A}_E''$ (1.3.14), then the automorphism τ does not depend on μ, so it is identical to the free time-evolution.

2. Free bosons. Let ω_φ be the equilibrium state of the field algebra of the free Bose gas at temperature $1/\beta$ and density ρ (see (2.5.51; 4)), which appears as the integrand in the decomposition of the canonical limiting state in (2.5.51; 1). (The decomposition is nontrivial iff $\rho > \rho_c(\beta)$—see also (2.5.33; 3).) The field algebra of the bosons is generated by the operators

$$W_f \equiv \exp[i(a_f^* + a_f)]; \qquad W_f W_g = \exp[-i \operatorname{Im}(f|g)] W_{f+g},$$

and the free time-evolution of the observables will be extended to the field algebra by $W_f \to W_{f_t}$,

$$\tilde{f}_t(\mathbf{k}) = \exp[it(|\mathbf{k}|^2 - \mu)]\tilde{f}(\mathbf{k}).$$

(The quantity $\mu = \mu(\rho)$ is a unique but not invertible function.) Then $A(f, g, t) \equiv \omega_\varphi(W_f W_{g_t})$ is the continuous boundary value of an analytic function of $z = t + iy$ on the strip $0 < y < \beta$, $t \in \mathbb{R}$, *viz.*,

$$\begin{aligned}\hat{A}(f, g, z) \equiv \exp\Bigg\{ -\int \frac{d^3k}{(2\pi)^3} \Bigg[(|\tilde{f}(\mathbf{k})|^2 + |\tilde{g}(\mathbf{k})|^2)\left(\frac{1}{2} + \frac{1}{\exp[\beta(|\mathbf{k}|^2 - \mu)] - 1}\right) \\ + \tilde{f}^*(\mathbf{k})\tilde{g}(\mathbf{k}) \exp[iz(|\mathbf{k}|^2 - \mu)]\left(1 + \frac{1}{\exp[\beta(|\mathbf{k}|^2 - \mu)] - 1}\right) \\ + \tilde{g}^*(\mathbf{k})\tilde{f}(\mathbf{k}) \exp[-iz(|\mathbf{k}|^2 - \mu)]\left(\frac{1}{\exp[\beta(|\mathbf{k}|^2 - \mu)] - 1}\right)\Bigg]\Bigg\} \\ \times \exp\{2i\sqrt{\rho - \rho_c(\beta)}\,\Theta(\rho - \rho_c(\beta)) \operatorname{Re}[(\tilde{f}(\mathbf{0}) + \tilde{g}(\mathbf{0})) \exp(i\varphi)]\},\end{aligned}$$

and the KMS condition is satisfied: $w(ab_{-t}) = w(a_t b) = w(ba_{t+i\beta})$,

$$\begin{aligned}\lim_{y \to +\beta} \hat{A}(f, g, t + iy) = \omega_\varphi(W_{g_t} W_f) = \omega_\varphi(W_g W_{f_{-t}}) = A(g, f, -t) \\ = \lim_{y \to +0} \hat{A}(g, f, -t + iy).\end{aligned}$$

It follows from $\rho < \rho_c(\beta)$ that $\mu(\rho) < 0$, so in this situation f and g can be arbitrary elements of L^2. However, $\mu(\rho) = 0$ for all $\rho \geq \rho_c(\beta)$, so ω_φ must be restricted, for example, to the algebra generated by the W_f with $f \in L^1 \cap L^2$. For general f and g it is not possible to extend $\hat{A}(f, g, z)$ analytically beyond the strip described above. However, if the support of either f or g is compact, then $\hat{A}(f, g, z)$ is an entire function of z.

Properties of a KMS state w (3.2.9)

1. *A* KMS *state w is invariant in time.*
2. *When extended to $\pi_w(\mathscr{A})''$, w remains* KMS.
3. *If w is faithful (as a positive functional), then π_w is faithful, and vice versa.*
4. *$\mathscr{Z} = \pi_w(\mathscr{A})' \cap \pi_w(\mathscr{A})''$ consists of time-invariant elements.*
5. *The* KMS *states for any fixed β form a weak-* compact, convex set.*
6. *If w is an extremal* KMS *state, then π_w is a factor.*
7. *For any w, there exists a unique time-evolution under which w is a* KMS *state.*

Remarks (3.2.10)

1. According to (1.3.5), if w is invariant in time, then on π_w we can write $a_{-t} = U_t a U_t^{-1}$, and the time evolution, when extended to $\pi_w(\mathscr{A})''$, transforms this algebra into itself: $a_n \to a \Rightarrow a_n(-t) = U_t a_n U_t^{-1} \to U_t a U_t^{-1} \in \pi_w(\mathscr{A})''$.
2. Of course, the extension of w to $\pi_w(\mathscr{A})''$ with cyclic vector $|\Omega\rangle$ is $w(a'') = \langle\Omega|a''|\Omega\rangle$ for all $a'' \in \pi_w(\mathscr{A})''$. Property 2 means that this state is KMS with respect to the time-evolution defined earlier on $\pi_w(\mathscr{A})''$.
3. According to (III: 2.3.10; 3),
$$\begin{aligned}\operatorname{Ker} w &= \{a \in \mathscr{A} : w(a) = 0\}\\ &\supset \mathscr{N} \equiv \{a \in \mathscr{A} : w(a^*a) = 0\}\\ &\supset \operatorname{Ker} \pi_w = \{a \in \mathscr{A} : w(b^*a^*ab) = 0 \quad \text{for all } b \in \mathscr{A}\},\end{aligned}$$
and the statement that w is faithful means that $\mathscr{N} = \{0\}$. Property 3 thus means that if $\operatorname{Ker} \pi_w = \{0\}$, then $\mathscr{N} = \{0\}$, so $|\Omega\rangle$ is a separating vector for $\pi_w(\mathscr{A})$: $\pi_w(a)|\Omega\rangle \neq 0$ for all $\pi_w(a) \neq 0$. (Speaking field-theoretically, no operator annihilates the vacuum.) If the algebra is simple, and hence has only faithful representations, then all KMS states are also faithful.
4. If the system is asymptotically Abelian, then $\mathscr{R}' = \mathscr{Z}$. The center $\mathscr{Z}$ contains the macroscopic observables, which are therefore constant in time in this case.
5. By Property 5, convex combinations and weak limits of KMS states (at a given β) are KMS states.
6. In a finite system, with $\mathscr{A} = \mathscr{B}(\mathscr{H})$, $U_t = \exp(iHt)$, there is only one normal KMS state. At $t = 0$ the condition is that
$$\operatorname{Tr} \rho ab = \operatorname{Tr} \rho b \exp(-\beta H) a \exp(\beta H) = \operatorname{Tr} \exp(-\beta H) a \exp(\beta H) \rho b$$
for all b, which means that $\rho a = \exp(-\beta H) a \exp(\beta H) \rho$ for all a, so $\exp(\beta H)\rho \in \mathscr{A}'$, and thus $\rho = \exp(-\beta H)$. Since the convex set of KMS states is compact, any KMS state may be decomposed into extremal KMS states. If the system is asymptotically Abelian, then according to Remark 6 a decomposition into extremal KMS states is the same as a decomposition into elements of the center (defined as a decomposition

into factors (1.4.9)), which is the same as a decomposition into extremal invariant states. In the characterization of ergodic states (3.1.22; 2) we learned that a factor state is not decomposable into invariant states, and thus *a fortiori* not decomposable into KMS states. Conversely, it is now being claimed that it is always possible to decompose a KMS state w further into other, extremal KMS states, if π_w is not a factor. This means that the extremal KMS states are ergodic and, as factors, even mixing. Since the decomposition by the center is unique, so is the decomposition into extremal KMS states. Hence the set of extremal KMS states is a simplex.

7. If the time-evolution is given, then there can be one or more KMS states (see Problem 2). In contrast, by Property 7, if w is given, then there is a unique time-evolution for which it is KMS.

Proof of (3.2.9)

1. Let $b = 1$; the function $\rho(a_t) = \rho(a_{t+i\beta})$ can be continued analytically to all of $\mathbb{C}$ and is periodic in Im t. Since it is bounded in a strip, it is bounded throughout $\mathbb{C}$ and therefore constant. It follows that ρ is time-invariant.
2. This proposition follows from a more general one to be stated later (3.2.13).
3. If $a \in \mathcal{N}$, then $w(a^*a) = 0$, which implies that for all b, $w(ba) = 0$ (by Cauchy–Schwarz), which means that for all b and c, $0 = w(c_{-i\beta}ba) = w(bac)$, and therefore $a \in \operatorname{Ker} \pi_w$.
4. Suppose $c \in \mathscr{Z}: w(a_t c) = w(a_{t+i\beta} c)$. As in Proposition 1, it can be concluded that $w(a_t c)$ is constant in t. If a is replaced with ab, it follows that $w(a_t c b_t) = \langle \Omega | a U_t c U_{-t} b | \Omega \rangle$ is constant for all a and b, so c is constant.
5. Convexity is trivial. If w_n converges in the weak-* sense to w, then for all $a \in \tilde{\mathscr{A}}$, $b \in \mathscr{A}$ and $t \in \mathbb{C}$, the quantities $w_n(a_t b)$ converge to $w(a_t b)$ and are dominated by $\pi\gamma \|a\| \, \|b\| \exp \alpha |\operatorname{Im} t|$. Consequently, the limit is holomorphic throughout $\mathbb{C}$ and satisfies $w(a_{t-i\beta} b) = w(ba_t)$. As in Problem 1, this relationship remains valid for norm-limits of a's in the strip $0 \leq \operatorname{Im} t \leq \beta$, and can thus be extended to all of $\mathscr{A}$ (and, by Property 2, to all of $\mathscr{A}''$).
6. Unless π_w is a factor, $\mathscr{Z}$ contains a nontrivial projection P. Therefore w can be decomposed into a combination of $w_1(a) = w(Pa)/w(P)$ and $w_2(a) = w((1 - P)a)/w(1 - P)$, and both w_i are KMS states: $w(Pa_t b) = w(a_t Pb) = w(Pba_{t+i\beta})$.
7. Suppose that τ_t and $\bar{\tau}_t$ are distinct automorphisms under which w is a KMS state. Then if a is entire with respect to τ, and b is entire with respect to $\bar{\tau}$, it follows that

$$F(t) \equiv w(\bar{\tau}_{-t}(\tau_t(a)) \cdot b) = w(\tau_t(a) \cdot \bar{\tau}_t(b)) = w(\bar{\tau}_t(b) \cdot \tau_{t+i\beta}(a))$$
$$= w(\tau_{t+i\beta}(a) \cdot \bar{\tau}_{t+i\beta}(b)) = F(t + i\beta).$$

This fact implies that F is constant, so τ and $\bar{\tau}$ have the same action on $\tilde{\mathscr{A}}$ and hence on $\mathscr{A}$. □

The foregoing conclusions suggest an interpretation of the decomposition into extremal KMS states as a decomposition of an equilibrium state into its pure phases. Yet it will be apparent from examples that these pure phases are not necessarily identical to physical phases. Property 6 together with Remarks (3.1.26) ensures that these states have mixing properties, meaning that local perturbations eventually die out, and equilibrium gets re-established. The canonical states were characterized earlier as the states of greatest entropy at a given energy, and the evolution towards them can be thought of as a tendency toward greater entropy. On the other hand, if the system is infinite, it is not the total entropy that is finite, but rather the average entropy, which is unaffected by local perturbations. If a state is normal when restricted to a local algebra (1.3.3; 6), then it is possible to define the local entropy, which will then tend to its equilibrium value. It is not, however, claimed that it increases monotonically to that value.

The diagram in Figure 27 collects together the various properties of asymptotically Abelian systems in invariant states and shows their connection with the time-evolution. It will be shown later (3.3.17) that the spectrum of H is ordinarily the whole real line $(-\infty, \infty)$. The spectral properties stated then include the supposition that the systems that we shall be concerned with have neither dense point spectrum nor singular continuous spectrum.

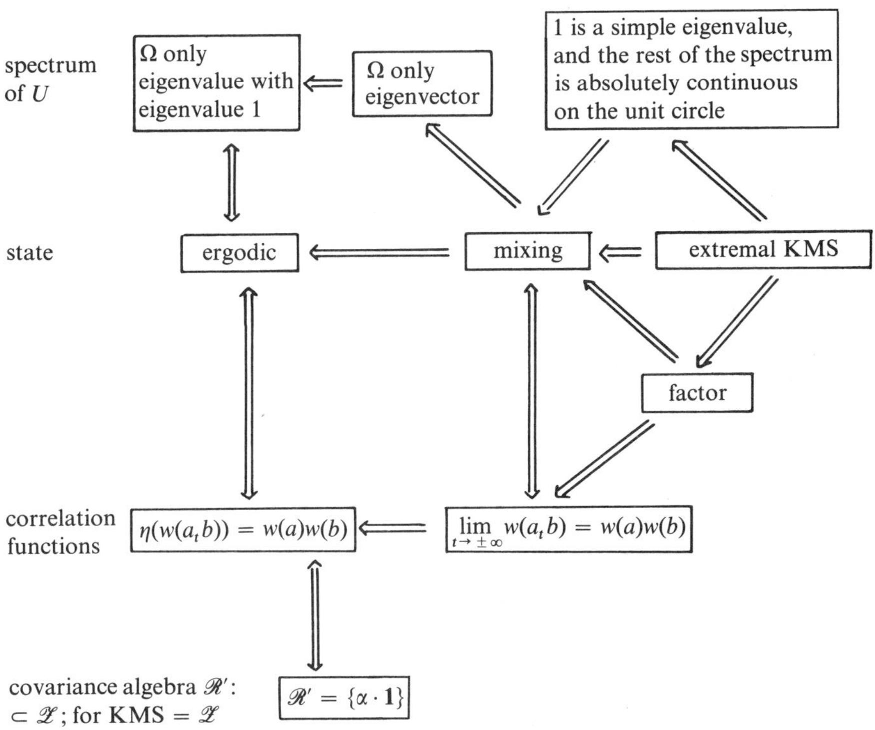

Figure 27 Implications among the ergodic properties.

Examples (3.2.11)

1. Free fermions. Consider a system of n kinds of free fermions, described by the field operators $a_{\alpha, f}$, $\alpha = 1, \ldots, n$. The algebra $\mathscr{A}_E$ of observables will be taken to consist only of polynomials containing an equal number of a_α and a_α^* for any α, in accordance with Definition (1.3.8). In other words it contains the densities and currents of the particles. The state is taken as the product of the grand canonical states (2.5.49), i.e.,

$$\langle a^*_{1, f_1^1} \cdots a^*_{1, f^1_{m_1}} a_{1, g_1^1} \cdots a_{1, g^1_{m_1}} a^*_{2, f_1^2} \cdots a^*_{2, f^2_{m_2}} a_{2, g_1^2} \cdots \\ a_{2, g^2_{m_2}} a^*_{3, f_1^3} \cdots a_{n, g^n_{m_n}} \rangle = \prod_\alpha \operatorname{Det}\langle g_i^\alpha | \rho_1^\alpha f_j^\alpha \rangle,$$

$$\langle f | \rho_1^\alpha g \rangle = \int \frac{d^3k}{(2\pi)^3} \frac{\tilde{f}^*(\mathbf{k})\tilde{g}(\mathbf{k})}{\exp[\beta((|\mathbf{k}|^2/2m_\alpha) - \mu_\alpha)] + 1}.$$

It is KMS with respect to the automorphism $a_{\alpha, f_\alpha} \to a_{\alpha, f_\alpha(t)}$,

$$\tilde{f}_\alpha(t) = \exp\left(\frac{it|\mathbf{k}|^2}{2m_\alpha}\right)\tilde{f}_\alpha.$$

Observe that for this automorphism of the algebra of observables there is an n-parameter family of KMS states. They can be parametrized by the chemical potentials μ_α, and, as factor states, they are extremal. A general KMS state at a given β is an integral over them with some probability measure on the μ_α, which corresponds to the mixture of phases posited in the usual procedure known as Gibbs's phase rule. As remarked in (2.3.41), with a variable β and n types of matter having only one phase, there is an $n + 1$-dimensional manifold of states.

2. Bose condensation. If $\rho > \rho_c(\beta)$, then the canonical state (2.5.51; 1) may be written as an integral $\int_0^{2\pi} (d\varphi/2\pi)w_\varphi$ over the factor states

$$w_\varphi(\exp(ia_f^*)\exp(ia_f)) \\ = \exp\left[-\int \frac{d^3k}{(2\pi)^3} \frac{|\tilde{f}(\mathbf{k})|^2}{\exp(\beta|\mathbf{k}|^2) - 1} + 2i\sqrt{\rho_0}\operatorname{Re}(\tilde{f}(\mathbf{0})\exp(i\varphi))\right].$$

These states are KMS with respect to the transformation $\tilde{f}(\mathbf{k}) \to \exp(i|\mathbf{k}|^2 t)\tilde{f}(\mathbf{k})$, and are consequently extremal KMS states. They describe the coexistence of two phases, the normal phase with particle density $\int d^3k[\exp(\beta|\mathbf{k}|^2) - 1]^{-1}(2\pi)^{-3}$ and a condensed phase of density ρ_0. The latter phase still depends parametrically on the argument φ of a_0, and so for fixed β there are two parameters, ρ_0 and φ, to specify the extremal KMS states. These extremal KMS states are not the same as the phases of Gibbs's phase rule. Although different phases of a substance are coexisting if $\mu = 0$ and $0 < T < T_c$, the condensed phase makes its appearance not as a single, pure phase, but rather as combination of infinitely many pure phases, differing in their values of the "hidden parameter" φ, which has no effect on the thermodynamic functions (2.5.33; 3). In this way the decomposition into extremal KMS states is finer than the phase decomposition of (2.3.39) into extremal points of the concave function $\sigma(\varepsilon, \rho)$. If the field algebra is confined to its even part $\mathscr{A}_E$ (in the Fock representation, $\mathscr{A}_E = \mathscr{A}_B \cap \{N\}'$), then all the w_φ become the same state. This is apparent when it is observed that gauge transformations τ_φ: $W_f \to W_{\exp(i\varphi)f}$ transform the w_φ into one another: $(w_\varphi \circ \tau_{\varphi'})(W_f) = w_{\varphi+\varphi'}(W_f)$. The restriction to $\mathscr{A}_E$ makes $\tau_{\varphi'}$ the identity, so $w_\varphi = w_{\varphi+\varphi'}$. Recall that for asymptotically Abelian systems the decomposition into extremal KMS states is unique according to (3.2.9; 6); the extremal states form a simplex. In contrast, we were not able to adduce any theoretical reasons for why the flat pieces of $\sigma(\varepsilon, \rho)$ had the structure of a simplex.

3. A model of a ferromagnet. The time-evolution of Example (2.3.33; 2) was investigated in (3.1.1; 4). We found that if $B = 0$ and $T < 2$, it was no longer an automorphism of the spin algebra $\mathscr{A} = \{\boldsymbol{\sigma}_i\}$, but rather of the strong closure $\pi(\mathscr{A})''$. The state

$$\langle \sigma_1^{\alpha_1} \cdots \sigma_m^{\alpha_m} \rangle = \int_{S_2} d\mathbf{n} s^m n_{\alpha_1} \cdots n_{\alpha_m}, \qquad s = \tanh(2\beta s),$$

is KMS with respect to this time-evolution. In each of the factors $\pi_{\mathbf{n}}$ it is a rotation about the axis $\mathbf{n}$ at angular velocity $4s$. For example, if $\mathbf{n}$ points in the z-direction, then $\sigma^+(t) = \exp(-4ist)\sigma^+$ and

$$\langle \sigma^+ \sigma^- \rangle = \frac{\langle 1 + \sigma \rangle}{2} = \frac{1 + s}{2} = \langle \sigma^- \sigma_{i\beta}^+ \rangle = \exp(4\beta s) \langle \sigma^- \sigma^+ \rangle$$
$$= \exp(4\beta s) \frac{1 - s}{2},$$

because $s(1 + \exp(4\beta s)) = \exp(4\beta s) - 1$. The individual factors $\pi_{\mathbf{n}}$ thus give rise to extremal KMS states, corresponding to spontaneous magnetization in the direction $\mathbf{n}$. Again, from the physical point of view this model would be described as having one magnetized phase, whereas the decomposition into extremal KMS states would distinguish among different directions of $\mathbf{n}$, and treat magnetization in each direction as a distinct phase. Notice that the phase transition at $T = 2$ is connected with a change of the type of factor; if $T < 2$ the integral runs over factors of type III, while if $T > 2$, the factors are of type II_1.

Remarks (3.2.12)

1. There are many different possible reasons for the existence of several KMS states. One is that the center of the algebra of observables $\mathscr{A}$ might be nontrivial. Unitary elements of the center generate transformations, which, like gauge transformations, leave each element of the algebra invariant. Therefore it is possible to combine the action of these transformations with that of time-evolution τ and study the KMS states with

respect to the resulting automorphisms. When restricted to $\mathscr{A}$, these automorphisms are identical to the time-evolution, so all such states are also τ-KMS for $\mathscr{A}$ (cf. Problem 2).

2. Many "degeneracies" of KMS states go away upon enlargement of the algebra of observables. If in Example 1 the particle number is also allowed to vary, for instance by a chemical reaction $(1) \rightleftharpoons (2) + (2)$, then noneven elements like $a_1^* a_2 a_2$ are introduced into the algebra of observables. They are not separately invariant under gauge transformations of the different types of particles, but are invariant only under certain combinations, e.g., if the generator of the transformation has the form $2N_1 + N_2$ in the Fock representation. Consequently, the KMS condition with the free time-evolution makes the chemical potentials satisfy a linear equation such as $\mu_1 - 2\mu_2 = 0$. Similarly, if two condensed Bose systems as in Example (3.2.11; 2) are coupled, the relative phase φ becomes observable (the Josephson effect).
3. It is possible that a symmetry is broken, which means that the extremal KMS states w are not invariant under some group σ of automorphisms that commute with τ. This is illustrated in Example (3.2.11; 2) with the gauge transformations and in (3.2.11; 3) with the rotations. If the symmetry is broken, then $w \circ \sigma_s$ is once again τ-KMS; thus with continuous groups there are even infinitely many KMS states.
4. The theoretical justification of Gibb's phase rule for continuous systems is still an open problem (cf. [20]).
5. So far we have been considering β as fixed. KMS states with different β's are disjoint, i.e., if $w = (w_{\beta_1} + w_{\beta_2})/2$, then $\pi_w = \pi_{\beta_1} \oplus \pi_{\beta_2}$. In this case the temperature β^{-1} becomes an observable belonging to the center of $\pi_w(\mathscr{A})$.

As discussed in §1.1, the ergodic property of a system has been an important ingredient of the justification of statistical mechanics throughout its history. Even though today ergodicity is no longer viewed as the central requirement, it can still be a noteworthy property of realistic systems, so it can still be valuable to have a formulation of ergodicity for infinite quantum systems. In a classical system, if there existed additional constants of the motion beyond H, it would be impossible for the trajectory of almost every point to wind densely throughout the energy shell. However, constants such as momentum or angular momentum are infinite for infinite systems, so ergodicity can not be defined as the absence of additional constants of the motion. But recall that classically constants of the motion also generate diffeomorphisms that commute with the flow of time (see I, §3.3). This property carries over to infinite systems, and even the notions of indecomposable time-invariant surfaces and of dense trajectories have analogies.

In order to characterize ergodic systems, it is only necessary to generalize (3.2.5) to infinite systems.

Modular Automorphisms of a von Neumann Algebra (3.2.13)

Let $\mathcal{M}$ be a von Neumann algebra of operators on a Hilbert space $\mathcal{H}$. For every vector $|\Omega\rangle$ that is both cyclic and separating (i.e., $\mathcal{M}|\Omega\rangle = \mathcal{H}$, and if $a|\Omega\rangle = 0$ for any $a \in \mathcal{M}$, then $a = 0$), there exists a unique one-parameter group of automorphisms $a \to \tau_t(a)$ and a conjugate-linear operator J such that

(i) *$w(a) \equiv \langle\Omega|a|\Omega\rangle$ is τ-KMS (with $\beta = 1$);*
(ii) *$J^2 = \mathbf{1}$, $J\mathcal{M}J = \mathcal{M}'$; and*
(iii) *$U_{-i/2}a|\Omega\rangle = Ja^*|\Omega\rangle$, where $\tau_t(a) = U_{-t}aU_t$.*

Remarks (3.2.14)

1. The idea of the proof follows that of (3.2.3), but with additional technical complications, for which reason the reader is referred to [21].
2. Properties (3.2.6) of the correlation functions hold also in the general case. Specifically, (iii) means that $\mathcal{A}|\Omega\rangle \subset D(\exp(-H/2))$, where $U_t = \exp(-iHt)$, from which it follows that $\mathcal{A}|\Omega\rangle \subset D(\exp(-yH)$ for $0 \leq y \leq \frac{1}{2}$, and $w(a^* \exp(-H)a) = w(aJ^2a^*) \leq \|a\|^2$. The proofs of the other properties can be repeated verbatim.
3. It is clear that a further generalization to arbitrary C^* algebras will not work. The state in Example (3.2.11; 3) is obviously faithful on the σ's, so it is a candidate for w. However, we have found that the related automorphism under which w is a KMS state maps the C^* algebra generated by the σ's out of itself, leaving only the von Neumann algebra $\pi_w(\mathcal{A})''$ invariant.
4. Suppose that w is a KMS state on the algebra $\mathcal{A}$ with respect to the time-evolution τ_t. By Property (3.2.9; 3) the vector $|\Omega\rangle$ given in the GNS representation π_w is cyclic and separates $\pi_w(\mathcal{A})''$, even if w fails to be faithful, and the representation of τ_t is identical to the modular automorphism.

Ergodic Quantum Systems (3.2.15)

Let τ be the time-evolution under which the C^* algebra $\mathcal{A}$ of observables is asymptotically Abelian, and let $\mathcal{T}$ be the set of faithful states w with the property that the normal extension of w to $\pi_w(\mathcal{A})''$ is also faithful. Then the following two properties are equivalent:

(i) *A state $w \in \mathcal{T}$ is ergodic if and only if it is an extremal* KMS *state; and*
(ii) *There is no $w \in \mathcal{T}$ such that its modular automorphism σ differs from τ, but $[\sigma, \tau] = 0$.*

If a system has these properties, we shall call it **ergodic**.

Proof that (i) $\Leftrightarrow$ (ii)

Not (ii) $\Rightarrow$ not (i). Let w be the σ-KMS state. Since σ and τ commute, $\rho \equiv \eta_t(w \circ \tau_t)$ is also σ-KMS, so our strategy will be to use it to construct a τ-ergodic state. Think of ρ as decomposed in two separate ways, on the one hand into τ-ergodic states and on the other into extremal σ-KMS states. By Remark (3.2.10; 6) the latter decomposition is the same as the decomposition into factors, whereas according to Remark (3.1.21; 3) the τ-ergodic decomposition is coarser than the factor decomposition. This means that the τ-ergodic components of ρ are combinations of extremal σ-KMS states, but not vice versa. Hence any such component is τ-ergodic but not τ-KMS, since it is not possible for it to be KMS with respect to σ and $\tau \neq \sigma$ at the same time.

Not (i) $\Rightarrow$ not (ii). Suppose that $w(a) = \langle \Omega | a | \Omega \rangle$ is τ-ergodic, and let σ denote the modular automorphism of $\pi_w(\mathscr{A})''$. Since w is invariant under τ and σ, both groups have unitary representations on π_w. Let $\exp(iHt)$ and $\exp(iGs)$ denote their representations. Since w is also σ-KMS, given any a and $b \in \mathscr{A}$,

$$\langle \Omega | \tau_t(a)\sigma_i(b) | \Omega \rangle = \langle \Omega | b\tau_t(a) | \Omega \rangle = \langle \Omega | \tau_{-t}(b)a | \Omega \rangle = \langle \Omega | \sigma_{-i}(a)\tau_{-t}(b) | \Omega \rangle,$$

so

$$\langle \Omega | a \exp(-iHt) \exp(-G) b | \Omega \rangle = \langle \Omega | a \exp(-G) \exp(-iHt) b | \Omega \rangle.$$

Since the vectors of the form $a|\Omega\rangle$ are dense, it follows that $[\exp(-G), \exp(-iHt)] = 0$, so $[\tau, \sigma] = 0$. However, if w is not KMS with respect to τ, then the groups of automorphisms must be different, since w is KMS with respect to σ. □

Remarks (3.2.16)

1. Unfortunately, no examples of ergodic quantum systems are known. Although the grand canonical state (2.5.49) of free particles is mixing, there are ergodic states that fail to be KMS: The momentum distribution $[\exp(\beta(|\mathbf{k}|^2 - \mu)) \pm 1]^{-1}$ would just have to be replaced with some other positive, integrable function. The state would then be time invariant and, as a factor state, ergodic, but not KMS. The hope is that when interactions are switched on, states of this kind will turn into equilibrium states (see §3.3).
2. Property (3.2.15(ii)) forbids the existence of additional constants of the motion. In finite quantum systems, in addition to the Hamiltonian H there are also the constants of the form $f(H)$. If H is nondegenerate, then this accounts for all the constants, because $\{H\}'$ is generated by $f(H)$ and the unitary transformations of the degeneracy space. If the system is infinite, then H exists only in representations π_w of invariant states w, and does not belong to $\pi_w(\mathscr{A})$. It can be shown [22] that only linear functions $f(H)$ produce automorphisms of $\pi_w(\mathscr{A})$. However, the function

$H \to cH$ does nothing more than change the scale of time, and we consider scaled time-evolutions as identical.

3. If particle numbers are conserved, then gauge transformations $a_f \to \exp(i\alpha)a_f$, $\alpha \in \mathbb{R}$, certainly commute with time-evolution, and the system is not ergodic as defined by (3.2.15). Yet the corresponding KMS states w are of the form (2.5.49) with infinite temperature but $\beta\mu = 1$,

$$w(a_f a_g^*) = \int \frac{d^3k}{(2\pi)^3} \frac{\tilde{f}^*(\mathbf{k})\tilde{g}(\mathbf{k})}{e + 1}.$$

The particle density in this state is infinite, $w(a(\mathbf{x})a^*(\mathbf{x})) = \delta(\mathbf{0})/(1 + e)$, however, so it is not of physical interest. This shows that in a nonergodic infinite system it may happen that the states that are ergodic but not KMS never actually occur, so the system behaves ergodically anyway. On the other hand, there is no similar objection to this state on a lattice system, for which $\mathbf{k}$ varies only over a compact region.

4. If an infinite system is homogeneous and isotropic, then translations and rotations commute with τ. The KMS states of these automorphisms have the same defect as that of Remark 3, that the local particle density is infinite.
5. Since under the measurability assumptions of (3.1.22; 3) ergodic states are time-averages of a pure state, the same will be true of the extremal KMS states of ergodic systems. This is the fulfillment of the hope of classical ergodic theory that the equilibrium state can be obtained as the closure of a single trajectory.

system / state	Finite, classical There are no additional constants of the motion	Finite, quantum-mechanical H is nondegenerate	Infinite, quantum-mechanical There exists no KMS σ such that $\sigma \neq \tau$, $[\sigma, \tau] = 0$
Microcanonical	Ergodic Time-average of pure states Not faithful	Ergodic Time-average of pure states Not faithful	
Canonical	Not ergodic Faithful	Not ergodic Time-average of pure states Faithful	
Extremal KMS			Ergodic Time-average of pure states Faithful

If we wish to conceive of ergodicity roughly as the absence of constants of motion other than $f(H)$, then it is useful to make a table of the implications of this for equilibrium states of systems of various types. As can be seen below, the KMS states of infinite quantum systems inherit the good properties of the canonical and microcanonical states of finite systems.

Problems (3.2.17)

1. Consider a sequence of states w_N on a C^* algebra $\mathscr{A}$ converging to w (in the weak-* sense). Show that if the modular automorphism $\tau_{N,t}(a)$ is a norm-convergent sequence in $\mathscr{A}$ for all $a \in \mathscr{A}$ and $t \in \mathbb{R}$, then the $\tau_{N,t}$ converge to the modular automorphism belonging to w.
2. Find an example of an algebra $\mathscr{A} \subset \mathscr{B}(\mathbb{C}^4)$ such that some nontrivial automorphism has many KMS states.
3. Construct the KMS states for translation and rotation of a system of free fermions.
4. In both classical and quantum mechanics, study the automorphisms of the anisotropic oscillator $H = \frac{1}{2}(p_1^2 + p_2^2 + \omega_1^2 q_1^2 + \omega_2^2 q_2^2)$, with ω_1/ω_2 irrational, that commute with the time-evolution. Is the system ergodic?

Solutions (3.2.18)

1. Consider the limits of the correlation functions $w_N(\tau_{N,t}(a(N, f))b)$, where

$$a(N, f) \equiv \int dt \tau_{N,t}(a) f(t),$$

and f is as in (3.2.6(v)), and let $\tau_t(\mathrm{a}) = \lim \tau_{N,t}(a)$. The norm-limit of $\tau_{N,t}(a(N, f))$ is $\tau_t(a(f))$ by the dominated convergence theorem, even for complex t, since $\int |f(t + iy)| dt \le \pi\gamma \exp(\alpha |y|)$. The first term of $[w(\tau_t(a(f))b) - w_N(\tau_t(a(f))b)] + w_N(\tau_t(a(f)) - \tau_{N,t}(a(N, f))b)$ goes to zero because of the weak-* convergence $w_N \to w$, and the second term goes to zero as a consequence of the norm-convergence of $a(N, f)$ to zero. Therefore, for all $a \in \mathscr{A}$ and $t \in \mathbb{C}$,

$$w_N(\tau_{N,t}(a(N, f))b) \to w(\tau_t(a(f))b).$$

These holomorphic functions converge pointwise and are uniformly bounded on every compact set in $\mathbb{C}$, because they are $\le \|a\| \, \|b\| \pi\gamma \exp(\alpha |y|)$; the limit is therefore holomorphic and identical to $w(b\tau_{t+i}(a(f)))$.

This means that the KMS condition holds for all $a \in \tilde{\mathscr{A}}$, and of course boundedness in the strip (3.2.6(ii)) is preserved in limits. Passing by norm-limits $a_n \to a$ to general $a \in \mathscr{A}$, if $-1 \le \mathrm{Im}\, t \le 0$, then $w(\tau_t(a_n)b)$ converges uniformly to $w(\tau_t(a)b)$, which is consequently continuous on the strip and holomorphic in its interior.

It is trivial to see that the identity $w(\tau_t(a)b) = w(b\tau_{t+i}(a))$ continues to hold for limits, as do the group property $\tau_{t+s} = \tau_t \circ \tau_s$ and the invariance of w: $w \circ \tau_t = w$. The GNS construction can now be carried out, so that τ_t is represented unitarily on π_w as U_t. If $\pi(a_n)$ converges weakly to $b \in \pi(\mathscr{A})''$, then $U_{-t}\pi(a_n)U_t$ converges weakly to $U_{-t}bU_t \equiv \tau_t(b)$. Therefore τ_t maps $\pi(\mathscr{A})''$ into itself, and is identical to the modular automorphism according to (3.2.9; 7) and (3.2.14; 4).

2. Let $\mathscr{A}$ be spanned by $(\mathbf{1}, \boldsymbol{\tau}) \otimes (\mathbf{1}, \sigma_3)$, and let the time-evolution be $\tau^{\pm}(t) = \exp(\pm i\omega t)\tau^{\pm}(0)$, with τ_3 and σ_3 constant. For a given β the density matrices of the form

$$\rho = \frac{\exp(-\beta\tau_3 - \alpha\sigma_3)}{\operatorname{Tr}\exp(-\beta\tau_3 - \alpha\sigma_3)}$$

yield KMS states for all real α.

3. They have the same structure as in (2.5.49), with

$$\langle f | \rho, g \rangle = \int \frac{d^3k}{(2\pi)^3} \frac{\tilde{f}^*(\mathbf{k})\tilde{g}(\mathbf{k})}{1 + \exp(k_1)}$$

for translations in the 1-direction, and

$$\int_0^{\infty} r^2\, dr \sum_{l,m} \frac{\hat{f}^*_{lm}(r)\hat{g}_{lm}(r)}{1 + \exp m}$$

for rotations about the 3-axis, where $\hat{f}_{lm}$ denote the expansion coefficients of f in spherical harmonics.

4. Classically, $H_i = \frac{1}{2}(p_i^2 + \omega_i^2 q_i^2)$ are two independent constants of the motion, and generate flows that commute with time-evolution. The system is not ergodic in the sense of Table I. Quantum mechanically, H has the eigenvalues $(n_1 + \frac{1}{2})\omega_1 + (n_2 + \frac{1}{2})\omega_2$ and is thus nondegenerate. All constants are of the form $f(H)$, and the system is ergodic in the sense of Table I.

3.3 Stability and Passivity

The distinguishing feature of the equilibrium state is that it does not change abruptly when subjected to a local perturbation. The second law of thermodynamics can be proved in a version stating that a system prevents energy from being extracted by a cyclic perturbation only if it is in equilibrium.

The final part of the general theory that will be investigated will be the influence of local perturbations on equilibrium. In the mathematical treatment local perturbations play the role of the speck of dust invoked in the traditional theory of statistical mechanics to convert stationary states, not yet in equilibrium, into equilibrium states. As a matter of fact, what makes the KMS states special in the mathematical theory is that they have certain stability properties—they change continuously when the Hamiltonian is perturbed slightly. This is certainly not true of all stationary states, and can even be used to characterize the extremal KMS states of an infinite system; they are precisely the set of states that turn continuously into the unperturbed states as a certain family of perturbations tends to zero. Mixed KMS states represent quantum-mechanical mixtures of phases, and lead to a nontrivial center of the algebra. If an observable from the center is added to H, the time-automorphism is unchanged, but the KMS states do change. Hence mixtures of KMS states exhibit a kind of instability in that they do not remain unchanged under the influence of a family of perturbations moving spatially off to infinity, and hence entering the center of the algebra.

A second important characteristic of KMS states is their passivity, which is the requirement that the energy of the system at time t can only have increased if the Hamiltonian depends on time and has returned to its initial form at time t. This condition also fixes the sign of β and means that no energy can be removed from a KMS state having $\beta > 0$, just as a periodic process can extract no energy from the ground state. This property does not constitute a kind of stability, and sheds no light on why Nature chiefly produces KMS states. However, it does show the most important empirically familiar feature of equilibrium.

As usual, the study of a finite system will provide us with a first exposure to the effects of perturbations. Its time-evolution will be caused by a self-adjoint operator, which also determines the equilibrium state w by $a_t = \exp(iHt)a\exp(-iHt)$, $w(a) = \text{Tr}\exp(-\beta H)a/\text{Tr}\exp(-\beta H)$. If H is subjected to a bounded, self-adjoint perturbation h, the effects can be written down as norm-convergent series. A simple generalization of (III: 3.4.10; 3) shows that

$$\exp(i(H + h)t)a\exp(-i(H + h)t)$$
$$= a_t + \sum_{n\geq 1} i^n \int_{0\leq t_1\leq t_2\cdots\leq t_n\leq t} dt_1\,dt_2\cdots dt_n[h_{t_1}, [h_{t_2}, \ldots, [h_{t_n}, a_t]\cdots]], \tag{3.3.1}$$

$$\exp(-H - h) = R_h\exp(-H), \qquad \exp(-(H + h)/2) = S_h\exp(-H/2),$$
$$R_h \equiv \mathbf{1} + \sum_{n\geq 1}(-1)^n \int_{0\leq s_1\leq\cdots\leq s_n\leq 1} ds_1\cdots ds_n h_{is_1}\cdots h_{is_n},$$
$$S_h \equiv \sum_{n\geq 0}(-1)^n \int_{0\leq s_1\leq\cdots\leq s_n\leq 1/2} ds_1\cdots ds_n h_{is_1}\cdots h_{is_n}. \tag{3.3.2}$$

Remarks (3.3.3)

1. Initially, h_{is} is well defined only if h is analytic in time (3.2.6(v)), but since such operators are dense in $\mathscr{A}$ in norm, the formulas it appears in extend to $\mathscr{A}$ by continuity. .
2. Inequalities (2.1.8: 3) and (2.1.8; 7) yield the estimates

$$\exp(-\|h\|) \leq \exp\left(\frac{-\text{Tr}\exp(-H)h}{\text{Tr}\exp(-H)}\right) \leq \frac{\text{Tr}\exp(-H - h)}{\text{Tr}\exp(-H)}$$
$$= \frac{\text{Tr}\,R_h\exp(-H)}{\text{Tr}\exp(-H)} \leq \min\{\|R_h\|, \|\exp(-h)\|\}.$$

Equation (3.3.1) can now be extended to cover infinite systems, for which H has continuous spectrum, as follows.

Perturbation of the Time–Evolution and KMS State (3.3.4)

Let $a \to a_t$ be an automorphism of a C^* algebra $\mathscr{A}$, and let $\tilde{\mathscr{A}}$ be the subalgebra that is analytic in time and w be a KMS state. Assume $\beta = 1$. If $h \in \tilde{\mathscr{A}}$ is self-adjoint, then a perturbed automorphism $a \to \tau_t^h(a)$ and perturbed state are defined by

$$\tau_t^h(a) = a_t + \sum_{n \geq 1} i^n \int_{0 \leq t_1 \leq \cdots \leq t_n \leq t} dt_1\, dt_2 \cdots dt_n [h_{t_1}, [h_{t_2}, \ldots, [h_{t_n}, a_t] \cdots]],$$

$$w_h(a) = \frac{w(aR_h)}{w(R_h)} = \frac{w(R_h^* a)}{w(R_h)} = \frac{w(S_h^* a S_h)}{w(R_h)},$$

where R_h and S_h are defined as in (3.3.2).

Remarks (3.3.5)

1. The operator h exists as a local perturbation on a purely algebraic level, whereas H exists only in certain representations. For that reason it is not possible to define $\tau_t^h(a)$ simply as $\exp(i(H + h)t)a \exp(-i(H + h)t)$. As in (3.3.2), for finite times the sums converge in norm.
2. If the system is asymptotically Abelian sufficiently strongly, then the limits as $t \to \pm\infty$ of $\tau_t^h \circ \tau_{-t}^0$ exist. However, such a limit may fail to be an automorphism; like the Møller transformations it might not be surjective. If it is surjective, its inverse transforms w into the perturbed state
$$w_h = \lim_{t \to \pm\infty} w \circ \tau_{-t}^0 \circ \tau_t^h.$$
3. See Problem 1 for the equivalence of the definitions of w_h.
4. $(\partial/\partial t)\tau_t^h(a) = \tau_t^h((\partial/\partial s)a_s|_{s=0}) + i\tau_t^h([h, a])$.
5. The function $\mathscr{A} \to \mathscr{A}: h \to \tau_t^h(a)$ is continuous for all $t \in \mathbb{R}$ and $a \in \mathscr{A}$, if $\mathscr{A}$ has either the strong or the norm topology.
6. The state w_h is KMS with respect to τ_t^h for $\beta = 1$: As shown by (3.3.1), $D(\exp(-H - h)) = D(\exp(-H))$ in the representation using π_w, and because $\exp(H) = \exp(H + h)R_h$, the domains of definition of $\exp(H + h)$ and $\exp(H)$ are also identical. Hence for all a and $b \in \mathscr{A}$,
$$w_h(\tau_{-i}^h(a)b) = \frac{w(R_h^* \exp(H + h)a \exp(-H - h)b)}{w(R_h)}$$
is well defined. From (3.3.1) and the KMS condition for w,
$$w_h(\tau_{-i}^h(a)b) = \frac{w(\tau_{-i}(aR_h)b)}{w(R_h)} = w_h(ba).$$

7. There is an analogue of the variational principle for the free energy, which generalizes (2.1.8; 3) for infinite systems. It is a consequence of the convexity of the function $h \to \ln w(R_h)$, which can be proved as follows: From Duhamel's formula (cf. the proof of (III: 3.3.15)),

$$\frac{d}{d\lambda}\exp(-(H+\lambda a))$$
$$= -\int_0^1 ds \exp(-s(H+\lambda a))a\exp(-(1-s)(H+\lambda a)),$$

it can be calculated that

$$\frac{d}{d\lambda} w(R_{h+\lambda a})|_{\lambda=0} = \int_0^1 w(\tau_{is}^h(a)R_h)\,ds = w(aR_h).$$

The second part of the equality makes use of the invariance of w_h under τ^h, which follows from the KMS condition shown above. Likewise,

$$\frac{d^2}{d\lambda^2} w(R_{h+\lambda a})|_{\lambda=0} = \int_0^1 ds w(a\tau_{is}^h(a)R_h),$$

and

$$\frac{d^2}{d\lambda^2}\log w(R_{h+\lambda a})|_{\lambda=0} = \frac{w(R_{h+\lambda a})''}{w(R_h)} - \left(\frac{w(R_{h+\lambda a})'}{w(R_h)}\right)^2$$
$$= \int_0^1 ds w_h((a - w_h(a))\tau_{is}^h(a - w_h(a))).$$

In (3.2.6(ii)) it was seen that the integrands are positive. As in (3.3.3; 2) this fact can be used to show that $w(R_h) \geq \exp(-w(h)) \geq \exp(-\|h\|)$.

If there is a bounded sequence of perturbations $h^{(n)}$ all the commutators of which with $\mathscr{A}$ tend to zero as $n \to \infty$, then the automorphism $\tau_t^{h^{(n)}}$ converges to the unperturbed automorphism because

$$\|\tau_t^h(a) - a_t\| \leq \exp(2\|h\|t)\int_0^t \|[h, a_{t-s}]\|\,ds.$$

This state of affairs can arise, for instance, if the algebra is asymptotically Abelian with respect to spatial translations. If Λ_n denotes the region Λ translated by $n\mathbf{a}$, $\mathbf{a} \in \mathbb{R}^3$, and $h^{(n)} \in \mathscr{A}_{\Lambda_n}$ is the corresponding translate of the operator h, then $\|[h^{(n)}, a]\| \to 0$, and consequently $\tau_t^h(a) \to a_t$. The question of whether the associated KMS states $w_{h^{(y)}}$ likewise converge to the unperturbed w depends on whether the KMS states are extremal. This is illustrated even in the finite-dimensional case by

Example (3.3.6)

With the notation of (1.1.1), let $\mathscr{A}$ be generated by $\{\mathbf{1}, \sigma_1, \sigma_1^\pm, \sigma_2\}$, and suppose that these observables evolve in time into $\{\mathbf{1}, \sigma_1, \exp(\mp 2it)\sigma_1^\pm, \sigma_2\}$. This time-evolution has a unitary representation as $U_t = \exp(it(\sigma_1 + c\sigma_2))$ for all $c \in \mathbb{R}$, so there is a one-parameter family of KMS states with density matrix $\rho = \exp(-\beta(\sigma_1 + \mu\sigma_2))$, which is not extremal, because

$$\exp(-\beta\mu\sigma_2) = \exp(-\beta\mu)\mathbf{1} \otimes \begin{vmatrix} 1 & 0 \\ 0 & 0 \end{vmatrix} + \exp(\beta\mu)\mathbf{1} \otimes \begin{vmatrix} 0 & 0 \\ 0 & 1 \end{vmatrix},$$

and $\exp(-\beta\sigma_1) \otimes \left(\begin{smallmatrix} 1 & 0 \\ 0 & 0 \end{smallmatrix}\right)$ provides a KMS state.

Although adding $h^{(n)} = (1/n)\sigma_1 + c'\sigma_2$ to the Hamiltonian leads to the same time-evolution as $n \to \infty$, the KMS state is different. Only the extremal KMS states provide two-dimensional representations, for which this can not happen.

Infinite systems generically have the property known as

Spatially Asymptotic Dynamical Stability (3.3.7)

Let $\mathscr{A}$ be a quasilocal algebra and w be a locally normal KMS *state on $\mathscr{A}$. The state w is an extremal* KMS *state iff for each sequence $h^{(n)}$ of perturbations such that $\|h^{(n)}\|$ and $\|h_i^{(n)}\|$ are bounded in n and $\tau^{h^{(n)}}(a) \to a_t$ for all $a \in \mathscr{A}$, the sequence $w^{(n)} \equiv w_{h^{(n)}} \to w$ converges in the weak-* sense to w.*

Remarks (3.3.8)

1. The assumption that $\mathscr{A}$ is quasilocal (1.3.3; 8) serves to guarantee the existence of suitable sequences $h^{(n)}$.
2. If $\mathscr{A}$ is also asymptotically Abelian in time, then the following propositions are equivalent for KMS states (recall Figure 27):

   (a) w is an extremal KMS state;
   (b) π_w is a factor;
   (c) $\lim_{t\to\infty} w(ab_t) = w(a)w(b)$;
   (d) $w_{h^{(n)}} \to w$ for all $h^{(n)}$ as described in (3.3.7).

Proof

1. If w is extremal, then $w^{(n)} \to w$: By assumption $\|h_i^{(n)}\|$ are bounded uniformly in n, so the same is true of the norms of $R_{h^{(n)}}$. Since, moreover, $w(R_{h^{(n)}}) \geq \exp(-\|h^{(n)}\|)$,

$$\rho_n = \frac{R_{h^{(n)}}}{w(R_{h^{(n)}})}$$

is a bounded sequence of operators. Bounded sequences of operators are weakly relatively compact ([33], VI; 9.6), and the set of states is weak-* compact (III: 2.1.23; 2), so there is a subsequence $h^{(k)}$, $k \in \mathbb{I} \subset \mathbb{N}$, such that $\bar{w} = \lim w^{(k)}$ and $\rho = \lim \rho_k$ exist, and $\bar{w}(a) = w(a\rho)$.
The automorphisms converge by assumption, and by Problem (3.2.17; 1) $\bar{w}$ is τ-KMS. But this means that ρ belongs not only to $\pi_w(\mathscr{A})''$ (by construction), but also to $\pi_w(\mathscr{A})'$ and thus belongs to the center:

$$w(a\rho b) = w(b_{-i}a\rho) = \bar{w}(b_{-i}a) = \bar{w}(ab) = w(ab\rho)$$

and

$$w(a\rho bc) = w(ab\rho c).$$

However, π_w is a factor, so $\rho = 1$, and since $\bar{w} = w$ is the only point of accumulation it is the limit of $w^{(n)}$.

2. Suppose now that w is not extremal. There is a nontrivial invariant element $z = z^*$ in the center of $\pi_w(\mathscr{A})''$. By Kaplansky's theorem [4] the unit ball of $\mathscr{A}$ is strongly dense in the unit ball of $\mathscr{A}''$, so z belongs to the closure of a bounded set of self-adjoint operators h of $\mathscr{A}$. Because of the locality assumption the closure of $\mathscr{A}_\Lambda|\Omega\rangle$ is a separable subspace of

$$\mathscr{H} = \overline{\mathscr{A}|\Omega\rangle} = \overline{\bigcup_n \mathscr{A}_{\Lambda(n)}|\Omega\rangle} \qquad (\Lambda(n) \to \mathbb{R}^3),$$

so $\mathscr{H}$ is also separable. As a consequence the strong topology on bounded sets of operators is metrizable, so z is actually the limit of some sequence $h^{(n)}$ in $\bigcup_n \mathscr{A}$. According to (3.3.4) τ_t^h converges to $\tau_t^z = \tau_t^0$. As in (3.2.6(v)) ρ_n can be constructed with the $h^{(n)}(f)$, as they converge to $z_t = z(f) = z$, just like $h_t^{(n)}(f)$ and $h_{is}^{(n)}(f)$. By the dominated convergence theorem it follows that

$$\lim_{n\to\infty} R_{h^{(n)}(f)} = R_z = \exp(-z),$$

and therefore

$$\lim w_{h^{(n)}(f)}(a) = \frac{w(\exp(-z)a)}{w(\exp(-z))}$$

is a KMS state different from w. □

The next topic is that of stability properties that can distinguish the extremal KMS states from other stationary states giving rise to factors. As shown by (3.3.4), if there is an extremal KMS state, then for all $h \in \mathscr{A}$ there exists a state that is stationary under the time-evolution including h as a perturbation, and which transforms continuously into the unperturbed state as $h \to 0$. It is not obvious that such a "linear-response theory" is possible. In fact, we learned (I, §3.3) that even in classical physics there are constants of motion that are not continuous in a parameter of the Hamiltonian. A density in phase space that is a function of such a constant will be unstable when perturbed, no matter by how little. This phenomenon is illustrated in quantum mechanics by the trivial

Example (3.3.9)

$\mathscr{H} = \mathbb{C}^2, H = 0 \in \mathscr{B}(\mathbb{C}^2)$. Every density matrix ρ corresponds to a stationary state, but with the perturbation $h = \mathbf{n} \cdot \boldsymbol{\sigma}$ the only stationary density matrices are $\rho = \mathbf{1}/2 + \lambda \mathbf{n} \cdot \boldsymbol{\sigma}$, $\lambda < |\mathbf{n}|/2$. This shows that only the density matrix $\rho = \mathbf{1}/2$ goes continuously into a density matrix that is stationary under all possible perturbed time-evolutions.

The example illustrates that only density matrices of the form $f(H)$, which are proportional to the identity in each degeneracy space of H, adapt themselves well to arbitrary perturbations. Despite the possibility of diagonalizing any stationary density matrix simultaneously with H, there is no telling from stationariness alone how it might vary within a degeneracy space. A requirement that two independent systems be stable would impose an additional restriction on the function f such that $w = f(H)$. The existence of two subsystems shows up mathematically as a tensor product, so if $H = H_1 \otimes \mathbf{1} + \mathbf{1} \otimes H_2$, then we would require that $f(H_1 \otimes \mathbf{1} + \mathbf{1} \otimes H_2) = f(H_1) \otimes f(H_2)$. Since H_1 and H_2 commute, both H_i may be regarded as ordinary numbers in their common spectral representation. Since the only reasonable functions satisfying $f(x + y) = f(x)f(y)$ are of the form $f(x) = \exp(-\beta x)$, we are led to the canonical density matrix, if the H_i may have arbitrary real spectral values. Since our infinite systems are asymptotically Abelian with respect to translations, and thus come to resemble tensor products of independent systems, it is a reasonable expectation that the condition of stability for such systems characterizes the KMS states. It will now be seen that this is the case, given some assumptions.

Local Dynamical Stability (3.3.10)

Suppose that the algebra $\mathscr{A}$ is asymptotically Abelian with respect to τ^0, and let w be a stationary factor state, and hence mixing. The question is whether for any perturbed automorphism τ^h it is possible for there to be a unique state w_h that is invariant under τ^h and turns into w as $h \to 0$. The states

$$w_{\pm} = \lim_{t \to \pm\infty} w \circ \tau_t^h$$

are reasonable candidates for w_h. If the limits exist, they would be invariant under τ^h, and the uniqueness of w_h means that the limits are equal. If τ^h is expanded as in (3.3.4) and we use the invariance of w under τ^0, we obtain the

Stability Condition to First Order in h (3.3.11)

If an invariant factor state w on an algebra $\mathscr{A}$ asymptotically Abelian in time is stable against arbitrary perturbations in the sense stated above, then for all h and $a \in \mathscr{A}$,

$$\int_{-\infty}^{\infty} dt w([h, a_t]) = 0.$$

Remarks (3.3.12)

1. The assumption that $h \in \mathscr{A}$ means that we consider only local perturbations. The requirement that $\mathscr{A}$ be asymptotically Abelian makes the commutator $[h, a_t]$ vanish as $t \to \pm\infty$. Condition (3.3.11) requires, roughly speaking, that $w(i[h, a_t])$ is equally often positive and negative.
2. The physical significance of (3.3.11) is that to first order in h the scattering transformation is the identity in the representation π_w. This can be interpreted as meaning that w is a locally perturbed equilibrium state with respect to the time-automorphism τ^h and should become the equilibrium state as $t \to \pm\infty$, so there is no net change between $t = -\infty$ and $t = +\infty$. In the kinetic theory of gases this is reflected in the argument that collisions do not alter the equilibrium distribution.

Let us introduce the abbreviations

$$F_{ab}(t) = w(ba_t) - w(a)w(b)$$

and

$$G_{ab}(t) = w(a_t b) - w(a)w(b) \tag{3.3.13}$$

in order to exploit (3.3.11) more fully.

Consequences for the Correlation Functions

Condition (3.3.11) makes

$$\int_{-\infty}^{\infty} dt(F_{ab}(t) - G_{ab}(t)) = 0.$$

Under the assumptions of (3.3.10) we know that F and G tend to zero as $t \to \pm\infty$. In order to ensure that this integral and others to follow make sense, it will be assumed that the correlation functions F and G are integrable in time from $-\infty$ to $+\infty$, at least for a dense set $\mathscr{S} \subset \mathscr{A}$. Since they are bounded, they belong to all $L^p(\mathbb{R})$ for $1 \leq p \leq \infty$. The assumption holds, for example, for free fermions. It will also be assumed that the higher correlation functions decrease rapidly enough for elements of $\mathcal{S}$ that integrals and limits may be interchanged.

If the state is a factor state, then as $u \to \pm\infty$, $w(ab_u c_t\, d_{t+u} - c_t\, d_{t+u} ab_u)$ tends to $w(ac_t)w(b\, d_t) - w(c_t a)w(d_t b)$. Therefore

$$\int_{-\infty}^{\infty} dt(F_{ca}(t)F_{db}(t) - G_{ca}(t)G_{db}(t)) = 0$$

for all a, b, c, and $d \in \mathcal{S}$. Similarly, from considering what happens to $w([ab_u c_v, d_t e_{u+t} f_{v+t}])$ as $u \to \infty$ and as $v \to \infty$,

$$\int_{-\infty}^{\infty} dt(F_{da}(t)F_{cf}(t)F_{be}(t) - G_{da}(t)G_{cf}(t)G_{be}(t)) = 0$$

for all a, b, c, d, e, and $f \in \mathcal{S}$. Because F and G belong to L^1, their Fourier transforms $\tilde{F}$ and $\tilde{G}$ exist and are continuous. Then if a, b, c, d, e, and $f \in \mathcal{S}$, the last three equations imply that

$$\tilde{F}_{ab}(0) = \tilde{G}_{ab}(0),$$

$$\int dE\tilde{F}_{ab}(E)\tilde{F}_{cd}(-E) = \int dE\tilde{G}_{ab}(E)\tilde{G}_{cd}(-E),$$

and

$$\int dE_1\, dE_2\, \tilde{F}_{ab}(E)\tilde{F}_{cd}(E' - E)\tilde{F}_{ef}(-E')$$
$$= \int dE_1\, dE_2\, \tilde{G}_{ab}(E)\tilde{G}_{cd}(E' - E)\tilde{G}_{ef}(-E'). \tag{3.3.14}$$

We shall now see that these equations imply the KMS condition.

In order to arrive at the KMS condition in Fourier-transformed space, $\tilde{F}_{ab}(E) = \exp(\beta E)\tilde{G}_{ab}(E)$, information about the supports of $\tilde{F}$ and $\tilde{G}$ is needed. It is at least clear that they are contained in the spectrum of H: Let $a_t = U_t^{-1} a U_t$, $U_t = \exp(-iHt)$, writing H as in (1.3.5) in the representation determined by w. Then

$$w(ba_t) = \langle b^*\Omega | U_t^{-1} | a\Omega\rangle,$$

so if $E \neq 0$, then

$$\tilde{F}_{ab}(E) = \tilde{F}_{b^*a^*}(E)^* = \tilde{G}_{ba}(-E) = \langle b^*\Omega | \delta(E - H) a\Omega\rangle. \tag{3.3.15}$$

This expression is to be interpreted in the spectral representation of H, in which the functions depend continuously on E when a and $b \in \mathcal{S}$.

In order to draw more far-reaching conclusions from these relationships, more information is needed about the energy spectrum. It would simply be additive if the Harmiltonian were the tensor product of Hamiltonians of independent systems: If H_1 and H_2 have eigenvalues $e_n^{(1)}$ and $e_n^{(2)}$, then

$H^{(1)} \otimes \mathbf{1} + \mathbf{1} \otimes H^{(2)}$ has eigenvalues $e_n^{(1)} + e_m^{(2)}$. This fact generalizes to an infinite system provided that the system is asymptotically Abelian with respect to an automorphism, such as the translations, that commutes with the time-evolution.

The Additivity of the Spectrum of H (3.3.16)

Let H generate a time-evolution τ on a factor state w, and suppose that the system is asymptotically Abelian with respect to an automorphism σ such that $[\sigma, \tau] = 0$ and $w \circ \sigma = w$. If H has the spectral values E_1 and E_2, then $E_1 + E_2$ also belongs to the spectrum of H.

Proof

Given any neighborhoods U_i of E_i, $i = 1, 2$, by assumption there exist f_i such that

$$a_{f_i}|\Omega\rangle \equiv \int_{-\infty}^{\infty} dt a_t f_i(t)|\Omega\rangle \neq 0,$$

where the Fourier transforms $\tilde{f}_i$ have their supports in U_i. Since by Property (3.1.18; 4) $\|\sigma_s(a_{f_1})a_{f_2}|\Omega\rangle\|^2$ approaches

$$\|a_{f_2}|\Omega\rangle\|^2\|a_{f_1}|\Omega\rangle\|^2 \neq 0$$

as $s \to \infty$, there must be a sufficiently large s that this vector is nonzero. Since the vector is supported in $E_1 + E_2 + U_1 + U_2$ in the spectral representation of H for all s, there are spectral values in every neighborhood of $E_1 + E_2$. Since the spectrum is closed, $E_1 + E_2$ itself belongs to the spectrum.

□

Remark (3.3.17)

If the system is asymptotically Abelian with respect to τ, then of course it is possible to take $\tau = \sigma$. Since w provides a factor, according to Table I in this case $|\Omega\rangle$ is the only eigenvector, and H has no eigenvalues other than 0. Since the spectrum is additive, it is either $0 \cup [\pm c, \pm\infty)$ for some $c \geq 0$, or else $(-\infty, \infty)$. In the first case there is a ground state; we shall be concerned only with the second possibility.

Derivation of the KMS Condition (3.3.18)

Let E_0 be in the spectrum of H and f be a function of the kind described in (3.2.6(v)) with $\tilde{f}(E_0) = 1$, supp $\tilde{f} \subset I \supset E_0$. Then $U_f \equiv \int dt f(t) U_t \neq 0$, and there exists an $a \in \mathscr{S}$ such that $U_f a\Omega = a_f \Omega \neq 0$. The operator a_f belongs to $\mathscr{S}$ whenever a does, and the functions $\tilde{F}$ and $\tilde{G}$ constructed with a_f are also supported in I, because

$$\tilde{F}_{a_f b}(E) = \tilde{f}(E)\tilde{F}_{ab}(E),$$
$$\tilde{G}_{a_f b}(E) = \tilde{f}(E)\tilde{G}_{ab}(E).$$

Let $b = a_f^*$ and shrink I down to E_0; this makes $\tilde{F}$ and $\tilde{G}$ proportional to $\delta(E - E_0)$. If we normalize so that

$$\int_{-\infty}^{\infty} dE\tilde{F}_{a_f a_f^*}(E) = w(a_f^* a_f) - |w(a_f)|^2 = 1,$$

and if

$$\int_{-\infty}^{\infty} dE\tilde{G}_{a_f a_f^*}(E) = w(a_f a_f^*) - |w(a_f)|^2 \geq 0$$

converges to some $\Phi \in \mathbb{R}^+$ (possibly after passage to some subsequence), then, because of the continuity of $\tilde{F}$ and $\tilde{G}$, (3.3.14) yields

$$\tilde{F}_{cd}(E_0) = \Phi\tilde{G}_{cd}(E_0) \quad \text{for all } c \text{ and } d \in \mathscr{S}.$$

This also proves that Φ may not be either 0 or ∞. Since this is true for all $E_0 \in \operatorname{Sp} H = \mathbb{R}$, there exists a universal function $\Phi(E)$ such that

$$\tilde{F}_{cd}(E) = \Phi(E)\tilde{G}_{cd}(E).$$

It follows from (3.3.15) that

$$\Phi(-E) = \Phi(E)^{-1} = \Phi^*(-E),$$

and the functional form then follows from the last equation of (3.3.14):

$$\int dE\, dE'(1 - \Phi(E)\Phi(E' - E)\Phi(-E'))\tilde{G}_{ab}(E)\tilde{G}_{cd}(E' - E)\tilde{G}_{ef}(-E') = 0$$

implies that

$$\Phi(E)\Phi(E' - E)\Phi(-E') = 1 \quad \text{for all } E \text{ and } E' \in \mathbb{R}.$$

Because of the equation derived above this,

$$\Phi(E)\Phi(-E') = \Phi(E - E'),$$

and since Φ is continuous it therefore has the functional form

$$\Phi(E) = \exp(\beta E) \quad \text{for some } \beta \in \mathbb{R}.$$

This shows the KMS condition for the dense set $\mathscr{S}$. However, since it can be written with the aid of (3.3.15) in the form

$$\langle b^*\Omega | f(-H)a\Omega\rangle = \langle a^*\Omega | f(H)\exp(-\beta H)b\Omega\rangle$$

for any bounded, continuous $f(H)$, it clearly suffices to derive it on a dense set.

In sum, the foregoing argument has shown the

Equivalence of Dynamical Stability and the KMS Condition (3.3.19)

Suppose that the algebra $\mathscr{A}$ is asymptotically Abelian with respect to the time-evolution and that w is a stationary state creating a factor representation. If for all $h \in \mathscr{A}$ there exists a normal state w_1 for $\pi_w(\mathscr{A})''$ to first order in h, such that w and w_1 are both stationary to first order under the perturbed time-evolution, and if w has an absolutely integrable correlation function, then either w is a KMS *state, or else the spectrum of H is* $\{0\} \cup [\pm c, \pm\infty)$, *in which case w is the ground state.*

Remarks (3.3.20)

1. It does not follow from this argument that $\beta > 0$. This fact did not even emerge from our argument with the tensor product of finite systems.
2. It is hard to tell how much the result suffers from the sharpening of the hypothesis of asymptotic commutativity. All the hypotheses are satisfied by a system of free fermions, but with a Coulomb interaction it is not even known if they hold in weakened forms. To a certain extent our assumptions about decrease at infinity and the interchangeability of limits belong to the realm of unproven hopes.
3. This shows that stability to first order in h implies KMS. Conversely, we have seen that KMS implies stability to every order in h, which means that the higher orders contribute no new information in this respect.

Whereas all the perturbations considered until now have been independent of time, we shall now turn our attention to perturbations $h(t)$ depending explicitly on time; they would be due to interference from outside the system. The time-evolution will not have the group property, but it will still be a one-parameter family of automorphisms. Let us, as usual, start by studying finite systems, for which the automorphisms are implemented by the unitary transformations

$$U_t = T\exp\left[-i\int_0^t dt'(H + h(t'))\right] \tag{3.3.21}$$

(cf. (III: 3.3.6)).

The most important quality of a passive state for our purposes will be that a system in a passive state will have gained energy when the perturbation has been switched off.

The Passivity of a State (3.3.22)

Let us suppose that a finite system evolves under the influence of $H + h(t)$, where by definition $h(0) = h(\tau) = 0$. The Hamiltonian generates a unitary time-evolution (3.3.21), so the change in energy from $t = 0$ to $t = \tau$ in the state w is given by $\operatorname{Tr} \rho(U_\tau H U_\tau^{-1} - H)$. A state is said to be **passive** if the change in energy is positive for all self-adjoint $h \in \mathscr{B}(\mathscr{H})$, in which case $\operatorname{Tr} \rho U H U^{-1} \geq \operatorname{Tr} \rho H$ for all $U = U^{*-1} \in \mathscr{B}(\mathscr{H})$.

Examples (3.3.23)

1. The canonical density matrix. Let $\rho = \exp(-\beta H)/\operatorname{Tr} \exp(-\beta H)$ and $\sigma = U^{-1} \rho U$. From (2.2.22(ii)) we know that

$$0 \leq \operatorname{Tr} \sigma(\ln \sigma - \ln \rho) = \operatorname{Tr}(\rho - \sigma)\ln \rho = -\beta \operatorname{Tr}(\rho - U^{-1}\rho U)H,$$

   so the system is passive.
2. Negative temperatures. Let ρ be as above, but $\beta < 0$. In order for $\operatorname{Tr} \exp(-\beta H)$ to be finite, H must be bounded from above; this would be realistic for a spin system. The inequality is then reversed, $\operatorname{Tr}(\rho - U^{-1}\rho U)H > 0$, so the system is not passive.

Remarks (3.3.24)

1. If it is desired to keep the energy $E = F + TS$ from increasing, the best tactic is to keep S constant (when $T > 0$). Our unitary time-evolution manages this automatically, and so the change in the energy E equals the change in the free energy F. Since the free energy is minimized with the canonical density matrix ρ, in the state ρ the only possibility is for E to increase, so ρ is passive.
2. Obviously, passivity requires the states of lower energy to be more densely occupied, so that the system is ready to gain energy. This is not the case when $\beta < 0$, in which circumstances the system would prefer to give energy away.

The General Form of Passive Density Matrices for Finite Systems (3.3.25)

A density matrix ρ on a finite system corresponds to a passive state if and only if

(i) $[\rho, H] = 0$; *and*
(ii) *if ρ_i and e_i designate respectively the ordered eigenvalues of ρ and H, then*

$$(e_i - e_k)(\rho_i - \rho_k) \leq 0.$$

Remarks (3.3.26)

1. The condition on the eigenvalues means that if the kth eigenvalue of H is greater than the ith, then the kth eigenvalue of ρ must be less than or equal to the ith. However, it is not necessary for ρ to be simply a function

of H, since in a degeneracy space for which $e_i = e_k$ it may happen that $\rho_i \neq \rho_k$.

2. The physical implication of the monotony is that lower-lying states are more densely occupied. On the other hand it implies nothing for the values of ρ where H does not vary:

$$H = \begin{pmatrix} 0 & & \\ & 0 & \\ & & 1 \end{pmatrix}, \quad \rho = \begin{pmatrix} \frac{1}{4} & & \\ & \frac{1}{2} & \\ & & \frac{1}{4} \end{pmatrix},$$

is passive.

Proof

(i) and (ii) $\Rightarrow$ passive $\Leftrightarrow \operatorname{Tr} \rho H \leq \operatorname{Tr} \rho U H U^{-1}$.
Let U be given in a matrix representation in the common eigenvectors of H and ρ as U_{ik}. The matrix $|U_{ik}|^2$ is doubly stochastic and therefore a convex combination of permutation matrices or a limit of such matrices (cf. (2.1.11; 4)). For any such matrix,

$$\operatorname{Tr} \rho U H U^{-1} = \sum_{i,k} e_i \rho_k \|U_{ik}\|^2 = \sum_P c_P \sum_i e_i \rho_{P_i},$$

where $\sum_P c_P = 1$, $c_P \geq 0$, and $\{P_i\}$ is a permutation of the $i \in \mathbb{Z}^+$. If $e_i < e_k$ implies that $\rho_i \geq \rho_k$, then for any permutation, $\sum_i e_i \rho_{P_i} \geq \sum_i e_i \rho_i = \operatorname{Tr} \rho H$.
Passive $\Rightarrow$ (i) and (ii). Suppose that $\operatorname{Tr} \rho U H U^{-1}$ has its minimum at $U = \mathbf{1}$, and write $U = \mathbf{1} + M_1 + M_2 + \cdots$, where $\|M_k\| < \varepsilon^k$ for sufficiently small ε. Then $\operatorname{Tr} \rho U H U^{-1} = \operatorname{Tr} \rho H + \operatorname{Tr}([H, \rho] M_1) + O(\varepsilon^2)$. The operator M_1 only needs to satisfy the condition that $M_1^* = -M_1$, and since $[\rho, H]$ is anti-Hermitian, it must equal zero, as otherwise the energy could be lowered. In order to prove (ii), choose U to have the form

$$U = \begin{pmatrix} \cos \varphi & \sin \varphi \\ -\sin \varphi & \cos \varphi \end{pmatrix}$$

on the subspace spanned by v_i and v_k, the eigenvectors with eigenvalues e_i, ρ_i and e_k, ρ_k. Then

$$\operatorname{Tr} \rho U H U^{-1} - \operatorname{Tr} \rho H = -(e_i - e_k)(\rho_i - \rho_k) \sin^2 \varphi,$$

which is positive only if $(e_i - e_k)(\rho_i - \rho_k) \leq 0$. □

In order to progress beyond the monotonic property to the statement that the function is exponential we must investigate infinite systems. We may either construct the infinite system by taking tensor products of copies of finite systems or go directly to the analysis of some asymptotically Abelian system. As before, the limiting case $\beta = \infty$, i.e., the ground state, would require a special treatment, which we shall not go into. Assuming therefore that β is finite, we can state the main proposition on the

Passivity of Infinite Systems (3.3.27)

Within the set of faithful factor states w on a C^* algebra with a time-automorphism τ and another automorphism commuting with τ and under which w is invariant and asymptotically Abelian, the passive states are precisely the KMS states, for any $\beta \geq 0$.

Remarks (3.3.28)

1. Translations of a homogeneous infinite system commute with the time-evolution. Since the local field algebra is asymptotically Abelian with respect to translations, this theorem can be used even if it is not known whether the time-evolution is asymptotically Abelian.
2. The sign of β is fixed by passivity, though of course its value is not.
3. To ensure that H is well-defined, assume that the time-evolution can be represented unitarily; then passivity is equivalent to the property that $w(U^{-1}HU - H) \geq 0$ for all unitary $U \in \mathscr{A}$.
4. Since the condition for passivity is linear in w, the passive states form a convex set. Passivity does not single out the extremal KMS states. We shall consider only factor states, which can not be decomposed further, as shown in §3.1.

Proof

Passive $\Rightarrow$ KMS. If the condition of passivity for an infinite system is written as $w(UHU^{-1}) \geq w(H)$, and we choose $U = \exp(i\varepsilon a)$ for a self-adjoint, then the first two terms of the expansion in powers of ε lead to

(i) $w([a, H]) = 0$ for all $a \in \mathscr{A}$, and
(ii) $w([a, [H, a]]) \geq 0$ for all $a \in \mathscr{A}$.

Equation (i) means that $(\partial/\partial t)w(a_t) = 0$, so w is stationary. In order to deduce the KMS condition from (ii) we employ the modular automorphism of w—call its generator $\bar{H}$. The KMS condition with respect to $\bar{H}$ can be used to write (ii) as

$$\begin{aligned} 0 &\leq \langle\Omega|2aHa - Ha^2 - a^2H|\Omega\rangle \\ &= \langle\Omega|2aHa - a\exp(-\bar{H})Ha - aH\exp(-\bar{H})a|\Omega\rangle \\ &= 2\langle\Omega|aH(\mathbf{1} - \exp(-\bar{H}))a|\Omega\rangle. \end{aligned}$$

In the last step we used the fact that $[H, \bar{H}] = 0$, in accordance with our assumption. Since the inequality holds for all $a = a^* \in \mathscr{A}$, it follows that $H(\mathbf{1} - \exp(-\bar{H})) \geq 0$. This means that in the common spectral representation of H and $\bar{H}$ the spectrum is restricted to the hatched region of the $(H, \bar{H})$-plane shown in Figure 28. Now the existence of the commuting,

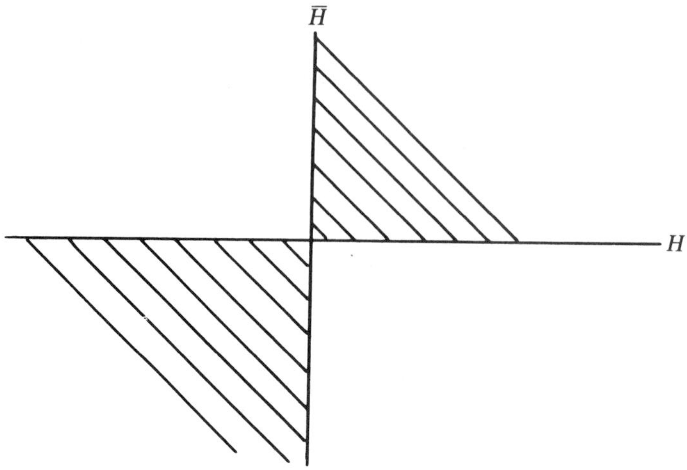

Figure 28 Possible location of the spectra of H and $\overline{H}$.

asymptotically Abelian automorphisms comes into play. According to (3.3.16), this implies that the spectrum is additive, i.e., if $(h_1, \overline{h_1})$ and $(h_2, \overline{h_2})$ are in the spectrum, then so is $(h_1 + h_2, \overline{h_1} + \overline{h_2})$. As a consequence the spectrum can at most be on a line through (0, 0), so $\overline{H} = \beta H$ for some $\beta > 0$.
KMS $\Rightarrow$ passive. Since $x \geq 1 - \exp(-x)$,

$$w(UHU^{-1}) \geq w(UU^{-1}) - w(U \exp(-H)U^{-1}) = w(UU^{-1}) - w(U^{-1}U) = 0. \qquad \square$$

Remarks (3.3.29)

1. The last inequality proved above is only the first of a whole family of inequalities that the expectation values in KMS states satisfy, and which completely characterize the KMS states [24]. They generalize trace inequalities, which are not directly applicable to infinite systems, since $\exp(-\beta H)$ is not trace-class.
2. Example (3.3.23; 1) showed that for finite systems, passivity follows from thermodynamic stability, or, in other words, from the minimum property of the free energy. This fact generalizes to infinite systems, for many of which the implication goes both ways, KMS $\Leftrightarrow$ thermodynamic stability, for instance for lattice systems with finite-range interactions. For these systems KMS is equivalent to global thermodynamic stability, provided that only translation-invariant states are considered, and that the free energy is interpreted as the free-energy density. However, for systems with long-range forces there exist KMS states that do not minimize the free energy; they are instead metastable, minimizing the free energy only one some reduced set of comparison states. Since the free energy is a convex functional on the states, it can not have a relative minimum on the set of all states that fails to be absolute.

3. The state $w_{\beta_1} \otimes w_{\beta_2}$ of two independent systems at different temperatures $T_1 > T_2$ is KMS with respect to the automorphism generated by $\beta_1 H_1 + \beta_2 H_2$. A perturbation $h(t)$ can cause the temperatures to equalize, and it may happen that the first system will have given up a positive amount of energy $\Delta E_1 \equiv E_1(0) - E_1(\tau) > 0$ by the end of the period. However, because the state is passive, $\beta_1 \Delta E_1 + \beta_2 \Delta E_2 \leq 0$, and the change in the total energy $\Delta E = \Delta E_1 + \Delta E_2$ is bounded by $\Delta E/\Delta E_1 \leq (T_1 - T_2)/T_1$. Since the total entropy remains constant under the unitary time-evolution, ΔE is the amount of energy provided by the total system, and this inequality is Carnot's classical bound on the thermal efficiency.

Another way to characterize the KMS states of an infinite system is known as reservoir stability, and it further justifies the physical interpretation of β as the reciprocal of the temperature. In outline it means that the KMS states are precisely the states that are suitable for thermal reservoirs, allowing the temperature $1/\beta$ to be defined. A more careful formulation states that if the reservoir is coupled to a finite system in the canonical state w, then in the weak-coupling limit w is invariant under the resulting semigroups (cf. (3.1.12)) for a reasonable class of couplings iff the reservoir is in a KMS state [24].

Problems (3.3.30)

1. Show that $w(R_h^* a) = w(aR_h) = w(S_h^* a S_h)$.
2. Estimate the length of time for which the "linear-response theory" remains valid; i.e., estimate

$$\left\| \tau_t^h(a) - a_t - i \int_0^t dt_1 [h_{t_1}, a_t] \right\|$$

3. Use the methods of §2.1 to conclude from $e_i > e_j \Rightarrow \rho_i \leq \rho_j$ that

$$\sum_i e_i \rho_i \leq \sum_i e_i \rho_{P_i}$$

for every permutation P.

Solutions (3.3.31)

1. Since H exists in the GNS representation with w, Equations (3.3.1) are applicable. The invariance of Ω holds also for complex z,

$$\exp(zH)|\Omega\rangle = |\Omega\rangle, \qquad R_h|\Omega\rangle = \exp(-H - h)|\Omega\rangle.$$

Now use the KMS condition for w in the form $w(ab) = \langle \Omega | b \exp(-H) a | \Omega \rangle$:

$$w(aR_h) = \langle \Omega | R_h \exp(-H) a | \Omega \rangle = \langle \Omega | \exp(-H - h) a | \Omega \rangle = w(R_h^* a).$$

It is also true in this representation that $S_h \exp(-H) S_h^* = \exp(-H - h)$, so

$$w(R_h^* a) = \langle \Omega | S_h \exp(-H) S_h^* a | \Omega \rangle = w(S_h^* a S_h).$$

2. Apply Taylor's formula $\|f(\alpha) - f(0) - \alpha f'(0)\| \leq \|\int_0^1 d\zeta(1 - \zeta) f''(\alpha\zeta)a^2\|$ to f: $[0, 1] \to \mathscr{B}(\mathscr{H})$, $\alpha \to \tau_t^{\alpha h}(a)$. According to (3.3.4),

$$\left\| \frac{\partial^2}{\partial\alpha^2} \tau_t^{\alpha h}(a)\bigg|_{\alpha=0} \right\| = \left\| 2\int_0^t dt_2 \int_0^{t_2} dt_1 [h_{t_1}, [h_{t_2}, a_t]] \right\| \leq 4t^2 \|h\|^2 \|a\|.$$

This is also true when $\alpha \in [0, 1]$; the only change when $\alpha > 0$ is that the time-evolution $a, h \to a_t, h_t$ becomes $a, h \to \tau_t^{\alpha h}(a), \tau_t^{\alpha h}(h)$, which does not affect the norms. Consequently the answer is that $\|\cdots\| \leq (2t\|h\|)^2\|a\|/2$. Recall that if $\|h\|$ is on the order of a Rydberg, then $t\|h\| \ll 1$ when $t \ll 10^{-15}$ sec. Therefore this *a priori* estimate guarantees only that the linear approximation remains valid for times on the atomic scale, and not for times measured in seconds. To go further would require knowing that the commutators go to zero for longer than macroscopic times.

3. Order e_i and ρ_i; then

$$\begin{aligned} e_1\rho_1 + e_2\rho_2 + e_3\rho_3 + \cdots = {} & (e_1 - e_2)\rho_1 + (e_2 - e_3)(\rho_1 + \rho_2) \\ & + (e_3 - e_4)(\rho_1 + \rho_2 + \rho_3) + \cdots. \end{aligned}$$

All the summands are positive, and permuting the ρ_i can at most make the summands larger.

Physical Systems 4

4.1 Thomas–Fermi Theory

Among the best examples of large quantum systems are atoms and molecules with highly charged nuclei. Classical features arise in the limit $Z \to \infty$, $N \to \infty$, except that the Fermi statistics continue to have an important effect.

Matter around us and within us consists of electrons and atomic nuclei, which are governed by the laws of quantum mechanics. Relativistic effects arise only in the fine details (cf. III, §1), so the forces of primary relevance are electrostatic and, for cosmic bodies, gravistatic (nonrelativistic). Moreover, the precise nature of the atomic nuclei is of little consequence on the macroscopic scale, so they can be considered as point charges. In order to understand the gross features of matter we shall study a Hamiltonian

$$H_{\text{mat}} = \sum_{i=1}^{M} \frac{|\mathbf{p}_i|^2}{2m_i} + \sum_{i>j} \frac{(e_i e_j - \kappa m_i m_j)}{|\mathbf{x}_i - \mathbf{x}_j|} \tag{4.1.1}$$

for ordinary matter. The first important issue to confront is that of why macroscopic bodies behave classically; in what sense is the thermodynamic limit $N \to \infty$ equivalent to the classical limit $\hbar \to 0$? There are a variety of ways to pass to the limit $N \to \infty$. In this section we begin by letting the nuclear charge Z and the nuclear masses both tend to infinity, while continuing to neglect gravity. This will permit a rather explicit mathematical treatment, as the action is determined by an average field, and the single-particle model becomes exact. The same will be true in §4.2 when we deal with cosmic bodies, for which gravitation predominates. However, macroscopic bodies on the scale of humans are far from these limits: nuclear charges are for the most part small, and yet gravitation is of little importance. In this intermediate range of normal matter it would be too much to hope for an explicit solution. Section 4.3 will discuss this case, but the results will be confined to general existence theorems and rather crude bounds on the values of observables of physical interest.

Let us consider now what happens to electrons in the field of fixed point charges. In order not to be distracted from the most important facts by physical constants, we shall use units in which $\hbar = 2m = e = k = 1$, so that (4.1.1) becomes

The Hamiltonian for Normal Matter (4.1.2)

$$H_N = \sum_{i=1}^{N} |\mathbf{p}_i|^2 - \sum_{i=1}^{N} \sum_{k=1}^{M} \frac{Z_k}{|\mathbf{x}_i - \mathbf{X}_k|} + \sum_{i>j} \frac{1}{|\mathbf{x}_i - \mathbf{x}_j|} + \sum_{k>l} \frac{Z_k Z_l}{|\mathbf{X}_k - \mathbf{X}_l|} + \sum_{i=1}^{N} W(\mathbf{x}_i).$$

Remarks (4.1.3)

1. The notation follows that of (III: 4.6.9), that is, $\mathbf{x}_i$ and $\mathbf{p}_i$ are the position and momentum of the ith electron, $\mathbf{X}_k$ and Z_k are the position and charge of the kth fixed nucleus, N is the number of electrons, and M is the number of nuclei.
2. The Hamiltonian H operates on an n-fold antisymmetrized tensor product of $L^2(\mathbb{R}^3) \otimes \mathbb{C}^2$ = configuration space $\otimes$ spin of a given electron. The nuclear coordinates $\mathbf{X}_i$ commute with everything, and are to be regarded as ordinary 3-vectors of numbers.
3. It is usually most convenient to study the many-particle system in the framework of the field algebra (1.3.2). If $a_\alpha(\mathbf{x})$, $\alpha = 1, 2$, denote the annihilation operators of electrons with spin up ($\alpha = 1$) and spin down ($\alpha = 2$), then (4.1.2) reads

$$H = \sum_\alpha \int d^3x \left[\nabla a_\alpha^*(\mathbf{x}) \cdot \nabla a_\alpha(\mathbf{x}) + \left(\sum_{k=1}^{M} \frac{-Z_k}{|\mathbf{x} - \mathbf{X}_k|} + W(\mathbf{x}) \right) a_\alpha^*(\mathbf{x}) a_\alpha(\mathbf{x}) \right] + \sum_{\alpha,\beta} \frac{1}{2} \int d^3x\, d^3x' \frac{a_\alpha^*(\mathbf{x}) a_\beta^*(\mathbf{x}') a_\beta(\mathbf{x}') a_\alpha(\mathbf{x})}{|\mathbf{x} - \mathbf{x}'|} + \sum_{k>l} \frac{Z_k Z_l}{|\mathbf{X}_k - \mathbf{X}_l|}.$$

4. If the temperature is finite, then the attraction of the nuclei is not strong enough to prevent the electrons from escaping to infinity, and the system must be imagined confined to a box. The box can be represented by a potential W, adding a term $\sum_\alpha \int d^3x W(\mathbf{x}) a_\alpha^*(\mathbf{x}) a_\alpha(\mathbf{x})$ to H. The wall potential W will be chosen to be the v_L of (2.5.23).

Most interesting systems are approximately neutral, so N is assumed to be about $\sum_{k=1}^{M} Z_k$. The thermodynamic limit $N \to \infty$ can consist either in

$M \to \infty$ or $Z_k \to \infty$. For the moment consider the latter case; the limit $M \to \infty$ will be studied in §4.3. The first step is to bound the grand canonical partition function in terms of the grand canonical partition function of a theory with free electrons in an external field. This means that the bounds of (III, §4.5) for the energies have to be generalized for arbitrarily complex systems at nonzero temperatures. After that we shall show that the upper and lower bounds coalesce (when properly scaled) as $Z_k \to \infty$, so the partition function can be calculated exactly in the thermodynamic limit. Finally, the limit of the grand canonical state will be analyzed.

Upper Bounds for the Partition Function (4.1.4)

These correspond to lower bounds for the Hamiltonian like those derived in (III: 4.5.20). The inequality (III: 4.5.24), though, is not well suited to our current purposes, and must be replaced with a variant, which will appear as a by-product of Thomas–Fermi theory in (4.1.46; 2). In it the Coulomb repulsion of the electrons is replaced by their energy in an external field:

$$\sum_{i>j=1}^{N} |\mathbf{x}_i - \mathbf{x}_j|^{-1} \geq \sum_{i=1}^{N} \int \frac{d^3x n(\mathbf{x})}{|\mathbf{x}_i - \mathbf{x}|} - \frac{1}{2} \int \frac{d^3x\, d^3x'}{|\mathbf{x} - \mathbf{x}'|} n(\mathbf{x})n(\mathbf{x}') - 3.68\gamma N$$
$$- \frac{3}{5\gamma} \int d^3x n^{5/3}(\mathbf{x})$$
$$\text{for all } \mathbf{x}_i \in \mathbb{R}^3, \quad \gamma > 0, \quad n \in L^1(\mathbb{R}^3) \cap L^{5/3}(\mathbb{R}^3). \qquad (4.1.5)$$

This yields a bound on the expression in (4.1.3; 3), which is quartic in the a's, in terms of a quadratic expression,

$$\frac{1}{2} \sum_{\alpha,\beta} \int \frac{d^3x\, d^3x'}{|\mathbf{x} - \mathbf{x}'|} a_\alpha^*(\mathbf{x})a_\beta^*(\mathbf{x}')a_\beta(\mathbf{x}')a_\alpha(\mathbf{x}) \geq \sum_\alpha \int d^3x a_\alpha^*(\mathbf{x})a_\alpha(\mathbf{x})$$
$$\times \left[\int \frac{d^3x' n(\mathbf{x}')}{|\mathbf{x} - \mathbf{x}'|} - 3.68\gamma\right] - \frac{1}{2} \int \frac{d^3x\, d^3x'}{|\mathbf{x} - \mathbf{x}'|} n(\mathbf{x}')n(\mathbf{x}) - \frac{3}{5\gamma} \int d^3x n^{5/3}(\mathbf{x}).$$

Consequently, H is bounded by a

Hamiltonian with an Effective Field (4.1.6)

$$H - \mu N \geq H_n - C_n + \sum_{k>l} \frac{Z_k Z_l}{|\mathbf{X}_k - \mathbf{X}_l|},$$

where

$$H_n \equiv \sum_\alpha \int d^3x \bigg\{ \nabla a_\alpha^*(\mathbf{x}) \cdot \nabla a_\alpha(\mathbf{x}) + a_\alpha^*(\mathbf{x})a_\alpha(\mathbf{x})$$
$$\times \left[-\sum_k \frac{Z_k}{|\mathbf{x} - \mathbf{X}_k|} + \int \frac{d^3x' n(\mathbf{x}')}{|\mathbf{x} - \mathbf{x}'|} + W(\mathbf{x}) - \mu - 3.68\gamma \right] \bigg\}$$
$$- \frac{1}{2} \int \frac{d^3x\, d^3x'}{|\mathbf{x} - \mathbf{x}'|} n(\mathbf{x})n(\mathbf{x}'),$$

and

$$C_n = \frac{3}{5\gamma}\int d^3x n^{5/3}(\mathbf{x}).$$

Remarks (4.1.7)

1. Although Inequality (4.1.5) holds for any $n(\mathbf{x})$, the optimal choice identifies it with the electron density. Thus the effective potential in the square brackets $[\cdots]$ consists of the attraction to the nuclei, the repulsion from other electrons, and the chemical potential. However, this interpretation counts the electron repulsion twice, as in $\sum_{i\neq k}|\mathbf{x}_i - \mathbf{x}_k|^{-1}$. The last term in H_n corrects this overcounting.
2. The correlations among the electrons due to their Fermi statistics have the effect of reducing their repulsion. Also, H_n contains the self-energy of the individual electrons. The constant C_n and $-3.68\gamma N$ serve to control any possible effect from these corrections.

The monotonic property (2.1.7; 4) translates (4.1.6) into an inequality for the partition function. Then with the aid of the maximum principle of (2.5.16; 2) the inequality can be expressed as the supremum of an expression linear in n.

The Partition Function with an Effective Field (4.1.8)

$$\Xi(H - \mu N) \equiv T \ln \operatorname{Tr} \exp[-\beta(H - \mu N)] \leq \Xi\left(H_n - C_n + \sum_{k>l}\frac{Z_k Z_l}{|\mathbf{X}_k - \mathbf{X}_l|}\right)$$

$$\leq \Xi(H_n) + C_n - \sum_{k>l}\frac{Z_k Z_l}{|\mathbf{X}_k - \mathbf{X}_l|},$$

$$\Xi(H_n) = \operatorname{tr} 2\ln(\mathbf{1} + \exp(-\beta h_n)) + \frac{1}{2}\int d^3x\, d^3x'\, \frac{n(\mathbf{x})n(\mathbf{x}')}{|\mathbf{x} - \mathbf{x}'|}$$

$$= \sup_{\rho_1} 2\operatorname{tr}\{T(-\rho_1 \ln \rho_1 - (\mathbf{1} - \rho_1)\ln(\mathbf{1} - \rho_1)) - \rho_1 h_n\}$$

$$+ \frac{1}{2}\int d^3x\, d^3x'\, \frac{n(\mathbf{x})n(\mathbf{x}')}{|\mathbf{x} - \mathbf{x}'|},$$

$$h_n = |\mathbf{p}|^2 - \sum_k \frac{Z_k}{|\mathbf{x} - \mathbf{X}_k|} + \int \frac{d^3x\, n(\mathbf{x}')}{|\mathbf{x} - \mathbf{x}'|} + W(\mathbf{x}) - \mu - 3.68\gamma.$$

Remarks (4.1.9)

1. The Hamiltonian h_n of one particle in the effective field acts on the space $\mathscr{H}_1 = L^2(\mathbb{R}^3)$. Spin is accounted for by the factor 2, and tr denotes the trace on $\mathscr{H}_1$.

2. As in Remark (2.5.16; 2), $\sup_{\rho_1}$ denotes the supremum over one-particle density matrices ρ_1 such that

$$0 \leq \rho_1 \leq 1, 2 \operatorname{tr} \rho_1 = N \equiv \left\langle \sum_\alpha \int d^3x a_\alpha^*(\mathbf{x}) a_\alpha(\mathbf{x}) \right\rangle.$$

3. There exist $c_i \geq 0$ such that $h_n \geq c_1 |\mathbf{p}|^2 + W(\mathbf{x}) - c_2$. This ensures that $\operatorname{tr} \ln(\mathbf{1} + \exp(-\beta h_n)) < \infty$.

The next task is to optimize the upper bound. The infimum over n of $\Xi(H_n)$ is in fact achieved. This is a consequence of the

Properties of the Functional $\Xi(H_n)$ (4.1.10)

The mapping $n \to \Xi(H_n)$ from $\mathcal{N}$ to $\mathbb{R}^+$, where $\mathcal{N}$ is the real Hilbert space of measurable functions $\mathbb{R}^3 \to \mathbb{R}$ finite in the norm

$$\|n\|_c^2 = \langle n | n \rangle_c = \int \frac{d^3x \, d^3x' n(\mathbf{x}) n(\mathbf{x}')}{|\mathbf{x} - \mathbf{x}'|}$$

is

(i) *weakly lower semicontinuous*;
(ii) *strictly convex*; *and*
(iii) *greater than* $\frac{1}{2}\|n\|_c^2$.

Proof

(i) In the second version $\Xi(H_n)$ depends on n through $\operatorname{tr} \rho_1 h_n$ and $\|n\|_c$. The norm is $\sup_{n' \in \mathcal{N}, \|n'\|_c \leq 1} \langle n' | n \rangle_c$, and $\operatorname{tr}(\rho_1 \int n(\mathbf{x}') \, d^3x'/|\mathbf{x} - \mathbf{x}'|)$ is weakly continuous for

$$\rho_1 \in C_M \equiv \left\{ \rho_1 : \int \frac{d^3x_1 \, d^3x_2}{|\mathbf{x}_1 - \mathbf{x}_2|} \langle \mathbf{x}_1 | \rho_1 | \mathbf{x}_1 \rangle \langle \mathbf{x}_2 | \rho_1 | \mathbf{x}_2 \rangle < M \in \mathbb{R}^+ \right\}.$$

The supremum is attained when $\rho_1 = (\exp(\beta h_n) + \mathbf{1})^{-1}$, which belongs to some C_M. Hence $\sup_{\rho_1}$ may be written as $\sup_{M \in \mathbb{R}^+} \sup_{\rho_1 \in C_M}$. In this way $\Xi(H_n)$ is expressed as the supremum over continuous functions, which is always lower semicontinuous.

(ii) This follows in the first version of $\Xi(H_n)$, when it is observed that $h \to \operatorname{tr} \ln(\mathbf{1} + \exp(-\beta h))$ is convex, $n \to h_n$ is linear, and $n \to \|n\|_c^2$ is strictly convex.

(iii) This follows in the first form of $\Xi(H_n)$, since $\operatorname{tr} \ln(\mathbf{1} + \exp(-\beta h)) \geq 0$.

□

Corollaries (4.1.11)

1. Because of Property (iii), the infimum over n lies in a compact region where $\|n\|_c < C$. Property (i) means that it is attained at some n_0, which is unique because of (ii).

2. Because of the convexity, we know that the function $\mathbb{R} \to \mathbb{R}^+ : t \to \Xi(H_{n_0 + tn_1})$ has a right derivative everywhere, and the minimum is attained at n_0 if and only if

$$\lim_{t \downarrow 0} t^{-1}(\Xi(H_{n_0 + tn_1}) - \Xi(H_{n_0})) \geq 0 \quad \text{for all } n_1 \in \mathcal{N}.$$

Although convexity does not imply the existence of a derivative, analyticity can be proved by a variant of Theorem (2.4.7). Granting that, the formal rules for differentiating tr ln(1 + A) are justified:

$$\frac{d}{dt} \operatorname{tr} \ln(\mathbf{1} + \exp(-\beta h_{n_0 + tn_1}))|_{t=0} = -\operatorname{tr} \int \frac{d^3x' n_1(\mathbf{x}')}{|\mathbf{x} - \mathbf{x}'|} \frac{\beta}{\exp(\beta h_{n_0}) + \mathbf{1}}.$$

Therefore the minimum at n_0 is characterized by

$$\int \frac{d^3x'\, d^3x}{|\mathbf{x} - \mathbf{x}'|} n_0(\mathbf{x}) n_1(\mathbf{x}') = 2 \operatorname{tr} \int \frac{d^3x' n_1(\mathbf{x}')}{|\mathbf{x} - \mathbf{x}'|} \frac{1}{\exp(\beta h_{n_0}) + \mathbf{1}} \quad \text{for all } n_1 \in \mathcal{N}.$$

If n_1 is made to tend to $\Delta\delta(\mathbf{x} - \mathbf{x}_0)$, then there results an equation for $n_0(\mathbf{x}_0)$. Since the integral kernel $K(\mathbf{x}, \mathbf{x}')$ of $(\exp(\beta h_n) + \mathbf{1})^{-1}$ is analytic for $\mathbf{x}, \mathbf{x}' \neq \mathbf{X}_k$ even though $\Delta\delta$ does not belong to $\mathcal{N}$, we have the

Existence of the Self-Consistent Field (4.1.12)

The equation

$$n_o(\mathbf{x}) = 2\langle \mathbf{x} | (\exp(\beta h_{n_0}) + 1)^{-1} | \mathbf{x} \rangle$$

has a unique solution, which minimizes $\Xi(H_n)$.

Remarks (4.1.13)

1. Since $2\langle \mathbf{x} | (\exp(\beta h_{n_0}) + \mathbf{1})^{-1} | \mathbf{x} \rangle$ equals $\sum_\alpha \langle a_\alpha^*(\mathbf{x}) a_\alpha(\mathbf{x}) \rangle_{n_0}$, it is the mean electron density in the state determined by the one-particle Hamiltonian h_{n_0}.
2. The ease with which the existence of the solution of the generalized Hartree equation (4.1.12) was proved depended on the wall potential W. In an infinite space without W there fails to be a solution when $N > \sum_k Z_k$, even at absolute zero temperature—the electrons escape to infinity, and the infimum is never attained. This is a reflection of the general mathematical fact that a strictly convex function need not achieve its infimum on a noncompact region; for example $1/x$ never reaches the value 0 on $[1, \infty)$.
3. A convex function on a finite-dimensional space is continuous on the interior of its domain of definition. This is not always the case when the dimension of the space is infinite, and $\|n\|_c^2$ is in fact not weakly continuous: The norms $\| \;\|_c$ of the charge distributions $n_R(\mathbf{x}) = R^{-5/2}\Theta(R - |\mathbf{x}|)$ are

all equal, but $\int d^3x n_R(\mathbf{x}) \to 0$ as $R \to 0$. Consequently $\langle n_R | n \rangle_c \to 0$ for all n, if

$$V_n(x) \equiv \int \frac{d^3x' n(\mathbf{x}')}{|\mathbf{x} - \mathbf{x}'|} \in L^\infty(\mathbb{R}^3).$$

Since the n's such that $V_n \in L^\infty$ are dense in $\mathcal{N}$, $n_R \rightharpoonup 0$, even though $\|n_R\|_c \not\to 0$. There even exist convex functions that fail to be lower semicontinuous, for example the functional of (III: 2.1.15; 2). Of course the function $n \to \|n\|_c^2$ is continuous in norm, but this finer norm topology can not be used, because we need the compactness of bounded sets.

4. At the minimum (4.1.12), it is indeed true that $n(\mathbf{x}) > 0$ and $\int d^3x n(\mathbf{x}) = N$.

Lower Bounds for $\Xi(H)$ (4.1.14)

In (III, §4.5) upper bounds on the energy were provided by the min–max principle, the generalization of which for nonzero temperatures is the Peierls–Bogoliubov inequality (2.1.8; 3) with $\Xi = -F$. Because

$$\left\langle H - \bar{\mu}N - \sum_{k>l} \frac{Z_k Z_l}{|\mathbf{X}_k - \mathbf{X}_l|} - H_{n_0} \right\rangle_{n_0}$$

$$= \sum_{\alpha,\beta} \frac{1}{2} \int \frac{d^3x\, d^3x'}{|\mathbf{x} - \mathbf{x}'|} (\langle a_\alpha^*(\mathbf{x}) a_\beta^*(\mathbf{x}') a_\beta(\mathbf{x}') a_\alpha(\mathbf{x}) \rangle_{n_0} - n_0(\mathbf{x}) n_0(\mathbf{x}'))$$

$$= -\sum_{\alpha,\beta} \frac{1}{2} \int \frac{d^3x\, d^3x'}{|\mathbf{x} - \mathbf{x}'|} |\langle a_\alpha^*(\mathbf{x}) a_\beta(\mathbf{x}') \rangle_{n_0}|^2 \equiv -A(n_0) < 0,$$

where $\bar{\mu} = \mu - 3.68\gamma$, it implies that

$$\Xi(H - \bar{\mu}N) \geq \Xi(H_{n_0}) + A(n_0) - \sum_{k>l} \frac{Z_k Z_l}{|\mathbf{X}_k - \mathbf{X}_l|}.$$

When this is combined with (4.1.8), it yields

Two-Sided Bounds for Ξ (4.1.15)

$$0 \leq A(n_0) \leq \Xi(H - \bar{\mu}N) + \sum_{k>l} \frac{Z_k Z_l}{|\mathbf{X}_k - \mathbf{X}_l|} - \Xi(H_{n_0}) \leq \frac{3}{5\gamma} \int d^3x\, n_0^{5/3}(\mathbf{x}).$$

Remarks (4.1.16)

1. This means that the true partition function exceeds the partition function with an effective field by more than A but less than C.
2. In particular (4.1.15) states for the exchange energy that $0 \leq A \leq C$. If Z is large, then n_0 approaches the electron density in Thomas–Fermi Theory, and we shall discover that $\int n^{5/3} \sim Z^{7/3}$. If γ is chosen $\sim (Z^{7/3}/N)^{1/2}$, then C and the additional term $3.68\gamma N$ in μ becomes $\sim N^{1/2} Z^{7/6}$. Since H goes as $Z^{7/3}$, if $N \sim Z$, then the relative error is $O(Z^{-2/3})$.

The Classical Limit (4.1.17)

The next topic of study is the way in which $\Xi(H_n)$ approaches the classical phase-space integral (2.5.17) as $Z \to \infty$. According to the general considerations of (1.2.4) the interesting limit would be expected to be that in which the system shrinks as $Z^{-1/3}$. Consider, therefore, a sequence of Hamiltonians H_Z in which not only do the nuclear charges increase as $Z_k = Zz_k$, $\sum_k z_k = 1$, z_k fixed, but also the nuclear coordinates are scaled by changing $\mathbf{X}_k$ into $Z^{-1/3}\mathbf{X}_k$ and the wall potential varies at the same time:

$$\begin{aligned} H_Z \equiv \sum_\alpha \int d^3x \biggl[\nabla a_\alpha^*(\mathbf{x}) \cdot \nabla a_\alpha(\mathbf{x}) \\ + a_\alpha^*(\mathbf{x}) a_\alpha(\mathbf{x}) \left(-Z \sum_{k=1}^{M} \frac{z_k}{|\mathbf{x} - \mathbf{X}_k Z^{-1/3}|} + Z^{4/3} W(Z^{1/3}\mathbf{x}) \right) \biggr] \\ + \frac{1}{2} \sum_{\alpha,\beta} \int a_\alpha^*(\mathbf{x}) a_\beta^*(\mathbf{x}) a_\beta(\mathbf{x}') a_\alpha(\mathbf{x}') \frac{d^3x\, d^3x'}{|\mathbf{x} - \mathbf{x}'|} + \sum_{k>l} \frac{z_k z_l}{|\mathbf{X}_k - \mathbf{X}_l|} Z^{7/3}, \end{aligned}$$

$$\begin{aligned} H_{Z,n} \equiv \sum_\alpha \int d^3x \biggl\{ \nabla a_\alpha^*(\mathbf{x}) \cdot \nabla a_\alpha(\mathbf{x}) \\ + a_\alpha^*(\mathbf{x}) a_\alpha(\mathbf{x}) \biggl[-Z \sum_{k=1}^{M} \frac{z_k}{|\mathbf{x} - \mathbf{X}_k Z^{-1/3}|} + \int \frac{d^3x' n^Z(\mathbf{x}')}{|\mathbf{x} - \mathbf{x}'|} \\ + Z^{4/3}(W(\mathbf{x}Z^{1/3}) - \mu) - 3.68\gamma \biggr] \biggr\} \\ - \frac{1}{2} \int \frac{d^3x\, d^3x'}{|\mathbf{x} - \mathbf{x}'|} n^Z(\mathbf{x}) n^Z(\mathbf{x}'); \end{aligned}$$

$$n^Z = Z^2 n(Z^{1/3}\mathbf{x}).$$

In order always to work in a fixed volume and see what happens in the limit $Z \to \infty$, use a canonical transformation to convert the electron coordinates $\mathbf{x}$ into $Z^{-1/3}\mathbf{x}$ and $\mathbf{p}$ into $Z^{1/3}\mathbf{p}$ at the same time—this entails $a(\mathbf{x}) \to Z^{-1/2} a(Z^{-1/3}\mathbf{x}))$ as well. Since the number of electrons also grows as Z, the mean momentum of the electrons grows as $Z^{2/3}$, and every kind of energy per particle, such as T or μ, will depend in the same way on Z. Thus if we calculate $\operatorname{Tr} \exp[-\beta_Z(H_Z - \mu_Z N)]$ with $\beta_Z = Z^{-4/3}\beta$, and $\mu_Z = Z^{4/3}\mu$, and scale n appropriately, we are led to $\operatorname{tr} \ln(\mathbf{1} + \exp(-\beta h_n))$ with

$$h_n = Z^{-2/3}|\mathbf{p}|^2 - \sum_j \frac{z_j}{|\mathbf{x} - \mathbf{X}_j|} + \int \frac{d^3x' n(\mathbf{x}')}{|\mathbf{x} - \mathbf{x}'|} + W(\mathbf{x}) - \mu_\gamma$$

and

$$\mu_\gamma = \mu + Z^{-4/3} \cdot 3.68\gamma.$$

Observe that $Z^{-1/3}$ occurs in the position of $\hbar$, making the limit $Z \to \infty$ equivalent to the classical limit $\hbar \to 0$. Now use the bound (2.5.17) with

$$u^2(\mathbf{x}) = \frac{\kappa^3}{8\pi} \exp(-\kappa r).$$

The Fourier transform of this density is

$$\widetilde{u^2}(\mathbf{k}) = \frac{\kappa^4}{(|\mathbf{k}|^2 + \kappa^2)^2}.$$

Consequently $\int |\nabla u|^2 \, d^3x = \kappa^2$, and if $v = 1/r$, then

$$v_u(q) \equiv \int d^3x \frac{1}{|\mathbf{x}|} |u(\mathbf{x} - \mathbf{q})|^2 = \frac{1}{q} - \frac{\exp(-\kappa q)}{q} - \frac{\kappa}{2} \exp(-\kappa q) \equiv \frac{1}{q} - v_s(q).$$

The Classical Upper Bound (4.1.18)

Since $1/r$ can not be represented as a smeared potential, v^u makes no sense. Thus it is first necessary to remove v_s, the short-range, singular part of $1/r$, and handle it separately. It can be neglected as $\kappa \to \infty$, and if the smeared remainder is unsmeared, we recover $1/r$:

$$\frac{1}{r} = v_u + v_s, \qquad (v_u)^u = \frac{1}{r}.$$

Let h_c be like the h_n of (4.1.17), but with v_u in place of $1/r$. Then

$$h_n = h_c + V_s,$$

$$h_c = \int \frac{d^3q \, d^3p}{(2\pi)^3} |\mathbf{q}, \mathbf{p}\rangle\langle\mathbf{q}, \mathbf{p}| \left(Z^{-2/3}(|\mathbf{p}|^2 - \kappa^2) - \sum_j \frac{z_j}{|\mathbf{q} - \mathbf{X}_j|} \right.$$

$$\left. + \int \frac{d^3x n(\mathbf{x})}{|\mathbf{q} - \mathbf{x}|} + W^u(\mathbf{q}) - \mu_\gamma \right),$$

$$V_s = -\sum_j z_j v_s(\mathbf{x} - \mathbf{X}_j) + \int d^3y n(\mathbf{y}) v_s(\mathbf{x} - \mathbf{y}).$$

In the x-representation, $|\mathbf{q}, \mathbf{p}\rangle$ is

$$\left(\frac{\kappa^3}{8\pi}\right)^{1/2} \exp(i\mathbf{p} \cdot \mathbf{x}) \exp(-\kappa|\mathbf{x} - \mathbf{q}|),$$

and we let $W^u(\mathbf{x})$ be the unsmeared wall potential W of (2.5.23). Convexity can be appealed to to bound the influence of V_s:

$$\begin{aligned} \operatorname{tr} \ln(\mathbf{1} + \exp(-\beta h_n)) \le {} & \operatorname{tr} \ln(\mathbf{1} + \exp(-\beta h_c)) \\ & + \alpha^{-1} \operatorname{tr}[\ln(\mathbf{1} + \exp(-\beta(h_c + \alpha V_s))) \\ & - \ln(\mathbf{1} + \exp(-\beta h_c))] \qquad \text{for all } \alpha \ge 1. \end{aligned}$$

The number α will be picked so large that the addition to the first term on the right side goes away in the limit $Z \to \infty$. By (2.5.17), the second term is bounded by

$$\operatorname{tr}\ln(\mathbf{1} + \exp(-\beta h_c)) \le \int \frac{d^3q\, d^3p}{(2\pi)^3} \ln\left(\mathbf{1} + \exp\left[-\beta\left(Z^{-2/3}(|\mathbf{p}|^2 - \kappa^2) - \sum_j \frac{z_j}{|\mathbf{q} - \mathbf{X}_j|} + \int \frac{d^3x n(\mathbf{x})}{|\mathbf{q} - \mathbf{x}|} + W^u(\mathbf{q}) - \mu_\gamma\right)\right]\right)$$
$$= Z \int \frac{d^3q\, d^3p}{(2\pi)^3} \ln\left(\mathbf{1} + \exp\left[-\beta\left(|\mathbf{p}|^2 - \sum_j \frac{z_j}{|\mathbf{q} - \mathbf{X}_j|} + \int \frac{d^3x n(\mathbf{x})}{|\mathbf{q} - \mathbf{x}|} + W^u(\mathbf{q}) - \mu_\gamma - Z^{-2/3}\kappa^2\right)\right]\right).$$

The additional part containing V_s can be taken care of because even for a singular potential $V(\mathbf{x}) \in L^{5/2}(\mathbb{R}^3)$ there is a bound of this form weakened by a factor $C \cong 7$:

$$\operatorname{tr}\ln(\mathbf{1} + \exp[-\beta(|\mathbf{p}|^2 + V(\mathbf{x}))])$$
$$\le c \int \frac{d^3p\, d^3q}{(2\pi)^3} \ln(1 + \exp[-\beta(|\mathbf{p}|^2 + V(\mathbf{q}))]). \tag{4.1.19}$$

The derivation of this formula is left for Problems 1 and 2. In this case it leaves us with

$$\operatorname{tr}\ln(\mathbf{1} + \exp[-\beta(h_c + \alpha V_s)])$$
$$\le cZ \int \frac{d^3q\, d^3p}{(2\pi)^3} \ln\left(1 + \exp\left[-\beta(|\mathbf{p}|^2 + W^u(\mathbf{q}) - \sum_j z_j\left(\frac{1}{|\mathbf{q} - \mathbf{X}_j|} + (\alpha - 1)v_s(\mathbf{q} - \mathbf{X}_j)\right) + \int d^3y n(\mathbf{y})\left(\frac{1}{|\mathbf{q} - \mathbf{y}|} + (\alpha - 1)v_s(\mathbf{q} - \mathbf{y})\right) - \mu_\gamma - Z^{-2/3}\kappa^2\right)\right]\right).$$

It remains to be shown that α and κ can be sent to infinity with Z in such a way that the additions to the classical one-particle potential in the effective field become negligible. To this end assume that W^u tends to infinity outside some compact set K containing the $\mathbf{X}_i$ so rapidly that the contribution to the integral over the complement CK is insignificant, that is, $\int_K d^3q \ln(\cdots) > \int_{CK} d^3q \ln(\cdots)$ for all $\alpha > 0$. Then it suffices to estimate the integral over K, which can be done in terms of the L^p norms of the potential on K, i.e., $\|V\|_p = (\int_K d^3q |V(\mathbf{q})|^p)^{1/p}$. If $|x|_- \equiv |x|\Theta(-x)$, then

$$\ln(1 + \exp(-x)) = |x|_- + \ln(1 + \exp(-|x|)) \le |x|_- + \exp(-|x|),$$

and if

$$\mathbf{q} \in K_- \equiv \{\mathbf{q} \in K \mid V(\mathbf{q}) < 0\},$$

then with $\varepsilon = |V(\mathbf{q})|\eta$,

$$I \equiv \int_0^\infty d\varepsilon\sqrt{\varepsilon}\ln(1 + \exp[-\beta(\varepsilon + V(\mathbf{q}))])$$

$$< \int_0^\infty d\eta\sqrt{\eta}(\beta|\eta - 1|_- |V(\mathbf{q})|^{5/2} + |V(\mathbf{q})|^{3/2}\exp[-\beta|V(\mathbf{q})||\eta - 1|]),$$

and if $\mathbf{q} \in K_+ \equiv \{\mathbf{q} \in K | V(\mathbf{q}) > 0\}$, then

$$I < \int_0^\infty d\eta\sqrt{\eta}\, V(\mathbf{q})^{3/2}\exp[-\beta V(\mathbf{q})(\eta + 1)].$$

Because $|\eta - 1| \leq \eta + 1$ for all $\eta \geq 0$ and

$$\int_0^\infty d\varepsilon\sqrt{\varepsilon}\ln(1 + \exp(-\beta\varepsilon)) < \beta^{-3/2}\sqrt{\pi}/2,$$

if $K' = K_+ \cup K_-$, then

$$\int_K d^3q \int_0^\infty d\varepsilon\sqrt{\varepsilon}\ln(1 + \exp[-\beta(\varepsilon + V(\mathbf{q}))])$$

$$< \int_{K'} d^3q \int_0^\infty d\eta\sqrt{\eta}(\beta|V(\mathbf{q})|_-^{5/2}|\eta - 1|_-$$

$$+ |V(\mathbf{q})|^{3/2}\exp[-\beta|V(\mathbf{q})||\eta - 1|]) + \beta^{-3/2}\frac{\sqrt{\pi}}{2}.$$

The required bound now follows from

$$\int_0^\infty d\varepsilon\sqrt{\varepsilon}\exp(-\gamma|\varepsilon - 1|) \leq \int_0^1 d\varepsilon\sqrt{\varepsilon} + \int_0^\infty d\varepsilon(\sqrt{\varepsilon} + 1)\exp(-\gamma\varepsilon)$$

$$= \tfrac{2}{3} + \gamma^{-1} + \frac{\sqrt{\pi}}{2}\gamma^{-3/2},$$

for

$$\int_0^\infty d\varepsilon\sqrt{\varepsilon}\int_K d^3q\ln(1 + \exp[-\beta(\varepsilon + V(\mathbf{q}))])$$

$$\leq \int_K d^3q\left[\frac{4\beta}{15}|V|_-^{5/2} + \tfrac{2}{3}|V|^{3/2} + \beta^{-1}|V|^{1/2} + \sqrt{\pi}\beta^{-3/2}\right].$$

In the case at hand, since $\|V_s\|_p \sim \kappa^{1-3/p}$ and

$$\int d^3x|V + (\alpha - 1)V_s|^p \leq (\|V\|_p + (\alpha - 1)\|V_s\|_p)^p, \qquad p = \tfrac{5}{2}, \tfrac{3}{2},$$

or, respectively,

$$\int d^3x|V + (\alpha - 1)V_s|^{1/2} \leq \|V\|_{1/2}^{1/2} + \sqrt{\alpha - 1}\,\|V_s\|_{1/2}^{1/2},$$

it follows that $\mathbf{1} + \exp[-\beta(h_c + \alpha V_s)]$ remains bounded in the limit as α and $\kappa \to \infty$ when $\alpha \sim \kappa^{1/5}$. If κ goes as $Z^{1/3-\varepsilon}$, $0 < \varepsilon < \frac{1}{3}$, then the correction $Z^{-2/3}\kappa^2$ to the kinetic energy tends to zero, and all corrections to the classical one-particle phase-space integral with the effective field are smaller than this quantity by a factor $Z^{-1/15+\varepsilon/5}$. The quantity μ_γ is no trouble at all, since it approaches μ, provided that $\gamma Z^{-4/3} \to 0$. Likewise, $W^u(\mathbf{q})$ and $W_u(\mathbf{q})$ approach $W(\mathbf{q})$ in the limit $\kappa \to \infty$.

The Classical Lower Bound (4.1.20) .

For the classical bound (2.5.17), the $1/r$ occurring in the classical phase-space integral has to be replaced with $v_u = 1/r - v_s$. As before, convexity is useful for estimating the influence of the v_s, except that this time the convexity of f for $\alpha > 0$,

$$f(-1) \geq f(0) + \frac{f(0) - f(\alpha)}{\alpha},$$

is used for the other side of the equation. The result is

$$\begin{aligned}
&\operatorname{tr} \ln(\mathbf{1} + \exp(-\beta h_n)) \\
&\quad \geq Z \int \frac{d^3q\, d^3p}{(2\pi)^3} \ln\Bigg(1 + \exp\Bigg[-\beta\Bigg\{|\mathbf{p}|^2 + Z^{-2/3}\kappa^2 - \mu_\gamma + W_u(\mathbf{q}) \\
&\qquad - \sum_j z_j\left(\frac{1}{|\mathbf{q} - \mathbf{X}_j|} - v_s(\mathbf{q} - \mathbf{X}_j)\right) \\
&\qquad + \int d^3x n(\mathbf{x})\left(\frac{1}{|\mathbf{q} - \mathbf{x}|} - v_s(\mathbf{q} - \mathbf{x})\right)\Bigg\}\Bigg]\Bigg) \\
&\quad \geq Z \int \frac{d^3q\, d^3p}{(2\pi)^3} \Bigg[\ln\Bigg(1 + \exp\Bigg[-\beta\Bigg\{|\mathbf{p}|^2 + Z^{-2/3}\kappa^2 - \sum_j \frac{z_j}{|\mathbf{q} - \mathbf{X}_j|} \\
&\qquad + \int \frac{d^3x n(\mathbf{x})}{|\mathbf{q} - \mathbf{x}|} + W(\mathbf{q}) - \mu_\gamma\Bigg\}\Bigg]\Bigg)\left(1 + \frac{1}{\alpha}\right) \\
&\qquad - \frac{1}{\alpha} \ln\Bigg(1 + \exp\Bigg[-\beta\Bigg\{|\mathbf{p}|^2 + Z^{-2/3}\kappa^2 \\
&\qquad - \sum_j z_j(|\mathbf{q} - \mathbf{X}_j|^{-1} + \alpha v_s(\mathbf{q} - \mathbf{X}_j)) \\
&\qquad + \int d^3x n(\mathbf{x})(|\mathbf{q} - \mathbf{x}|^{-1} + \alpha v_s(\mathbf{q} - \mathbf{x})) + W_u(\mathbf{q}) - \mu_\gamma\Bigg\}\Bigg]\Bigg)\Bigg].
\end{aligned}$$

The integrals that show up are the same as for the upper bounds, so with $\alpha = \kappa^{1/5}$, $\kappa = Z^{1/3-\varepsilon}$, $0 < \varepsilon < \frac{1}{3}$, the corrections to the classical expression vanish as $Z \to \infty$. The $n(\mathbf{x})$ considered earlier was constant, while that defined by (4.1.12) depends on Z. However, it is shown in Problem 4 that the minimum values also converge, so our bounds prove the

Classical Limit of the Partition Function (4.1.21)

$$\begin{aligned}
&\lim_{Z\to\infty} Z^{-1} \ln \operatorname{Tr} \exp[-\beta(Z^{-4/3}H_Z - \mu N)] \\
&\quad = \lim_{Z\to\infty} Z^{-1} \ln \operatorname{Tr} \exp(-\beta Z^{-4/3} H_{Z,n}) - \beta \sum_{k>l} \frac{z_k z_l}{|\mathbf{X}_k - \mathbf{X}_l|} \\
&\quad = 2 \int \frac{d^3p\, d^3q}{(2\pi)^3} \ln\left(1 + \exp\left[-\beta\left(|\mathbf{p}|^2 - \sum_j \frac{z_j}{|\mathbf{q} - \mathbf{X}_j|}\right.\right.\right. \\
&\qquad \left.\left.\left. + \int d^3y \frac{n(\mathbf{y})}{|\mathbf{q} - \mathbf{y}|} + W(\mathbf{q}) - \mu\right)\right]\right) \\
&\qquad - \beta \sum_{k>l} \frac{z_k z_l}{|\mathbf{X}_k - \mathbf{X}_l|} + \frac{\beta}{2} \int \frac{d^3x\, d^3y}{|\mathbf{x} - \mathbf{y}|} n(\mathbf{x}) n(\mathbf{y}).
\end{aligned}$$

According to Remark (2.5.18; 4), the optimal density for this formula satisfies

$$\begin{aligned}
n(\mathbf{x}) = 2 \int \frac{d^3p}{(2\pi)^3} \left[\exp\left(\beta\left(|\mathbf{p}|^2 - \sum_j \frac{z_j}{|\mathbf{x} - \mathbf{X}_j|}\right.\right.\right. \\
\left.\left.\left. + \int d^3y \frac{n(\mathbf{y})}{|\mathbf{q} - \mathbf{y}|} + W(\mathbf{x}) - \mu\right) + 1\right)\right]^{-1}.
\end{aligned} \tag{4.1.22}$$

Remarks (4.1.23)

1. The classical functional also has Properties (4.1.10), which ensure the existence and uniqueness of a solution of (4.1.22).
2. As yet unproved conjectures [11] imply that Equation (4.1.19) holds even with $c = 1$. If that turns out to be true, then many of the proofs given here can be simplified.

The Density in Phase Space (4.1.24)

Now that Ξ has been shown to converge, we can study the limiting behavior of the expectation values of a suitable subalgebra of observables. The densities on classical phase space would seem to be an appropriate subalgebra, since in the classical limit $Z \to \infty$ it ought to make sense to speak of position and momentum simultaneously. As mentioned above (cf. (1.2.4)) position goes as $Z^{-1/3}$ while momentum goes as $Z^{2/3}$, so the product of their relative mean-square deviations would be expected to go as $Z^{-2/3}$, and as $Z \to \infty$ the physics should become classical. This rather airy argument can be made mathematically substantial, and we shall discover that in convenient units, fermions distribute themselves in phase space according to

$$\rho(\mathbf{q}, \mathbf{p}) = \left[\exp \beta\left\{|\mathbf{p}|^2 - \sum_j \frac{z_j}{|\mathbf{q} - \mathbf{X}_j|} + \int \frac{d^3x n(\mathbf{x})}{|\mathbf{x} - \mathbf{q}|} + W(\mathbf{q}) - \mu\right\} + 1\right]^{-1}.$$

Particularly interesting is the observation that fermions behave more classically than bosons. The latter have a -1 in the denominator, so $\rho(\mathbf{q}, \mathbf{p})$ becomes negative when $\mathbf{q} = \mathbf{X}_j$, and thus can not turn out to be a probability density on phase space.

To make the connection with (2.2.10; 5) we define creation and annihilation operators at the point $(\mathbf{q}, \mathbf{p})$ in phase space, and choose u as a sufficiently smooth, decreasing function such that $\|u\|_2 = 1$, like the function of (4.1.17):

The Field Algebra on Phase Space (4.1.25)

The operators

$$a_{\mathbf{q}, \mathbf{p}; \alpha} = Z^{3\varepsilon/2} \int d^3x a_\alpha(\mathbf{x}) \exp(iZ^{2/3}\mathbf{p} \cdot \mathbf{x}) u(Z^\varepsilon(\mathbf{x} - Z^{-1/3}\mathbf{p})),$$

$$\tfrac{1}{3} < \varepsilon < \tfrac{2}{3}, u^* = u,$$

satisfy the commutation relations

$$[a_{\mathbf{q}, \mathbf{p}; \alpha}, a^*_{\mathbf{q}', \mathbf{p}'; \beta}]_+ = \delta_{\alpha\beta} \int d^3x \exp(iZ^{2/3-\varepsilon}\mathbf{x} \cdot (\mathbf{p} - \mathbf{p}')) u(\mathbf{x} - Z^{\varepsilon-1/3}\mathbf{q})$$
$$\times u(\mathbf{x} - Z^{\varepsilon-1/3}\mathbf{q}').$$

If $\mathbf{q} = \mathbf{q}'$ and $\mathbf{p} = \mathbf{p}'$, then the right side is $\delta_{\alpha\beta}$, and otherwise it goes to zero as $Z \to \infty$. Hence $\rho_{\mathbf{q}, \mathbf{p}} = \sum_\alpha a^*_{\mathbf{q}, \mathbf{p}; \alpha} a_{\mathbf{q}, \mathbf{p}; \alpha}$ are bounded above and below by $0 \leq \rho_{\mathbf{q}, \mathbf{p}} \leq 2$, and generate an algebra that is Abelian in the limit $Z \to \infty$.

Defining $d\Omega \equiv d^3q\, d^3p/(2\pi)^3$, we calculate

$$\int d\Omega \rho_{\mathbf{q}, \mathbf{p}} F(\mathbf{q}) = Z^{-1} \sum_\alpha \int d^3x a^*_\alpha(\mathbf{x}) a_\alpha(\mathbf{x}) Z^{3\varepsilon} |u(Z^3(\mathbf{x} - \mathbf{x}'))|^2$$
$$\times F(Z^{1/3}\mathbf{x}')\, d^3x',$$

$$\int d\Omega \rho_{\mathbf{q}, \mathbf{p}} |\mathbf{p}|^2 = Z^{-7/3} \sum_\alpha \int d^3x (\nabla a^*_\alpha(\mathbf{x}) \cdot \nabla a_\alpha(\mathbf{x})$$
$$+ a^*_\alpha(\mathbf{x}) a_\alpha(\mathbf{x}) Z^{2\varepsilon} \int d^3y |\nabla u(\mathbf{y})|^2)$$

$$\int d\Omega\, d\Omega' \frac{\rho_{\mathbf{q}, \mathbf{p}} \rho_{\mathbf{q}', \mathbf{p}'}}{|\mathbf{q} - \mathbf{q}'|} = Z^{-7/3} \sum_{\alpha, \beta} \int d^3x\, d^3x' a^*_\alpha(\mathbf{x}) a^*_\beta(\mathbf{x}') a_\beta(\mathbf{x}') a_\alpha(\mathbf{x}) v_{uu}(\mathbf{x} - \mathbf{x}')$$
$$+ Z^{-7/3} \sum_\alpha \int d^3x a^*_\alpha(\mathbf{x}) a_\alpha(\mathbf{x}) v_{uu}(\mathbf{0}), \qquad (4.1.26)$$

where $F \in L^\infty(\mathbb{R}^3)$ and

$$0 \leq v_{uu}(\mathbf{x} - \mathbf{x}') \equiv \int d^3q\, d^3q' \frac{|u(Z^\varepsilon(\mathbf{x} - \mathbf{q}))|^2 |u(Z^\varepsilon(\mathbf{x}' - \mathbf{q}'))|^2}{|\mathbf{q} - \mathbf{q}'|} Z^{6\varepsilon} < \frac{1}{|\mathbf{x} - \mathbf{x}'|}.$$

Remarks (4.1.27)

1. As $Z \to \infty$, $Z^{3\varepsilon}|u(Z^\varepsilon(\mathbf{x} - \mathbf{x}'))|^2$ approaches $\delta(\mathbf{x} - \mathbf{x}')$. It is not hard to convince oneself that when the classical Hamiltonian with ρ or $\rho \cdot \rho$ is integrated, the result is H to order $Z^{7/3}$.
2. For neutral states, i.e., $\sum_\alpha \langle N_\alpha \rangle = Z$, it follows that $\langle \int d\Omega \rho_{\mathbf{q},\mathbf{p}} \rangle = 1$.

The convexity of the partition function (2.4.7) can be used to calculate an expectation value by allowing it to be written as the derivative of the partition function by a perturbation parameter. We shall show that the perturbed Ξ still converges as $Z \to \infty$, which will simultaneously prove that the foregoing results are stable against small variations in H. The limit will turn out to be likewise convex and differentiable in the perturbation parameters, so by Problem (2.4.18; 3) the limit of the derivative is the derivative of the limit. Since our real aim is to prove that the expectation value of $\rho_{\mathbf{q},\mathbf{p}}$ approaches the Thomas–Fermi density and that the deviations of $\rho_{\mathbf{q},\mathbf{p}}$ vanish, we will perturb H both linearly and quadratically in ρ. To an accuracy of $Z^{-2/3}$ we can by-pass the intermediate steps (4.1.15), so we shall not require the more refined inequality (4.1.5). Thus we get by with a somewhat simpler effective Hamiltonian.

The Perturbed Hamiltonian (4.1.28)

$$H_\lambda \equiv H_Z + \lambda_1 Z^{7/3} \int d\Omega \rho_{\mathbf{q},\mathbf{p}} f(\mathbf{q}, \mathbf{p}) + \lambda_2 \frac{Z^{7/3}}{2} \left(\int d\Omega \rho_{\mathbf{q},\mathbf{p}} f(\mathbf{q}, \mathbf{p}) \right)^2,$$

$$H_{\lambda,n} \equiv \int d^3x \sum_\alpha \Bigg\{ \nabla a_\alpha^*(\mathbf{x}) \cdot \nabla a_\alpha(\mathbf{x}) + a_\alpha^*(\mathbf{x}) a_\alpha(\mathbf{x})$$

$$\times \left[-Z \sum_{k=1}^{M} \frac{z_k}{|\mathbf{x} - Z^{-1/3}\mathbf{X}_k|} + Z^{4/3}(W(Z^{1/3}\mathbf{x}) - \mu) \right] \Bigg\}$$

$$+ Z^{7/3} \int \frac{d\Omega\, d\Omega'}{|\mathbf{q} - \mathbf{q}'|} (\rho_{\mathbf{q},\mathbf{p}} - \tfrac{1}{2} n(\mathbf{q}, \mathbf{p})) n(\mathbf{q}', \mathbf{p}')$$

$$+ Z^{7/3}(\lambda_1 + \lambda_2 g) \int d\Omega \rho_{\mathbf{q},\mathbf{p}} f(\mathbf{q}, \mathbf{p}) - Z^{7/3} \lambda_2 g^2/2,$$

where $\lambda_i \in \mathbb{R}$ and $f \in C_0^\infty$. We shall choose $n(\mathbf{q}, \mathbf{p})$ as $\langle \rho_{\mathbf{q},\mathbf{p}} \rangle$ and let $g \equiv \int d\Omega n(\mathbf{q}, \mathbf{p}) f(\mathbf{q}, \mathbf{p})$. With the idea of (4.1.24), because $0 \leq v_{uu}(\mathbf{x}) \leq 1/|\mathbf{x}|$,

$$\begin{aligned}
H_\lambda - Z^{4/3}\mu N - Z^{7/3} \sum_{k>l} \frac{z_k z_l}{|\mathbf{X}_k - \mathbf{X}_l|} - H_{\lambda, n}
&= \frac{1}{2} \int \frac{d^3x\, d^3x'}{|\mathbf{x} - \mathbf{x}'|} a_\alpha^*(\mathbf{x}) a_\beta^*(\mathbf{x}') a_\beta(\mathbf{x}') a_\alpha(\mathbf{x}) \\
&\quad - Z^{7/3} \int \frac{d\Omega\, d\Omega'}{|\mathbf{q} - \mathbf{q}'|} (\rho_{\mathbf{q},\mathbf{p}} - \tfrac{1}{2} n(\mathbf{q}, \mathbf{p})) n(\mathbf{q}', \mathbf{p}') \\
&\quad + Z^{7/3} \frac{\lambda_2}{2} \left(\int d\Omega \rho_{\mathbf{q},\mathbf{p}} f(\mathbf{q}, \mathbf{p}) - g \right)^2 \\
&\geq \frac{Z^{7/3}}{2} \int d\Omega\, d\Omega' (\rho_{\mathbf{q},\mathbf{p}} - n(\mathbf{q}, \mathbf{p}))(\rho_{\mathbf{q}',\mathbf{p}'} - n(\mathbf{q}', \mathbf{p}')) \\
&\quad \times \left[\frac{1}{|\mathbf{q} - \mathbf{q}'|} + \lambda_2 f(\mathbf{q}, \mathbf{p}) f(\mathbf{q}', \mathbf{p}') \right] - \frac{N}{2} v_{uu}(\mathbf{0}).
\end{aligned}$$

Remarks (4.1.29)

1. Since the Fourier transform in the q variables, $\tilde{f}(\mathbf{k}, \mathbf{p})$, decreases in $\mathbf{k}$ faster than any power, $|\mathbf{k}|^2 + \lambda_2 \tilde{f}(\mathbf{k}, \mathbf{p}) \tilde{f}(\mathbf{k}, \mathbf{p}')$ is positive for sufficiently small $|\lambda_2|$. The expression in square brackets $[\cdots]$ is then of positive type, and the inequality extends to the statement that

$$H_\lambda - Z^{4/3}\mu N - Z^{7/3} \sum_{k>m} \frac{z_k z_m}{|\mathbf{X}_k - \mathbf{X}_m|} - H_{\lambda, n} \geq -\frac{N}{2} v_{uu}(\mathbf{0}).$$

   It is easy to calculate that $v_{uu}(\mathbf{0}) \sim Z^\varepsilon$, so the right side is dominated by $Z^{7/3}$, and in the limit as $Z \to \infty$,

$$Z^{-1}\Xi(Z^{-4/3}H_\lambda - \mu N) \leq Z^{-1}\Xi(Z^{-4/3}H_{\lambda, n}) - \sum_{k>m} \frac{z_k z_m}{|\mathbf{X}_k - \mathbf{X}_m|}.$$

2. According to (4.1.24),

$$Z^{7/3} \int d\Omega\, d\Omega' \frac{\rho_{\mathbf{q},\mathbf{p}} n(\mathbf{q}', \mathbf{p}')}{|\mathbf{q} - \mathbf{q}'|} = \sum_\alpha \int d^3x\, d^3x' a_\alpha^*(\mathbf{x}) a_\alpha(\mathbf{x}) v_u(\mathbf{x} - \mathbf{x}') n(\mathbf{x}'),$$

   where

$$n(\mathbf{x}) = \int \frac{d^3p}{(2\pi)^3} n(\mathbf{x}, \mathbf{p}).$$

   Therefore the Coulomb repulsion of the electrons in the Hamiltonian $H_{\lambda, n}$ of (4.1.28) is reduced by $v_s = 1/r - v_u$. As in (4.1.14) the Hamiltonian $H_{\lambda, n}$ with v_u in place of $1/r$ furnishes a lower bound for Ξ. On the other hand, it was shown in (4.1.18) and (4.1.20) that the effect of v_s on $\Xi(H_{\lambda,n})$ was negligible as $Z \to \infty$. Moreover, $\int d\Omega \rho_{\mathbf{q},\mathbf{p}} f(\mathbf{q},\mathbf{p})$ is the second quantization of the one-particle operator $\int d\Omega |\mathbf{q},\mathbf{p}\rangle\langle\mathbf{q},\mathbf{p}| f(\mathbf{q},\mathbf{p})$, $|\mathbf{q},\mathbf{p}\rangle = \exp(i\mathbf{p}\cdot\mathbf{x})u(\mathbf{x}-\mathbf{q})$, the expectation value of which in the state $|\mathbf{q}',\mathbf{p}'\rangle$ reduces to $f(\mathbf{q}',\mathbf{p}')$ in the limit $Z \to \infty$. The generalization of (4.1.21) is consequently

$$\lim_{Z\to\infty} Z^{-1} \ln \operatorname{Tr} \exp[-\beta(Z^{-4/3}H_\lambda - \mu N)]$$
$$= 2\int \frac{d^3q\,d^3p}{(2\pi)^3} \ln\left\{1 + \exp\left[-\beta\left(|\mathbf{p}|^2 - \sum_k \frac{z_k}{|\mathbf{q}-\mathbf{X}_k|}\right.\right.\right.$$
$$\left.\left.\left. + \int \frac{d^3q' n_\lambda(\mathbf{q}')}{|\mathbf{q}-\mathbf{q}'|} + W(\mathbf{q}) + f(\mathbf{q},\mathbf{p})(\lambda_1 + \lambda_2 g_\lambda) - \frac{\lambda_2}{2} g_\lambda^2 - \mu\right)\right]\right\}$$
$$+ \beta\left(\frac{1}{2}\int \frac{d^3q\,d^3q'}{|\mathbf{q}-\mathbf{q}'|} n_\lambda(\mathbf{q}) n_\lambda(\mathbf{q}') - \sum_{k>l} \frac{z_k z_l}{|\mathbf{X}_k - \mathbf{X}_l|}\right), \tag{4.1.30}$$

where

$$n_\lambda(\mathbf{q}) = \int \frac{d^3p}{(2\pi)^3} n_\lambda(\mathbf{q},\mathbf{p}),$$
$$n_\lambda(\mathbf{q},\mathbf{p}) = 2\left\{\exp \beta\left[|\mathbf{p}|^2 - \sum_k \frac{z_k}{|\mathbf{q}-\mathbf{X}_k|} + \int \frac{d^3q' n_\lambda(\mathbf{q}')}{|\mathbf{q}-\mathbf{q}'|}\right.\right.$$
$$\left.\left. + W(\mathbf{q}) + f(\mathbf{q},\mathbf{p})(\lambda_1 + \lambda_2 g_\lambda) - \frac{\lambda_2}{2} g_\lambda^2 - \mu\right] + 1\right\}^{-1},$$
$$g_\lambda = \int d\Omega\, n_\lambda(\mathbf{q},\mathbf{p}) f(\mathbf{q},\mathbf{p}),$$

and $|\lambda_2|$ is sufficiently small.

Differentiation by λ_1 and λ_2 at $\lambda_1 = \lambda_2 = 0$ and an optimization of $f \in C_0^2(\mathbb{R}^6)$ reveal the

Convergence of the Expectation Values (4.1.31)

$$\lim_{Z\to\infty} \langle \rho_{\mathbf{q},\mathbf{p}} \rangle_Z \equiv \lim_{Z\to\infty} \frac{\operatorname{Tr}(\rho_{\mathbf{q},\mathbf{p}} \exp[-\beta(Z^{-4/3}H_Z - \mu N)])}{\operatorname{Tr} \exp[Z^{-4/3}H_Z - \mu N]}$$
$$= 2\left\{\exp\left[\beta\left(|\mathbf{p}|^2 - \sum_k \frac{z_k}{|\mathbf{q}-\mathbf{X}_k|}\right.\right.\right.$$
$$\left.\left.\left. + \int \frac{d^3q' n_0(\mathbf{q}')}{|\mathbf{q}-\mathbf{q}'|} + W(\mathbf{q}) - \mu\right)\right] + 1\right\}^{-1} = n_0(\mathbf{q},\mathbf{p}),$$

$$\lim_{Z\to\infty} \langle \rho_{\mathbf{q},\mathbf{p}}, \rho_{\mathbf{q}',\mathbf{p}'} \rangle_Z = n_0(\mathbf{q},\mathbf{p}) n_0(\mathbf{q}',\mathbf{p}').$$

Remarks (4.1.32)

1. Since f is not arbitrary, but assumed in $C_0^2(\mathbb{R}^6)$, the limit converges only in the sense of distributions. The C^* algebra $\mathscr{A}_Z$ generated by the "smeared" densities on phase space, $\rho_g \equiv \int d\Omega g(\mathbf{q}, \mathbf{p})\rho_{\mathbf{q},\mathbf{p}}$, together with the identity becomes Abelian in the "weak" limit $Z \to \infty$. Hence, according to the Gel'fand isomorphism (III: 2.2.28), if $Z = \infty$, then $\mathscr{A}_Z$ can be represented as the set of continuous functions on a compact Hausdorff space. The space of characters of an Abelian C^* algebra $\mathscr{A}$, i.e., *-homomorphisms from $\mathscr{A}$ to $\mathbb{C}$, is the same as the set $\mathscr{E}$ of pure states and is a compact Hausdorff space in the (relative) weak-* topology. With the identification $[a](\omega) = \omega(a) \in \mathbb{C}$ for all $a \in \mathscr{A}$ and $\omega \in \mathscr{E}$, $\mathscr{A}$ is equivalent to the C^* algebra of the continuous functions with the supremum norm on the set $\mathscr{E}$, given the weak-* topology. In our case, $\mathscr{E} = \{n \in L^\infty(\mathbb{R}^6) | n \geq 0 \text{ a.e.}, \|n\|_\infty \leq 2\}$, with the weak-* topology with respect to the linear functionals belonging to the predual $L^1(\mathbb{R}^6)$. (Since $C_0^2(\mathbb{R}^6)$ is dense in $L^1(\mathbb{R}^6)$ in norm, the corresponding weak-* topologies agree on $L^\infty(\mathbb{R}^6)$.) Since $\mathscr{E}$ is the intersection of the cone of the functions that are non-negative a.e., which is a weak-* closed set, with a multiple of the unit cube of L^∞, it is weak-* compact. The Gel'fand isomorphism correlates ρ_g with the mapping $[\rho_g](n) = \int ng\, d\Omega$, and since $\|[\rho_g] - [\rho_{g'}]\|_\infty \leq 2\|g - g'\|_1$, the completion contains for instance all ρ_g such that $g \in L^1(\mathbb{R}^6)$. The set of all states on the algebra is the weak-* closure of the convex combinations of characters and can be represented as a set of probability measures; pure states correspond to point measures. If the state is mixed, $\alpha\langle\ \rangle_{n_1} + (1 - \alpha)\langle\ \rangle_{n_2}$, then the two-point function can not be factorized:

$$\begin{aligned}
\alpha\langle\rho_{z_1}\rho_{z_2}\rangle_{n_1} + (1-\alpha)\langle\rho_{z_1}\rho_{z_2}\rangle_{n_2} &= \alpha n_1(z_1)n_1(z_2) + (1-\alpha)n_2(z_1)n_2(z_2) \\
&= (\alpha\langle\rho_{z_1}\rangle_{n_1} + (1-\alpha)\langle\rho_{z_1}\rangle_{n_2})(\alpha\langle\rho_{z_2}\rangle_{n_1} \\
&\quad + (1-\alpha)\langle\rho_{z_2}\rangle_{n_2}) \quad \text{for all } z_1, z_2 \\
&\Rightarrow n_1(z) = n_2(z) \quad \text{for all } z = (\mathbf{q}, \mathbf{p}).
\end{aligned}$$

Hence it follows from (4.1.31) that the limiting state is a character, and consequently pure.

2. Although the system acts classically on a distance scale $\sim Z^{-1/3}$, it would be expected to behave like a free Fermi gas on the scale $Z^{-2/3} \sim$ the average distance between particles $\sim$ reciprocal of momentum. If the microscopic field operators

$$a_{\mathbf{q}}(\xi) = Z^{-1}a(Z^{-1/3}\mathbf{q} + Z^{-2/3}\xi), \qquad [a_q(\xi), a_q^*(\xi')]_+ = \delta(\xi - \xi')$$

are introduced, it can be seen from (4.1.31) that its expectation value for free Fermions is

$$\int \frac{d^3p}{(2\pi)^3} \exp(i\mathbf{p}\cdot\xi)\rho_{\mathbf{q},\mathbf{p}}$$
$$= \int \frac{d^3p}{(2\pi)^3} \exp[i\mathbf{p}\cdot(Z^{2/3}(\mathbf{x}-\mathbf{x}')+\xi)]\, a^*(\mathbf{x})a(\mathbf{x}')\frac{Z^{3/2}}{\pi^{3/2}}$$
$$\times \exp\left[-\frac{Z}{2}(|\mathbf{x}-Z^{-1/3}\mathbf{q}|^2+|\mathbf{x}'-Z^{-1/3}\mathbf{q}|^2)\right] d^3x\, d^3x'$$
$$= \int d^3z a^*_{\mathbf{q}+\mathbf{x}}\left(-\frac{\xi}{2}\right)a_{\mathbf{q}+\mathbf{x}}\left(\frac{\xi}{2}\right)\exp\left[-Z^{1/3}|\mathbf{x}|^2-\frac{Z^{-1/3}|\xi|^2}{4}\right]\frac{Z^{1/2}}{\pi^{3/2}},$$

where the chemical potential is determined by the potential $V(\mathbf{q})$ at the point $\mathbf{q}$, and we set $\varepsilon = \frac{1}{2}$, $u = \pi^{-3/4}\exp(-|\mathbf{x}|^2/2)$,

$$\int \frac{d^3p}{(2\pi)^3} \exp(i\mathbf{p}\cdot\xi)\rho_{\mathbf{q},\mathbf{p}}$$
$$= \int \frac{d^3p}{(2\pi)^3} \exp[i\mathbf{p}\cdot(Z^{2/3}(\mathbf{x}-\mathbf{x}')+\xi)]a^*(\mathbf{x})a(\mathbf{x}')\frac{Z^{3/2}}{\pi^{3/2}}$$
$$\times \exp\left[-\frac{Z}{2}(|\mathbf{x}-Z^{-1/3}\mathbf{q}|^2+|\mathbf{x}'-Z^{-1/3}\mathbf{q}|^2)\right] d^3x\, d^3x'$$
$$= \int d^3x a^*_{\mathbf{q}+\mathbf{x}}\left(-\frac{\xi}{2}\right)a_{\mathbf{q}+\mathbf{x}}\left(\frac{\xi}{2}\right)\exp\left[-Z^{1/3}|\mathbf{x}|^2-\frac{Z^{-1/3}|\xi|^2}{4}\right]\frac{Z^{1/2}}{\pi^{3/2}}.$$

Therefore

$$\int \frac{d^3p \exp(i\mathbf{p}\cdot\xi)(2\pi)^{-3}}{\exp[\beta(|\mathbf{p}|^2-V(\mathbf{q}))]+1}$$
$$= \lim_{Z\to\infty}\int d^3x\, \frac{Z^{1/2}}{\pi^{2/3}}\exp\left[-Z^{1/3}|\mathbf{x}|^2-\frac{Z^{-1/3}|\xi|^2}{2}\right]$$
$$\times \left\langle a^*_{\mathbf{q}+\mathbf{x}}\left(-\frac{\xi}{2}\right)a_{\mathbf{q}+\mathbf{x}}\left(\frac{\xi}{2}\right)\right\rangle_{n_0} \equiv \left\langle a^*_{\mathbf{q}}\left(-\frac{\xi}{2}\right)a_{\mathbf{q}}\left(\frac{\xi}{2}\right)\right\rangle_{\infty},$$

$$V(\mathbf{q}) = -\sum_k \frac{z_k}{|\mathbf{q}-\mathbf{X}_k|} + \int\frac{d^3x n_0(\mathbf{x})}{|\mathbf{q}-\mathbf{x}|} + W(\mathbf{q}) - \mu.$$

3. Results have also been obtained concerning the time-evolution in the limit $Z\to\infty$ [26], but they have only been proved for regularized potentials v_u and not for $1/r$, so they will not be presented here. At any rate the time-evolution of $\omega(a_t)$, where the nonstationary state ω has the

same scaling properties as the grand canonical state ρ, is the free time-evolution, as is that of $\rho(a_t b)$, when only the microscopic observables (4.1.32; 2) are considered. The equation for the expectation values of the macroscopic observables $\rho_{\mathbf{q},\mathbf{p}}$ is known as the **Vlasov equation**; it describes a classical time-evolution according to

$$\frac{dn}{dt} = \mathbf{p} \cdot \frac{\partial n}{\partial \mathbf{q}} - \frac{\partial V}{\partial \mathbf{q}} \cdot \frac{\partial n}{\partial \mathbf{p}},$$

where the potential itself depends on the particle density,

$$V(\mathbf{q}) = -\sum_k \frac{Z_k}{|\mathbf{q} - \mathbf{X}_k|} + \int \frac{d^3q'\, d^3p'}{(2\pi)^3} \frac{n(\mathbf{q}', \mathbf{p}')}{|\mathbf{q} - \mathbf{q}'|}.$$

Thomas–Fermi theory thus reduces the quantum-mechanical many-body problem to the solution of the integral equation (4.1.22). Although (4.1.22) is much simpler than the original Schrödinger equation, it can still be solved with reasonable numerical effort and skill only in the radially symmetric case. Despite that, some valuable relationships and properties can be obtained just from the maximum property.

The Relationships among the Contributions to Ξ (4.1.33)

Write

$$-\lim Z^{-1}\Xi(Z^{-4/3}H_Z - \mu N) - \sum_{k>l} \frac{z_k z_l}{|\mathbf{X}_k - \mathbf{X}_l|}$$

$$= \inf_{0 \le n \le 2} \int \frac{d^3q\, d^3p}{(2\pi)^3} \left\{ 2T\left[\frac{n(\mathbf{q}, \mathbf{p})}{2} \ln \frac{n(\mathbf{q}, \mathbf{p})}{2} + \left(1 - \frac{n(\mathbf{q}, \mathbf{p})}{2}\right) \ln\left(1 - \frac{n(\mathbf{q}, \mathbf{p})}{2}\right)\right]\right.$$

$$+ n(\mathbf{q}, \mathbf{p})\left(-\mu + |\mathbf{p}|^2 - \sum_{j=1}^{M} \frac{z_j}{|\mathbf{q} - \mathbf{X}_j|}\right.$$

$$\left.\left. + \frac{1}{2} \int \frac{d^3q'\, d^3p'}{2(\pi)^3} \frac{n(\mathbf{q}', \mathbf{p}')}{|\mathbf{q} - \mathbf{q}'|} + W(\mathbf{q})\right)\right\}$$

$$= -TS - \mu\lambda + K - A + R + W,$$

where

$$\lambda = \int \frac{d^3q\, d^3p}{(2\pi)^3} n(\mathbf{q}, \mathbf{p}) = \lim_{Z \to \infty} \frac{N}{Z},$$

K is the kinetic energy of the electrons, A is the potential attracting the electrons to the nuclei, and R is the interelectronic repulsion. Then for the values of μ

at which the infimum is attained as a minimum (at a given phase-space density n_0*),*

(i) $-3(TS + \mu\lambda) + 5K - 3A + 6R + 3W = 0$; *and*
(ii) *if an atom is isolated and in the ground state, i.e.,* $M = 1$, $X_1 = 0$, $W = 0$, $T = 0$, *then*

$$-3\mu\lambda + 3K - 2A + 5R = 0.$$

Proof

(i) Take the infimum over n' of the form $n_0(q, \gamma^{-1}p)$. A change of the variables of integration $\mathbf{p} \to \gamma_1 \mathbf{p}$ converts (4.1.33) into

$$-\gamma_1^3(TS + \mu\lambda + A - W) + \gamma_1^5 K + \gamma_1^6 R.$$

This has its minimum at $\gamma_1 = 1$ when condition (i) holds.
(ii) Now dilate $\mathbf{q}$ so that $n(\mathbf{q}, \mathbf{p}) = n_0(\gamma_2^{-1}\mathbf{q}, \mathbf{p})$, and proceed as before; then

$$\frac{d}{d\gamma_2}[\gamma_2^3(K - \mu\lambda) - \gamma_2^2 A + \gamma_2^5 R]|_{\gamma_2=1} = 0$$

yields Relationship (ii). □

Corollary (4.1.34)

In case (ii) with $\mu = 0$, the three contributions to the energy stand in the ratio

$$K:A:R = 3:7:1.$$

Remarks (4.1.35)

1. The dilatation required for (ii) affects the nuclear coordinates (other than $\mathbf{X}_1 = \mathbf{0}$) and the wall. The reason for setting $T = 0$ was to avoid problems connected with the latter.
2. Since A, K, and R are positive, the second derivatives at $\gamma = 1$ are automatically positive.
3. If $T = \mu = 0$, then $-\Xi$ becomes the minimum of the energy without fixed particle number. We shall learn that the minimum is achieved by a neutral system in Thomas–Fermi theory, and that in case (ii)

$$\int \frac{d^3q\, d^3p}{(2\pi)^3} n_0(\mathbf{q}, \mathbf{p}) = z_1.$$

The comparison densities $n(\gamma^{-1}\mathbf{q}, \mathbf{p})$ and $n(\mathbf{q}, \gamma^{-1}\mathbf{p})$ correspond to different numbers of particles, and the mystical numbers in (4.1.34) reflect the stability of neutral atoms against spontaneous ionization.

In the limit $T \to 0$, the quantity $(\exp[\beta(\varepsilon - a)] + 1)^{-1}$ approaches $\Theta(a - \varepsilon)$. In that case W may be chosen identically zero, and the integration over $\mathbf{p}$ becomes elementary. The computation yields

The Electron Density in Configuration Space (4.1.36)

$$\rho(\mathbf{x}) \equiv \int \frac{d^3p}{(2\pi)^3} n_0(\mathbf{x}, \mathbf{p}) = \frac{1}{3\pi^2} |\Phi(\mathbf{x}) + \mu|_+^{3/2}, \qquad |z|_+ \equiv |z|\Theta(z),$$

$$\Phi(\mathbf{x}) \equiv \sum_j \frac{z_j}{|\mathbf{x} - \mathbf{X}_j|} - \int \frac{d^3x' \, \rho(\mathbf{x}')}{|\mathbf{x} - \mathbf{x}'|}.$$

The kinetic-energy density is

$$\int \frac{d^3p}{(2\pi)^3} |\mathbf{p}|^2 n_0(\mathbf{x}, \mathbf{p}) = \tfrac{3}{5}(3\pi^2)^{2/3} \rho^{5/3}(\mathbf{x}).$$

(Since the particles have spin 1/2, the factor $(6\pi^2)^{2/3}$ of (2.5.32) has become $(3\pi^2)^{2/3}$.)

This reveals

The Range of Values of μ and $\Phi(x)$ (4.1.37)

(i) *μ takes on the values* $-\infty < \mu \leq 0$; *and*
(ii) *Φ takes on the values* $0 \leq \Phi < \infty$.

Proof

We shall only demonstrate the impossibility of $\mu > 0$ and $\Phi < 0$; Problem 3 will assure us that a minimizing ρ exists for all $\mu \leq 0$, and it can be seen directly that $\Phi(\mathbf{x})$ ranges over $[0, \infty)$ as $\mathbf{x}$ ranges over $\mathbb{R}^3$.

(i) Since $\rho(\mathbf{x})$ must be integrable, $\Phi(\mathbf{x}) \to 0$ as $|\mathbf{x}| \to \infty$. If $\mu > 0$, then $\rho(\mathbf{x})$ would have to approach $\mu^{3/2}/3\pi^2$ as $|\mathbf{x}| \to \infty$, which would contradict integrability.
(ii) The set $A \equiv \{\mathbf{x} \in \mathbb{R}^3 : \Phi(\mathbf{x}) < 0\}$ is open and does not contain $\mathbf{x}_i$. Because $\mu \leq 0$, the density ρ vanishes identically on A, so $\Delta\Phi(\mathbf{x}) = 0$ holds throughout A. Since Φ equals zero on the boundary of A and at infinity and is harmonic, it would have to equal zero on A, because its maximum would be attained either on ∂A or at infinity. However, this contradicts the definition of A, so A must be empty. □

The quantity $\lambda \equiv \int d^3x\rho(\mathbf{x}) = \lim_{Z\to\infty} N/Z$, where N is the number of electrons and Z is the sum of the nuclear charges, is more intuitively understandable than μ. By expressing the energy as a function of λ, we can find the limits of the observables studied in (III: §4.5).

Properties of the Thomas–Fermi Functional at $T = 0$ (4.1.38)

Let

$$K(\rho) = \tfrac{3}{5}(3\pi^2)^{2/3} \int d^3x \rho^{5/3}(\mathbf{x}),$$

$$A(\rho) = \sum_j z_j \int \frac{d^3x \rho(\mathbf{x})}{|\mathbf{x} - \mathbf{X}_j|},$$

$$R(\rho) = \frac{1}{2} \int \frac{d^3x\, d^3x'}{|\mathbf{x} - \mathbf{x}'|} \rho(\mathbf{x})\rho(\mathbf{x}'),$$

$$E(\rho) \equiv K(\rho) - A(\rho) + R(\rho),$$

and

$$S_\lambda = \left\{\rho\colon \rho(\mathbf{x}) \geq 0, \int d^3x \rho(\mathbf{x}) = \lambda\right\}, \qquad \sum_j z_j = 1.$$

Then $E[\lambda] = \inf_{\rho \in S_\lambda} E(\rho)$ *satisfies*

(i) $$E[\lambda] = -\inf_\mu \left(\Xi_\infty(\mu, 0) - \mu\lambda + \sum_{k>l} \frac{z_k z_l}{|\mathbf{X}_k - \mathbf{X}_l|}\right),$$

$$\Xi_\infty(\mu, T) = \lim_{Z \to \infty} Z^{-1} T \ln \operatorname{Tr} \exp[-\beta(Z^{-4/3}H - \mu N)];$$

(ii) $\partial E/\partial\lambda = \mu$ *if* $\lambda \leq 1$, *and* $= 0$ *if* $\lambda > 1$;
(iii) $E[\lambda]$ *is a nonpositive, convex, decreasing function of* λ; *and*
(iv) *in the atomic case* $z_1 = 1$, *all other* $z_i = 0$, $-\lambda^{-1/6}(-E[\lambda])^{1/2}$ *is a concave, increasing function of* λ

Proof

(i) Observe first that $E[\lambda]$ is convex, since the convexity of $E(\rho)$ as a function of ρ means that $E[\alpha\lambda_1 + (1 - \alpha)\lambda_2] \leq E(\alpha\rho_1 + (1 - \alpha)\rho_2) \leq \alpha E[\lambda_1] + (1 - \alpha)E[\lambda_2]$, in which $E[\lambda_i] = E(\rho_i)$, because $\alpha\rho_1 + (1 - \alpha)\rho_2 \in S_{\alpha\lambda_1 + (1-\alpha)\lambda_2}$. As remarked in (2.4.15; 2(i)), the Legendre transformation

$$-\Xi_\infty(\mu, 0) = \inf_\lambda \inf_{\rho \in S_\lambda} \left(E(\rho) - \mu\lambda + \sum_{k>m} \frac{z_k z_m}{|\mathbf{X}_k - \mathbf{X}_m|}\right)$$

can be inverted for the concave function $-E[\lambda]$, yielding (i).

(ii) The formula $dE/d\lambda = \mu$ will follow from Property (i) once $E[\lambda]$ has been shown to be differentiable. Let ρ_λ denote the minimizing ρ (4.1.36). A calculation shows that

$$\frac{\partial}{\partial t} E(\rho_\lambda(1 + t))|_{t=0} = \mu\lambda,$$

so $E[(1+t)\lambda] - E[\lambda] \le t\mu\lambda + o(t)$ and $E[(1-t)\lambda] - E[\lambda] \le -t\mu\lambda + o(t)$. In the limit $t \to 0$, this becomes $dE/d\lambda = \mu$. It remains to show that $\lambda < 1 \Leftrightarrow \mu < 0$ and $\lambda = 1 \Rightarrow \mu = 0$, which $\Rightarrow \lambda \ge 1$. Note that Φ goes asymptotically as $(1-\lambda)/r$. If μ were 0, then

$$\rho \overset{r\to\infty}{\sim} \left(\frac{1-\lambda}{r}\right)^{3/2},$$

which would not be integrable; thus μ must be negative when $\lambda < 1$. When $\lambda > 1$, there is no minimum, since if there were, then Φ would be negative as $r \to \infty$, which is impossible because of (4.1.37). However, the infimum has to be $E(1)$, since for $\lambda > 1$ and for any $\varepsilon > 0$ a ρ can be constructed such that $E(\rho) < E(1) + \varepsilon$; start with a ρ_1 with $\lambda = 1$ and compact support, and such that $E(\rho_1) \le E(1) + \varepsilon/2$, and then let

$$\rho_k = \rho_1 + \frac{1}{k}\chi_k, \qquad k \in \mathbb{N},$$

where the characteristic functions χ_k satisfy $\chi_k \rho_1 \equiv 0$ and $\|\chi_k\|_1 = k(\lambda - 1)$ to ensure that $\rho_k \in S_\lambda$. Then $\|\rho_1 - \rho_k\|_p \to 0$ for all $p > 1$, and it is easy to verify that $E(\rho_k) \to E(\rho_1)$. This accords with the intuitive feeling that a thin electron cloud at a great distance affects the energy only slightly. It means that $E[\lambda]$ decreases while $0 < \lambda < 1$, and becomes constant thereafter.

(iii) This follows from the proofs of (i) and (ii), since $\mu \le 0$.

(iv) Make both of the scaling transformations of (4.1.33) simultaneously and define

$$\inf_{\rho \in S_1} (K(\rho) - ZA(\rho) + \alpha R(\rho)) = Z^2 \inf_{\rho \in S_1} \left(K(\rho) - A(\rho) + \frac{\alpha}{Z} R(\rho)\right)$$

$$\equiv Z^2 f\left(\frac{\alpha}{Z}\right).$$

This is the infimum of a set of linear functions and consequently concave in (Z, α). The condition that

$$\frac{\partial^2}{\partial Z^2}\frac{\partial^2}{\partial \alpha^2} \le \left(\frac{\partial^2}{\partial Z \alpha}\right)^2$$

implies that $2f'' \le f'^2/f$, so $-\sqrt{-f}$ is concave. Because $f' = R(\rho) > 0$, the function f is increasing. With still another scaling transformation, with $\rho(\mathbf{x}) = \lambda\bar{\rho}(\lambda^{2/3}\mathbf{x})$,

$$f(\lambda) = \inf_{\rho \in S_1} (K(\rho) - A(\rho) + \lambda R(\rho)) = \lambda^{-1/3} \inf_{\bar{\rho} \in S^\lambda} (K(\bar{\rho}) - A(\bar{\rho}) + R(\bar{\rho}))$$

$$= \lambda^{-1/3} E[\lambda]. \qquad \square$$

The at first sight contradictory properties (iii) and (iv) determine the form of $E[\lambda]$ rather narrowly for an atom, making it almost linear:

Properties of $f(\lambda) = \lambda^{-1/3}E[\lambda]$ for an Atom (4.1.39)

(i) $0 \le f' \le -\dfrac{1}{3\lambda} f$;

(ii) $\dfrac{2f}{9\lambda^2} - \dfrac{2f'}{3\lambda} \le f'' \le \dfrac{f'^2}{2f}$.

Proof

(i) This follows from $E' < 0$ and $f' = \lambda^{-4/3}R(\rho_\lambda) = R(\rho_1) > 0$, where ρ_λ and ρ_1 are the minimizing densities of S_λ and S_1.

(ii) This follows from $E'' \ge 0$ and the concavity of $-\sqrt{-f}$. □

Consequence (4.1.40)

1. With the aid of the virial theorem, $2K = A - R$, which follows from (4.1.33) for any μ, Property (i) may be rewritten as $7R(\rho_\lambda) < A(\rho_\lambda)$, $0 \le \lambda < 1$. This generalizes Corollary (4.1.34), which held for $\lambda = 1 \Rightarrow \mu = 0$, to the statement that $7R = A$, provided that $0 \le \lambda < 1$.
2. It is not hard to calculate analytically that $f(0) = -0.572$ and $f'(0) = 0.2424$ (Problem 4); computer analysis of the Thomas–Fermi equation has shown that $f(1)$ is -0.384, and by (4.1.38(ii)) and (4.1.34), $f'(1) = -f(1)/3$. Integrating Property (ii) leads to the bounds

$$\max\{-\lambda^{-1/6}|f(1)|^{1/2}, -\lambda|f(1)|^{1/2} - (1-\lambda)|f(0)|^{1/2}\} \le -|f(\lambda)|^{1/2}$$
$$\le \min\left\{-|f(0)|^{1/2}\left(1 + \frac{\lambda}{2}\frac{f'(0)}{f(0)}\right), -|f(1)|^{1/2}\frac{7-\lambda}{6}\right\}$$

(cf. (III: 4.3.21)). The concave hull of the left side can be taken, in which case the greatest difference between the bounds is <2% (see Figure 29). Since this is already better accuracy than that of the Thomas–Fermi theory itself, there is no point in making fancy numerical calculations of $E[\lambda]$.

If from (4.1.36) we now deduce

The Thomas–Fermi Equation (4.1.41)

in the form

$$\Delta\Phi(\mathbf{x}) = -4\pi\delta^3(\mathbf{x}) + 4\pi\rho(\mathbf{x}) = -4\pi\delta^3(\mathbf{x}) + \frac{4}{3\pi}|\mu + \Phi(\mathbf{x})|_+^{3/2},$$

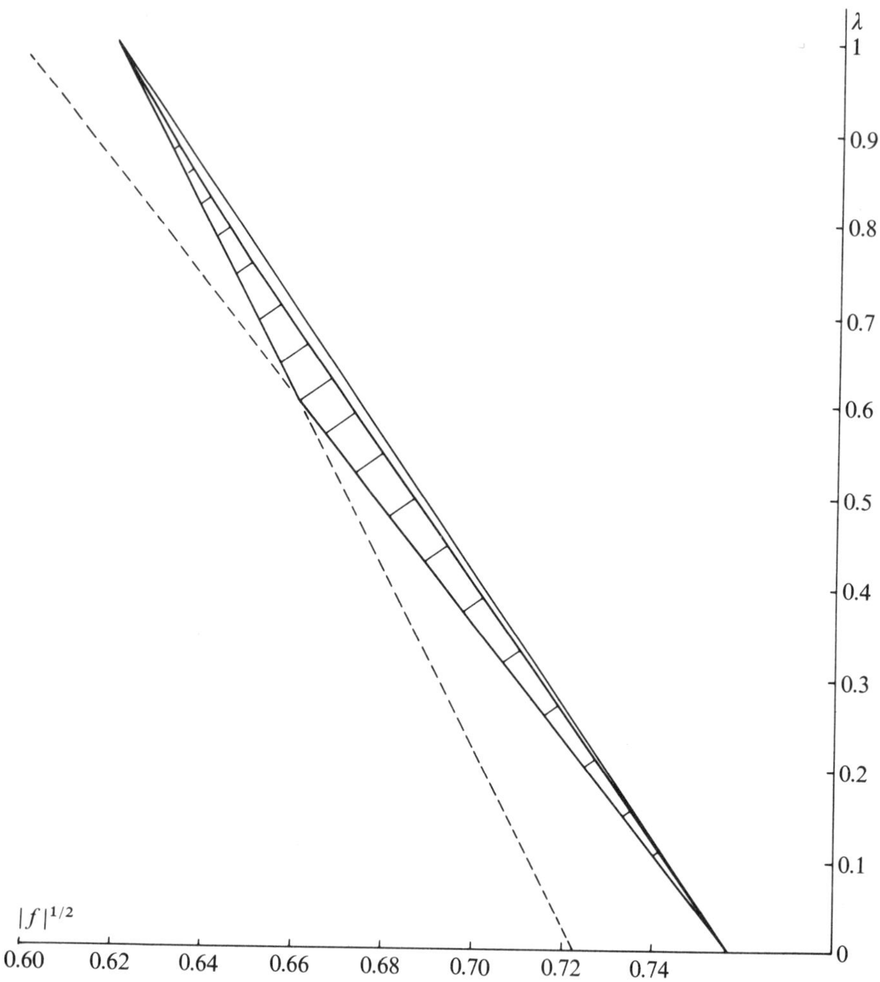

Figure 29 The bounds (4.1.40; 2) from the concavity of $f(\lambda) = \lambda^{-1/3}E(\lambda)$. The hatched region is allowed.

then it reduces to $\sqrt{\zeta}\chi'' = \chi^{3/2}\Theta(\chi)$ for spherically symmetric densities, with the substitution $|\mathbf{x}| = r = \zeta(3\pi/4)^{2/3}$, $\Phi(\mathbf{x}) + \mu = \chi(\zeta)/r$. The delta function is taken care of by the boundary condition $\chi(0) = 1$. The second boundary condition, required to make the solution unique, is $\chi'(\infty) = \mu$, which follows from $\int \rho \leq 1$ with Gauss's theorem. The function χ is concave and decreasing, and has the limiting forms

$$\chi(\zeta) \begin{array}{l} \xrightarrow{\zeta \to 0} 1 - 1.59\zeta \\ \xrightarrow[\zeta \to \infty]{} 144/\zeta^3 \end{array}$$

for $\mu = 0$. This means that for neutral atoms ρ behaves like $r^{-3/2}$ at small r, and like r^{-6} at large r. A numerical solution is shown in Figure 30. A compu-

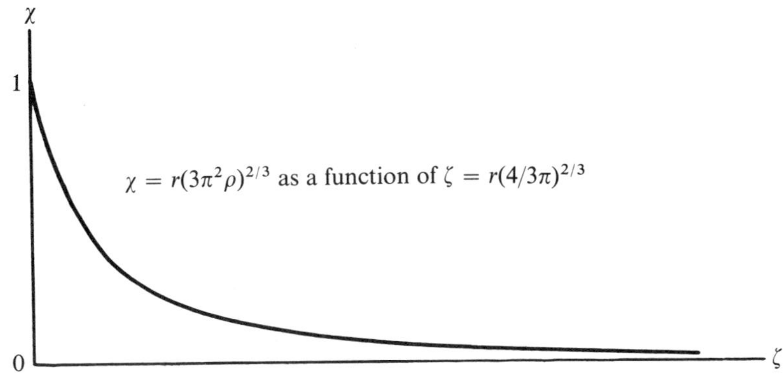

Figure 30 The Thomas–Fermi density of an atom.

tation of the energy of the solution yields the value $E(1) = -0.384$, i.e., -0.77 atomic units, or -20.7 eV.

The final proposition deduced from Thomas–Fermi theory will be that there is no chemical binding, which means that actual chemical binding energies must be smaller than the errors in the theory. In §4.3 it will be learned that this theory with some constants changed gives a lower bound for quantum-mechanical energies even for finite Z, and thereby leads to a simple proof of the stability of matter. Finally, we shall obtain the long-deferred proof of Inequality (4.1.5).

Monotony of the Thomas–Fermi Potential with Respect to the Nuclear Charges (4.1.42)

Let $\Phi_{1,2}$ and $\rho_{1,2}$ be the solutions of the Thomas–Fermi equation with $\mu = 0$ and nuclear charges $z_k^{(1,2)}$. If $z_k^{(1)} \geq z_k^{(2)}$ for all k, then $\Phi_1(\mathbf{x}) \geq \Phi_2(\mathbf{x})$ and $\rho_1(\mathbf{x}) \geq \rho_2(\mathbf{x})$ for all $\mathbf{x}$.

Remarks (4.1.43)

1. The normalization $\sum_k z_k^{(i)} = 1$ has of course been dropped.
2. The condition $\mu = 0$ means $\int d^3x\rho_1(\mathbf{x}) = \sum_k z_k^{(1)} \geq \int d^3x\rho_2(\mathbf{x}) = \sum_k z_k^{(2)}$.
3. This can be interpreted as showing how increasing all the nuclear charges causes the configuration with lower energy to have a higher electron density.

Proof

As in the proof of (4.1.37(ii)), let $A \equiv \{\mathbf{x} \in \mathbb{R}^3 : \Phi_1(\mathbf{x}) < \Phi_2(\mathbf{x})\}$. Then A is open and contains none of the $\mathbf{X}_k$, and on it $\psi(\mathbf{x}) \equiv \Phi_1(\mathbf{x}) - \Phi_2(\mathbf{x})$ is negative, continuous, and satisfies

$$\Delta\psi|_A = (\Phi_1^{3/2} - \Phi_2^{3/2})|_A < 0.$$

Hence ψ approaches its infimum on A either on the boundary or at infinity. Since it then vanishes throughout A, the set A must be empty. □

The next fact to show is that molecular energies are always greater than those of the isolated atoms. This will require the

Feynman–Hellmann Formula of Thomas–Fermi Theory (4.1.44)

Let $E(Z) = \inf_\rho (K(\rho) - ZA(\rho) + R(\rho))$. *Then* $\partial E/\partial Z = -A(\rho_Z)$, *where* ρ_Z *is the density that minimizes* $E(Z)$.

Proof

The function $E(Z)$ is concave, and its right and left derivatives are $\lim_{\varepsilon \downarrow 0} (-A(\rho_{Z \pm \varepsilon}))$, another consequence of the interplay between the concavity of $E(Z)$ and the convexity of the functional in the variable ρ as in (4.1.38(ii)). Since, as shown in Problem 3, for any Z there exists a unique minimizing ρ_Z on a certain compact set, the densities ρ_Z depend continuously on Z. In fact the individual contributions to $E(Z)$ are continuous in Z as well as $E(Z)$ itself. Therefore both the right and the left derivative coincide with $-A(\rho_Z)$. □

Let us now start treating E as a function of each of the nuclear charges, so

$$E(z_1, \ldots, z_M) = \inf_{\rho \in S} \left\{ \tfrac{3}{5}(3\pi^2)^{2/3} \int \rho^{5/3} - \sum_{k=1}^{M} z_k \int \frac{\rho(\mathbf{x})}{|\mathbf{x} - \mathbf{X}_k|} + \frac{1}{2} \int \frac{\rho(\mathbf{x})\rho(\mathbf{y})}{|\mathbf{x} - \mathbf{y}|} + \sum_{k>i} \frac{z_k z_i}{|\mathbf{X}_k - \mathbf{X}_i|} \right\},$$

and define

$$\begin{aligned} E(Z) &= E(Zz_1, Zz_2, \ldots, Zz_M), \\ E_1(Z) &= E(Zz_1, \ldots, Zz_j, 0, 0, \ldots), \\ E_2(Z) &= E(0, \ldots, 0, Zz_{j+1}, \ldots, Zz_M). \end{aligned}$$

Let $\rho \geq \rho_{1,2}$ and $\Phi \geq \Phi_{1,2}$ be the solutions of the appropriately subscripted Thomas–Fermi equations. Then

$$\begin{aligned} \frac{\partial E_1}{\partial Z} &= \sum_{k=1}^{j} z_k \left\{ Z \sum_{i \neq k} \frac{z_i}{|\mathbf{X}_i - \mathbf{X}_k|} - \int \frac{d^3x \rho(\mathbf{x})}{|\mathbf{x} - \mathbf{X}_k|} \right\} \\ &= \sum_{k=1}^{j} z_k \lim_{\mathbf{x} \to \mathbf{X}_k} \left(\Phi_1(\mathbf{x}) - \frac{Zz_k}{|\mathbf{x} - \mathbf{X}_k|} \right), \end{aligned}$$

and likewise for E_2. The difference between the energy of the total system and the sum of the energies of the subsystems is easily found to satisfy

$$\frac{\partial E}{\partial Z} - \frac{\partial E_1}{\partial Z} - \frac{\partial E_2}{\partial Z} = \sum_{k=1}^{j} z_k(\Phi(\mathbf{x}_k) - \Phi_1(\mathbf{x}_k)) + \sum_{k=j+1}^{M} z_k(\Phi(\mathbf{x}_k) - \Phi_2(\mathbf{x}_k)) \geq 0.$$

Since E and $E_{1,2}$ become zero when $Z = 0$, this calculation proves the

Instability of Molecules in Thomas–Fermi Theory (4.1.45)

$$E(z_1, \ldots, z_M) \geq E(z_1, \ldots, z_j) + E(z_{j+1}, \ldots, z_M).$$

Remarks (4.1.46)

1. In the absence of nuclear repulsion the inequality is reversed; in that case

$$\frac{\partial E}{\partial Z} - \frac{\partial E_1}{\partial Z} - \frac{\partial E_2}{\partial Z} = \sum_{k=1}^{j} z_k \int \frac{\rho_1(\mathbf{x}) - \rho(\mathbf{x})}{|\mathbf{x} - \mathbf{X}_k|} + \sum_{k=j+1}^{M} \int \frac{\rho_2(\mathbf{x}) - \rho(\mathbf{x})}{|\mathbf{x} - \mathbf{X}_k|} \leq 0.$$

Although Thomas–Fermi theory predicts some attraction between the nuclei, it is weaker than their Coulomb repulsion. It can even be shown that if the nuclear coordinates are scaled by $X_k \to RX_k$, then E is a convex, decreasing function of R. Thus Thomas–Fermi theory predicts positive pressure and compressibility. However, the molecular energy is not a sum of pair potentials, but contains many-body potentials with alternating signs [34].

2. An alternative version of this theorem reads

$$\sum_{k>i} \frac{z_k z_i}{|\mathbf{X}_k - \mathbf{X}_i|} \geq \sum_{k=1}^{M} z_k \int \frac{\rho(\mathbf{x})}{|\mathbf{x} - \mathbf{X}_k|} - \frac{1}{2} \int \frac{\rho(\mathbf{x})\rho(\mathbf{y})}{|\mathbf{x} - \mathbf{y}|}$$
$$- \frac{3}{5}(3\pi^2)^{2/3} \int \rho^{5/3} + \sum_k E(z_k)$$

for all $\mathbf{X}_k \in \mathbb{R}^3$ and $\rho \in S$. If $K(\rho)$ is replaced with $(1/\gamma)K(\rho)$, then, because of the way dilatations affect single atoms, $E(z_k)$ becomes $\gamma E(z_k)$. The computed value $E(1) = -0.384$ then leads to Equation (4.1.5), provided that $\mathbf{X}_k$ are interpreted as the coordinates of the electrons.

3. The proof of (4.1.45) works the same way for a Yukawa potential $\exp(-\mu r)/r$ in place of $1/r$. Because $\Delta \exp(-\mu r)/r + 4\pi\delta^3(\mathbf{x}) = \mu^2 \exp(-\mu r)/r > 0$, the argument with subharmonicity likewise works: $\Delta\psi|_A = \Phi_1^{3/2} - \Phi_2^{3/2} + \mu^2(\Phi_1 - \Phi_2) < 0$, which implies that A must be empty.

Problems (4.1.47)

1. Let $H = |\mathbf{p}|^2 + V(\mathbf{x})$ act on $L^2(\mathbb{R}^3)$, and assume that $|V|_- \in L^{5/2}(\mathbb{R}^3)$ and let e_i be the negative eigenvalues of H. Use the bound of Ghirardi and Rimini (III: 3.5.37; 2) to show that

$$\sum_i |e_i| \leq \frac{4}{15\pi} \int d^3x |V(\mathbf{x})|_-^{5/2}$$

and derive Inequality (4.1.19) from this fact.

2. Use Problem 1 to prove the inequality

$$T \equiv \langle\psi| - \sum_{i=1}^{N} \Delta_i |\psi\rangle \geq \frac{3}{5}\left(\frac{3\pi}{4}\right)^{2/3} \int d^3x \rho^{5/3}(\mathbf{x})$$

for spin $-\frac{1}{2}$ fermions, where

$$\rho(\mathbf{x}_1) = N \sum_{\alpha_i} \int d^3x_2 \cdots d^3x_N |\psi(\mathbf{x}_1, \mathbf{x}_2, \ldots, \mathbf{x}_N; \alpha_2, \ldots, \alpha_N)|^2,$$

α being the spin index. (Hint: use $\rho^{2/3}$ as the potential in Problem 1.)

3. Show that the sets $S \equiv \{\rho \in L^1 \cap L^{5/3}: \rho \geq 0, \|\rho\|_1 \leq N, \|\rho\|_{5/3} \leq K\}$ are compact in the weak $L^{5/3}$ topology, and that the functional $S \to \mathbb{R}$:

$$\varepsilon(\rho) = \tfrac{3}{5}(3\pi^2)^{2/3} \int d^3x \rho^{5/3}(\mathbf{x}) - \int d^3x \rho(\mathbf{x}) \left(\sum_k \frac{z_k}{|\mathbf{x} - \mathbf{X}_k|} + \mu\right)$$
$$+ \frac{1}{2} \int \frac{d^3x\, d^3x'}{|\mathbf{x} - \mathbf{x}'|} \rho(\mathbf{x})\rho(\mathbf{x}') + \sum_{k>j} \frac{z_k z_j}{|\mathbf{X}_k - \mathbf{X}_j|}$$

has Properties (4.1.10) if $\mu \leq 0$: It is

(i) weakly $L^{5/3}$ lower semicontinuous;
(ii) strictly convex; and
(iii) $\geq \frac{3}{5}(3\pi^2)^{2/3} \|\rho\|_{5/3}^{5/3} - 3(\frac{2}{3})^{5/6}(8\pi)^{1/3} \|\rho\|_{5/3}^{5/6} + |\mu| \|\rho\|_1$.

Conclude that the infimum is attained, and in fact precisely with the ρ of (4.1.36).

4. Solve the Thomas–Fermi equation without Coulomb repulsion, compare with (III: 4.5.9), and conclude that the next correction is $0(N^{6/3})$. Use the solution to calculate $f(0)$ and $f'(0)$ of (4.1.40; 2).

5. Minimize the functional

$$E(\rho) = \int d^3x \left(\frac{\rho^2(\mathbf{x})2\pi}{\mu^2} - \sum_i \frac{z_i}{|\mathbf{x} - \mathbf{X}_i|} \rho(\mathbf{x}) + \frac{1}{2} \int \frac{d^3y \rho(\mathbf{y})}{|\mathbf{x} - \mathbf{y}|}\right) + \sum_{i>j} \frac{z_i z_j}{|\mathbf{X}_i - \mathbf{X}_j|},$$

and use the result for a new derivation of (III: 4.5.24):

$$\sum_{i>j} |\mathbf{X}_i - \mathbf{X}_j|^{-1} \geq \sum_{j=1}^{N} \int \frac{\rho(\mathbf{x})}{|\mathbf{x} - \mathbf{X}_i|} - \frac{1}{2} \int \frac{\rho(\mathbf{x})\rho(\mathbf{y})}{|\mathbf{x} - \mathbf{y}|} - \frac{2\pi}{\mu^2} \int \rho^2 - \frac{\mu N}{2}$$

for all $\mathbf{X}_i \in \mathbb{R}^3$, $\rho \in L^1 \cap L^2$.

Solutions (4.1.48)

1. Let $N_E(V)$ denote the number of eigenvalues less than or equal to E. According to (III: 3.5.37; 2), for all $\alpha > 0$,

$$N_{-\alpha}(V) \le N_{-\alpha/2}\left(\left|V + \frac{\alpha}{2}\right|_-\right) \le \operatorname{tr}\left[\left(|\mathbf{p}|^2 + \frac{\alpha}{2}\right)^{-1/2}\left|V + \frac{\alpha}{2}\right|_-\left(|\mathbf{p}|^2 + \frac{\alpha}{2}\right)^{-1/2}\right]^2$$

$$= \frac{1}{(4\pi)^2}\int d^3x\, d^3y \left|V(\mathbf{x}) + \frac{\alpha}{2}\right|_- \frac{\exp(-\sqrt{2\alpha}|\mathbf{x} - \mathbf{y}|)}{|\mathbf{x} - \mathbf{y}|^2}\left|V(\mathbf{y}) + \frac{\alpha}{2}\right|_-$$

$$\le \frac{1}{4\pi\sqrt{2\alpha}}\int d^3x \left|V(\mathbf{x}) + \frac{\alpha}{2}\right|_-^2.$$

The last step used Young's inequality, $\|f \cdot (v * g)\|_1 \le \|v\|_1 \|f\|_2 \|g\|_2$. Now simply think about what $N_E(V)$ means (see Figure 31).

$$\sum_j |e_j(V)| = \int_0^\infty d\alpha N_{-\alpha}(V) \le \frac{\sqrt{2}}{8\pi}\int d^3x \int_0^{2|V(\mathbf{x})|_-} \frac{d\alpha}{\sqrt{\alpha}}\left(V(\mathbf{x}) + \frac{\alpha}{2}\right)^2$$

$$= \frac{4}{15\pi}\int d^3x |V(\mathbf{x})|_-^{5/2}.$$

If $|V|_- \in L^{5/2}$, then the negative part of the spectrum of H is discrete, and we may also write

$$\operatorname{Tr}||\mathbf{p}|^2 + V(\mathbf{x})|_- \le 4\pi \int \frac{d^3x\, d^3p}{(2\pi)^3} ||\mathbf{p}|^2 + V(\mathbf{x})|_-.$$

The partition function can be bounded with the observation that

$$\ln(1 + \exp(-\beta H)) = \int_{-\infty}^{\infty} dE |H - E|_- \beta(1 + \exp(\beta E))^{-1} \Rightarrow$$

$$\operatorname{tr}\ln(1 + \exp[-\beta(|\mathbf{p}|^2 + V(\mathbf{x}))]) \le 4\pi \int \frac{d^3x\, d^3p}{(2\pi)^3} \ln(1 + \exp[-\beta(|\mathbf{p}|^2 + V(\mathbf{x}))]).$$

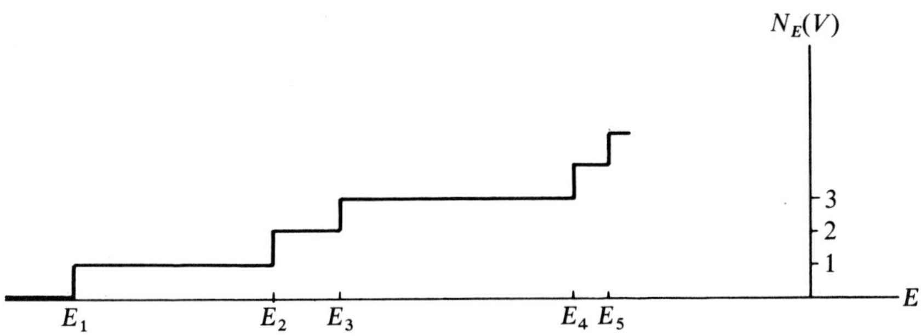

Figure 31 The dependence of the function $N_E(V)$ on E.

2. Let

$$\rho_\pm(\mathbf{x}_1) = N \sum_{\alpha_2,\ldots,\alpha_N} \int d^3x_2 \cdots d^3x_N |\psi(\mathbf{x}_1, \mathbf{x}_2, \ldots, \mathbf{x}_N; \pm, \alpha_2, \ldots, \alpha_N)|^2$$

be the densities of the electrons with spin $\pm\frac{1}{2}$ and $K = T(\int \rho_+^{5/3} + \int \rho_-^{5/3})^{-1}$. Because of Problem 1 and the min–max principle the lowest energy E_0 of the Hamiltonian

$$H = \sum_i \left(|\mathbf{p}_i|^2 - \frac{5K}{3} [\pi_{i,+} \rho_+^{2/3}(\mathbf{x}_i) + \pi_{i,-} \rho_-^{2/3}(\mathbf{x}_i)] \right),$$

where $\pi_{i,\pm}$ are the spin projections, satisfies the inequalities

$$-\frac{4}{15\pi}\left(\frac{5K}{3}\right)^{5/2} \left(\int \rho_+^{5/3} + \int \rho_-^{5/3}\right) \leq E_0 \leq \langle \psi | H | \psi \rangle = T - \frac{5K}{3}\left(\int \rho_+^{5/3} + \int \rho_-^{5/3}\right).$$

This implies that $K \geq \frac{3}{5}(3\pi/2)^{2/3}$, and then the convexity of the function $x \to x^{5/3}$ yields the inequality for $\rho = \rho_+ + \rho_-$.

3. Since

$$\|\rho\|_{5/3} = \sup_{\|\rho'\|_{5/2} = 1} |\langle \rho' | \rho \rangle| \quad \text{and} \quad \|\rho\|_1 = \sup_{\substack{\rho' \in L^{5/2} \cap L^\infty \\ \|\rho'\|_\infty = 1}} |\langle \rho' | \rho \rangle|$$

are suprema over weakly continuous functions, they are weakly lower semicontinuous, so S is weakly compact.

(i) This proposition is equivalent to the statement that $\rho_n \rightharpoonup \rho \Rightarrow \underline{\lim}\, \varepsilon(\rho_n) \geq \varepsilon(\rho)$. First note that $\|\rho\|_{5/3}$ is weakly lower semicontinuous, i.e., $\lim \|\rho_n\|_{5/3} \geq \|\rho\|_{5/3}$. Moreover, $\lim_{n\to\infty} \int \rho_n(1/|\mathbf{x}|) = \int \rho(1/|\mathbf{x}|)$. If the potential $1/|\mathbf{x}|$ is broken up as $1/|\mathbf{x}| = V_1 + V_2$, where $V_1 \in L^{5/2}$, $V_2 \in L^p$, $3 < p \leq \infty$, then by assumption $\int \rho_n V_1$ converges to $\int \rho V_1$. Since $\sup_n \|\rho_n\|_1 < \infty$ (by assumption $\{\rho_n\}$ is bounded in L^1), $\rho_n \to \rho$ in the weak topologies of all L^q spaces with $1 < q \leq \frac{5}{3}$. This follows from the density of $L^{5/2} \cap L^s$ in L^s for $s \geq \frac{5}{2}$, $1/s + 1/q = 1$, and $\sup_n \|\rho_n\|_q < \infty$, because $\|\rho\|_q \leq \|\rho\|_p^\alpha \|\rho\|_r^{1-\alpha}$ for $1/q = \alpha/p + (1-\alpha)/r$. Hence also $\int \rho_n V_2 \to \int \rho V_2$, proving the convergence of the nuclear attraction. Finally, for the repulsion of the electrons we can write

$$\left\| \left(\rho_n * \frac{1}{|\mathbf{x}|}\right)\rho_n \right\|_1 = \left\| \left((\rho_n - \rho) * \frac{1}{|\mathbf{x}|}\right)(\rho_n - \rho) \right\|_1 + 2\left\| \left(\rho * \frac{1}{|\mathbf{x}|}\right)\rho_n \right\|_1 - \left\| \left(\rho * \frac{1}{|\mathbf{x}|}\right)\rho \right\|_1$$

By Young's inequality, if V is broken up as above and $\rho \in L^1$, then $\rho * V_1 \in L^{5/2}$, $\rho * V_2 \in L^p$, $3 < p \leq \infty$, so the mixed term on the right converges to $2\|(\rho * 1/|\mathbf{x}|)\rho\|_1$, while the first term is positive. Therefore

$$\underline{\lim} \left\| \left(\rho_n * \frac{1}{|\mathbf{x}|}\right)\rho_n \right\| \geq \left\| \left(\rho * \frac{1}{|\mathbf{x}|}\right)\rho \right\|.$$

(ii) $\rho^{5/3}$ is strictly convex, $\int \rho(1/|\mathbf{x}|)$ is linear, and

$$\int \frac{d^3x \rho(\mathbf{x})\rho(\mathbf{y})}{|\mathbf{x} - \mathbf{y}|} = c \int \frac{d^3k}{|\mathbf{k}|^2} |\tilde{\rho}(\mathbf{k})|^2,$$

$c > 0$, is strictly convex.

(iii) The proof of semiboundedness on S will require the following refinements of our earlier estimates. Let $R > 0$,

$$I_+ \equiv \int_{|\mathbf{x}| \geq R} d^3x \frac{\rho(\mathbf{x})}{|\mathbf{x}|}, \qquad I_- \equiv \int_{|\mathbf{x}| \leq R} d^3x \frac{\rho(\mathbf{x})}{|\mathbf{x}|} \quad \text{and} \quad f(\mathbf{x}) = \frac{\delta(|\mathbf{x}| - R)}{4\pi}.$$

It follows from

$$\int \frac{d^3x\, d^3y}{|\mathbf{x} - \mathbf{y}|} (\rho(\mathbf{x})\Theta(|\mathbf{x}| - R) - f(\mathbf{x}))(\rho(\mathbf{y})\Theta(|\mathbf{y}| - R) - f(\mathbf{y})) \geq 0,$$

$$I_+ = \int_{\|\mathbf{y}\| \geq R} \frac{d^3x\, d^3y}{|\mathbf{x} - \mathbf{y}|} f(\mathbf{x})\rho(\mathbf{y}),$$

and

$$\int \frac{d^3x\, d^3y}{|\mathbf{x} - \mathbf{y}|} f(\mathbf{x}) f(\mathbf{y}) = \frac{1}{R}$$

that

$$\frac{1}{2} \int_{\substack{|\mathbf{x}| \geq R \\ |\mathbf{y}| \geq R}} \frac{d^3x\, d^3y}{|\mathbf{x} - \mathbf{y}|} \rho(\mathbf{x})\rho(\mathbf{y}) - I_+ \geq -\frac{1}{2R},$$

and by Hölder's inequality,

$$|I_-| \leq \left\| \frac{\Theta(R - |\mathbf{x}|)}{|\mathbf{x}|} \right\|_{5/2} \|\rho\|_{5/3} = (64\pi^2 R)^{1/5} \|\rho\|_{5/3}.$$

If R is chosen as $R = \frac{5}{2}((8\pi)^{2/5} \|\rho\|_{5/3})^{-1}$, then with $\sum_k z_k \leq 1$,

$$\varepsilon(\rho) \geq \tfrac{3}{5}(3\pi^2)^{2/3} \|\rho\|_{5/3}^{5/3} - 3(\tfrac{2}{5})^{5/6}(8\pi)^{1/3} \|\rho\|_{5/3}^{5/6} + \sum_{k>j} \frac{z_k z_j}{|\mathbf{X}_k - \mathbf{X}_j|},$$

and the function $ax^2 - bx + c$ is bounded below on $\mathbb{R}$ for non-negative a, b, and c.

If $\mu < 0$, then because of (iii) the infimum is attained for a ρ in the interior of one of the compact sets S, and ρ must satisfy the Thomas–Fermi equation (4.1.36) by the same argument as in (4.1.12). If $\mu = 0$ then there is also the possibility that the infimum is attained on the boundary $\|\rho\|_1 = N$ of every set S. In that event it would still satisfy the Thomas–Fermi equation with some μ as the Lagrange multiplier for the constraint $\|\rho\|_1 = N$. However, if $N > \sum_i z_i \leq 1$, then there is no such solution, as otherwise $\Phi(\mathbf{x})$ would be negative for large $|\mathbf{x}|$, contradicting (4.1.37(ii)). Therefore, if $\mu = 0$, then the infimum still lies in the interior of some set S.

4. Use units such that $e = \hbar = 2m = 1$, and suppose there is spin. Then

$$E = \int d^3x \left[\frac{3}{5}(3\pi^2)^{2/3} \rho^{5/3} - \frac{Z}{r} \rho \right].$$

From the Thomas–Fermi equations,

$$(3\pi^2\rho)^{2/3} - \frac{Z}{r} + \mu = 0 \Rightarrow \rho = \frac{Z^{3/2}}{3\pi^2}\left(\frac{1}{r} - \frac{1}{R}\right)^{3/2}, \qquad \mu = Z/R.$$

$$N = \frac{Z^{3/2}}{3\pi^2} 4\pi \int_0^R r^2\,dr\left(\frac{1}{r} - \frac{1}{R}\right)^{3/2} = \frac{(ZR)^{3/2}}{12} \Rightarrow R = N^{-1/3}\frac{N}{Z}4(\tfrac{3}{2})^{2/3},$$

$$-V = Z^{5/2}\frac{4\pi}{3\pi^2}\int_0^R r\,dr\left(\frac{1}{r} - \frac{1}{R}\right)^{3/2} = \frac{6NZ}{R},$$

$$T = \tfrac{3}{5}Z^{5/3}\frac{4\pi}{3\pi^2}\int_0^R r^2\,dr\left(\frac{1}{r} - \frac{1}{R}\right)^{5/2} = -\frac{V}{2},$$

and

$$E = -T = \frac{3NZ}{R} = -\tfrac{1}{2}(\tfrac{3}{2})^{1/3}Z^2N^{1/3} \quad \text{(in units with } 2m = 1\text{; twice this if } m = 1\text{)},$$

so

$$f(0) = -\tfrac{1}{2}(\tfrac{3}{2})^{1/3} = -0.572,$$

$$f'(0) = \frac{1}{2}\int\frac{d^3x\,d^3y}{|\mathbf{x} - \mathbf{y}|}\rho(\mathbf{x})\rho(\mathbf{y}) = \left(\frac{4}{3\pi}\right)^2\int_0^1\frac{dr}{\sqrt{r}}(1 - r)^{3/2}\int_0^r dr'\sqrt{r'}(1 - r')^{3/2}\frac{(ZR)^3}{R}$$

$$= 0.24244, \quad \text{if } Z = N = 1.$$

If we read the exact ground-state energy off from (III: 4.5.15), then to $o(N^{-1/3})$,

$$E_{\text{exact}}/E_{\text{Thomas–Fermi}} = 1 - N^{-1/3}\tfrac{1}{2}(\tfrac{3}{2})^{1/3}.$$

Thus the Thomas–Fermi energy is below the actual ground-state energy.

5. The density that minimizes E is

$$\rho_0(\mathbf{x}) = \frac{\mu^2}{4\pi}\sum_j z_j\frac{\exp(-\mu|\mathbf{x} - \mathbf{X}_j|)}{|\mathbf{x} - \mathbf{X}_j|},$$

with which

$$E(\rho_0) = -\frac{N\mu}{2} + \sum_{i>j} z_i z_j\left(\frac{1}{|\mathbf{X}_i - \mathbf{X}_j|} - \mu\exp[-\mu|\mathbf{X}_i - \mathbf{X}_j|]\right) > -\frac{N\mu}{2}$$

for all $\mathbf{X}_i \in \mathbb{R}^3$.

If $z_i = 1$, this reduces to (III: 4.5.24). In this variant of Thomas–Fermi theory the electron cloud creates an attractive potential $-\mu\exp(-\mu r)$ between the nuclei, which is also weaker than their $1/r$ Coulomb repulsion.

4.2 Cosmic Bodies

The Thomas–Fermi theory of stars is thermodynamically more interesting than that of atoms, since it predicts an unusual phase transition

In the year 1926 great discoveries about the laws of matter appeared in rapid succession. Shortly after E. Schrödinger published the equation named after him, E. Fermi discovered the distribution law (2.5.22; 1) governing particles that satisfy the exclusion principle. This inspired L. Thomas's ingenious idea that the electron cloud of a large atom should satisfy equation (4.1.36) at $T = 0$. Then, still in the year 1926, R. Fowler realized that the stability of cosmic matter is ensured by the zero-point energy of the electrons, and that a cosmic body is closely analogous to a "gigantic molecule in the ground state." Yet it has taken considerably longer to found this vision in mathematics and derive everything from the Schrödinger equation. Today, however, the derivation goes through without gaps, and the Thomas–Fermi theory of atoms and stars is the only many-body problem with realistic forces to have succumbed, in the appropriate thermodynamic limit, to mankind's attempts at calculation.

Yet the zero-point energy guarantees stability only in so far as the speeds of the electrons remain slow in comparison with light. If they enter the regime of relativistic kinematics, for which the kinetic energy $\sim c|\mathbf{p}|$, then the zero-point energy goes as $N(N/V)^{1/3}$, whereas the gravitational energy goes as $-\kappa N^2/V^{1/3}$. If $N > (\kappa m_p^2)^{-3/2} \sim 10^{57}$, then the latter predominates, and as V becomes smaller and smaller, the total energy goes to $-\infty$. We shall avoid this instability by remaining within the framework of nonrelativistic kinematics, considering only stars of masses somewhat smaller than that of the sun. Then, according to the estimates (1.2.23; 3), if $N > 10^{54}$, the minimum energy occurs when $V \sim N^{-1}$. The situation is again like that of Thomas–Fermi theory, which leads to the hope that the many-body problem can be solved in the limit $N \to \infty$ with the Hamiltonian

$$H = \sum_{i=1}^{N} \frac{|\mathbf{p}_i|^2}{2m_i} + \sum_{i>j} \frac{e_i e_j - \kappa m_i m_j}{|\mathbf{x}_i - \mathbf{x}_j|}. \tag{4.2.1}$$

In this limit the system becomes a highly compressed plasma, so the average gravitational field would be expected to be so dominant that the Thomas–Fermi equation is valid. Of course, the total charge of the system must be zero, or, more exactly, the possible excess charge ΔQ is bounded by $(\Delta Q)^2 \leq \kappa m_p^2 N_p^2$, so for gravity to predominate, $\Delta Q < 10^{-19} N_p$. Indeed, these conjectures can be derived mathematically for all three ensembles:

The Asymptotic Forms of the State Functions (4.2.2)

Let $H_{N,V}$ be the Hamiltonian (4.2.1) *for N_1 positive and N_2 negative fermions of masses $M_{1,2}$, charges e_1 and $e_2 = -e_1$, and spin $\frac{1}{2}$ in a volume V. Let N denote the pair (N_1, N_2). Then the limits*

$$E(N, S, V) = \lim_{\lambda\to\infty} \lambda^{-7/3} \inf_{\mathscr{H}_{\lambda S}} \operatorname{Tr}_{\mathscr{H}_{\lambda S}} H_{\lambda N, \lambda^{-1}V},$$

$$F(N, \beta, V) = -\lim_{\lambda\to\infty} \beta^{-1}\lambda^{-1} \ln \operatorname{Tr} \exp(-\beta\lambda^{-4/3} H_{\lambda N, \lambda^{-1}V}). \tag{4.2.3}$$

exist. The grand-canonical function Ξ is not defined as in (4.1.8), as now the finiteness of the sum $\sum_N$ requires a factor $N^{-2/3}$ in the interaction and $V \sim N$ [27] (see (4.2.10; 4)).

With the solution of the Thomas–Fermi equation

$$\rho_\alpha(\mathbf{x}) = 2 \int \frac{d^3p}{(2\pi)^3} \left[1 + \exp\left(\beta\left(\frac{|\mathbf{p}|^2}{2M_\alpha} + W_\alpha(\mathbf{x}) - \mu_\alpha\right)\right)\right]^{-1}, \tag{4.2.4}$$

$$W_\alpha(\mathbf{x}) = \sum_\beta \int_V d^3x' \frac{e_\alpha e_\beta + \kappa M_\alpha M_\beta}{|\mathbf{x} - \mathbf{x}'|} \rho_\beta(\mathbf{x}'), \qquad \alpha, \beta = 1, 2, \tag{4.2.5}$$

$$\int_V d^3x \rho_\alpha(\mathbf{x}) = N_\alpha, \tag{4.2.6}$$

these quantities are found to be

$$E(N, S, V) = \sum_\alpha \int_V d^3x \left\{\tfrac{1}{2}\rho_\alpha(\mathbf{x})W_\alpha(\mathbf{x}) + 2\int \frac{d^3p}{(2\pi)^3} \frac{|\mathbf{p}|^2/2M_\alpha}{1 + \exp[\beta(|\mathbf{p}|^2/2M_\alpha + W_\alpha(\mathbf{x}) - \mu_\alpha)]}\right\}, \tag{4.2.7}$$

$$F(N, \beta, V) = \sum_\alpha \left\{-\int_V d^3x \tfrac{1}{2}\rho_\alpha(\mathbf{x})W_\alpha(\mathbf{x}) + N_\alpha\mu_\alpha - 2T \int_V d^3x \int \frac{d^3p}{(2\pi)^3} \times \ln\left(1 + \exp\left[-\beta\left(\frac{|\mathbf{p}|^2}{2M_\alpha} + W_\alpha(\mathbf{x}) - \mu_\alpha\right)\right]\right)\right\}, \tag{4.2.8}$$

and

$$\Xi(\mu_1, \mu_2, \beta, V) = \sum_\alpha \left\{\int_V d^3x \tfrac{1}{2}\rho_\alpha(\mathbf{x})W_\alpha(\mathbf{x}) + 2T \int_V d^3x \int \frac{d^3p}{(2\pi)^3} \times \ln\left(1 + \exp\left[-\beta\left(\frac{|\mathbf{p}|^2}{2M_\alpha} + W_\alpha(\mathbf{x}) - \mu_\alpha\right)\right]\right)\right\}. \tag{4.2.9}$$

Gloss (4.2.10)

1. For $\lambda S \in \ln \mathbb{Z}^+$, $\mathscr{H}_{\lambda S}$ is an $\exp(\lambda S)$-dimensional subspace of $\mathscr{H}$.
2. The thermodynamic limit has been taken in the sense discussed in (1.2.19), i.e., $E \sim N^{7/3}$, $V \sim N^{-1}$, $S \sim N$, and $T \sim E/S \sim N^{4/3}$. The energies E and F are accordingly neither per particle nor per volume; these specific energies and energy densities do not have thermodynamic limits.
3. The quantity $S = \beta(E - F)$ is extensive for $\beta \sim N^{-4/3}$ and $E - F \sim N^{7/3}$.
4. If one insists on the usual relationships $E \sim N$, $V \sim N$, $S \sim N$, with T constant, then according to (1.2.19) the interaction has to be taken as

$$N^{-2/3} \sum_{i>j} \frac{e_i e_j - \kappa m_i m_j}{|\mathbf{x}_i - \mathbf{x}_j|}.$$

This means that the system is imagined as getting larger and larger with an ever weaker interaction; all such problems are mathematically equivalent because of the scaling law of (1.2.1). Physically relevant systems are large but finite and have weak, but still nonzero, gravitational interaction. The question of how reasonable the thermodynamic limit is depends only on whether the physical object is sufficiently like the limiting system. If so, the convergence of the thermodynamic quantities (4.2.2) guarantees that the relevant observables of the finite system will have values fairly near those of the infinite idealization.

5. Since ρ_α is a strictly monotonic function of μ_α, the normalization (4.2.6) is an implicit equation for μ_α.
6. We shall soon discover that for certain values of β, N, and V there is more than one solution of the Thomas–Fermi equations. The question of which solutions are the correct limits (4.2.3) is decided by the minimum principles for the thermodynamic potentials (2.3.3; 4), (2.2.23; 1), and (2.5.3), which survive the limit $\lambda \to \infty$ in the following manner (cf. (4.1.21)): The functionals for energy, entropy, and the phase-space densities n_α are

$$E(n) = -\frac{1}{2}\sum_{\alpha,\beta}\int d^3x\, d^3x' \frac{d^3p\, d^3p'}{(2\pi)^6} n_\alpha(\mathbf{x}, \mathbf{p}) n_\beta(\mathbf{x}', \mathbf{p}') \frac{e_\alpha e_\beta - \kappa M_\alpha M_\beta}{|\mathbf{x} - \mathbf{x}'|}$$

$$+ \sum_\alpha \int d^3x \frac{d^3p}{(2\pi)^3} \frac{|\mathbf{p}|^2}{2M_\alpha} n_\alpha(\mathbf{x}, \mathbf{p}),$$

$$S(n) = -2\sum_\alpha \int d^3x \frac{d^3p}{(2\pi)^3}\left[\frac{n_\alpha}{2}\ln\frac{n_\alpha}{2} + \left(1 - \frac{n_\alpha}{2}\right)\ln\left(1 - \frac{n_\alpha}{2}\right)\right],$$

$$N_\alpha(n) = \int d^3x \frac{d^3p}{(2\pi)^3} n_\alpha(\mathbf{x}, \mathbf{p}).$$

The correct Thomas–Fermi densities are those that minimize the energy for given N_α and S. The variational derivative with T and μ_α as Lagrange multipliers leads to the Thomas–Fermi equations (4.2.4)–(4.2.7) again, with

$$\rho_\alpha(\mathbf{x}) = \int \frac{d^3p}{(2\pi)^3} n_\alpha(\mathbf{x}, \mathbf{p}),$$

for the solution of

$$\frac{\delta}{\delta n_\alpha(\mathbf{x}, \mathbf{p})}(E - TS + \mu_1 N_1 + \mu_2 N_2) = 0.$$

However, this equation is also satisfied by merely local extrema and by saddle points. At the minimizing density, $E(n) = E(N_1, N_2, S, V)$.

7. The ensembles are equivalent only in the region where the convex hull of the function $E(S)$ is the same as $E(S)$.

8. We have written E and F as functions of three variables, but it is clear from the definition that they depend on only two ratios. This is reflected in the Thomas–Fermi equation by its scaling behavior when $\mathbf{x} \to \lambda\mathbf{x}$.

Proof

If the only force is gravitation, as in a neutron star ($e = 0$), the methods of §4.1 are applicable; the lower bound for Ξ is trivial, and Inequality (2.1.8; 3) can be used for the upper bound. However, since it requires knowledge of the expectation value of H, it is necessary to estimate the norm of the quantum fluctuations. If e and κ differ from zero the estimate is much more difficult than that of §4.1, and can not be given in detail here. The strategy is as follows.

1. Regularization of the potential
   Since one expects the motion of the particles to be determined by an average field, the singular part of the $1/r$ potential should first be cut off, so that the influence of a near-by particle will not be stronger than that of the average field. There are also good physical grounds to insist that the important part of the potential is its long range rather than the singularity, as in reality the singularity is smoothed out with some form factor. By "long range" is meant a length comparable to the diameter of the star, which shrinks to zero as $\lambda \to \infty$. Hence the cut-off length has to be reduced while λ increases, or alternatively one can work in the scaled system (4.2.10; 4). It is thus useful to show that changing the potential by, say, $1/r \to (1 - \exp(-\lambda^{1/3}sr))/r$ makes little difference for large s in comparison with the main contribution to the energy, which is $\sim N^{7/3}$. This fact can be shown by an argument similar to the estimate (1.2.21) and making use of the bound (III: 4.5.15) on the number of bound states of a short-range potential.
2. Replacing the potential with a step function
   Since Thomas–Fermi theory is oriented toward free particles in a box, it is useful to divide the volume V into cells inside of which the potential is made constant. The proof that changing the potential to a step function has only a slight effect is trivial, since the continuous function $(1 - \exp(-sr))/r$ can be approximated uniformly on any compact set by a step function.
3. Insertion of walls
   In each of the cells of constant potential the Schrödinger equation reduces to the force-free equation, if they are separated by impenetrable walls. It is thus useful to show that inserting walls will not alter the result much. It is clear that the effect will be to raise all the energy levels. The min–max principle can be called upon to show that they do not rise by too much. The presence of the walls means that the wave-function vanishes at their positions, which costs kinetic energy. It is possible to patch together wave-functions for the system without walls so that they vanish at the

positions of the walls, and the expectation value of the kinetic energy in such a state is not increased by too much. It is important that the number of walls in this procedure remain constant in the limit $N \to \infty$ so that their effect can be neglected in comparison with $N^{7/3}$.

4. Filling the boxes
The foregoing manipulations leave the particles in separated boxes moving in constant potentials, which, however, depend on the distribution of the particles. One now finds that the thermodynamic functions of (4.2.2) are dominated by the contribution from a certain distribution of the particles among the boxes, which is determined by a self-consistent equation, namely the Thomas–Fermi equation for the step potential with walls.
5. Continuity of the Thomas–Fermi equation
Since we wish to end up with the Thomas–Fermi equation for a $1/r$ potential, we still need to show that the approximations made above for the $1/r$ potential do not change the energy of the solution much. Otherwise, if the solution depended discontinuously on the potential, it would be worthless; the Thomas–Fermi equations can not be solved analytically, and a numerical solution on a computer approximates the potential by a step function. It is thus indispensible, but fortunately also possible, to show that the Thomas–Fermi energy has the required continuity with respect to the potential. □

The structure of the Thomas–Fermi equation is different for stars than for atoms. The energy loses the properties of convexity and weak lower semicontinuity. Consequently the solution is not guaranteed to be unique and there is a possibility of a phase transition, which will be discussed at the conclusion of this section. Meanwhile, we prepare by proving a general

Virial Theorem (4.2.11)

The pressure

$$P \equiv -\frac{\partial}{\partial V} F(N, \beta, V),$$

kinetic energy

$$E_k = \sum_\alpha 2 \int_V d^3x \int \frac{d^3p}{(2\pi)^3} \frac{|\mathbf{p}|^2/2M_\alpha}{1 + \exp[\beta(|\mathbf{p}|^2/2M_\alpha + W_\alpha(\mathbf{x}) - \mu_\alpha)]},$$

and potential energy

$$E_p = \sum_\alpha \int d^3x \tfrac{1}{2}\rho_\alpha(\mathbf{x}) W_\alpha(\mathbf{x})$$

are connected by

$$3PV = 2E_k + E_p.$$

Proof

We start by convincing ourselves of the usual thermodynamic relationships

$$\frac{\partial F}{\partial N_\alpha} = \mu_\alpha \quad \text{and} \quad \beta \frac{\partial F}{\partial \beta} = E - F, \tag{4.2.12}$$

which follow directly from differentiating (4.2.8). For this purpose note that ρ depends on β and N, and thus implicitly so does W, but that this dependence does not show up when the Thomas–Fermi equations are satisfied. Next rewrite (4.2.8) by integrating by parts in the variable $\mathbf{p}$. Then

$$\frac{|\mathbf{p}|^2}{2M_\alpha} = \varepsilon, \qquad \int_0^\infty d\varepsilon \sqrt{\varepsilon}\ln(1 + \exp[-\beta(\varepsilon + c)]) = \tfrac{2}{3}\beta \int_0^\infty \frac{d\varepsilon\, \varepsilon^{3/2}}{1 + \exp[\beta(\varepsilon + c)]},$$

and we conclude that

$$F = \sum_\alpha N_\alpha \mu_\alpha - \tfrac{2}{3}E_k - E_p. \tag{4.2.13}$$

Finally, the dilatation relationship mentioned earlier implies that $F(N, \beta, V) = \lambda^{-7/3}F(\lambda N, \lambda^{-4/3}\beta, \lambda^{-1}V)$ for all $\lambda \in \mathbb{R}^+$.

With reference to (4.2.12), the derivative by λ produces

$$0 = -\tfrac{7}{3}F + \sum_\alpha N_\alpha \mu_\alpha - \tfrac{4}{3}(E - F) + PV,$$

which concludes the proof of the theorem when combined with (4.2.13). □

The local densities in phase space,

$$n_\alpha(\mathbf{x}, \mathbf{p}) = 2\left[\exp\left(\beta\left(\frac{|\mathbf{p}|^2}{2m} + W_\alpha(\mathbf{x}) - \mu_\alpha\right)\right) + 1\right]^{-1},$$

have the same significance as in §4.1. They are stationary solutions of the Vlasov equation (4.1.32; 3),

$$\sum_j \frac{\mathbf{p}_j}{M_\alpha} \cdot \frac{\partial}{\partial \mathbf{x}_j} n_\alpha(\mathbf{x}, \mathbf{p}) - \frac{\partial}{\partial \mathbf{p}_j} n_\alpha(\mathbf{x}, \mathbf{p}) \cdot \frac{\partial}{\partial \mathbf{x}_j} W_\alpha(\mathbf{x}) = 0. \tag{4.2.14}$$

In this equation quantum mechanics enters only through the initial condition $|n_\alpha(\mathbf{x}, \mathbf{p})| \leq 1$. In fact, as a classical equation it is the basis of stellar dynamics [35]. When reduced to configuration space, the local densities describe the hydrostatic equilibrium between the pressure of the matter and of gravitation, in the spherically symmetric case. Since the fermions behave like free particles on the microscopic level, one would expect from (2.5.32) that

$$P(\mathbf{x}) = \tfrac{2}{3}E_k(\mathbf{x}) \equiv \frac{2}{3}\sum_\alpha 2 \int \frac{d^3p}{(2\pi)^3} \frac{|\mathbf{p}|^2/2M_\alpha}{1 + \exp[\beta(|\mathbf{p}|^2/2M_\alpha + W_\alpha(\mathbf{x}) - \mu_\alpha)]} \tag{4.2.15}$$

functions as the pressure, and in fact if (4.2.14) is multiplied by $\mathbf{p}_i$, integrated by d^3p by parts, and one replaces $\mathbf{p}_i \cdot \mathbf{p}_j \to (|\mathbf{p}|^2/3)\delta_{ij}$, then

$$\nabla P(\mathbf{x}) = -\sum_\alpha \rho_\alpha(\mathbf{x})\nabla W_\alpha(\mathbf{x}), \tag{4.2.16}$$

which is the equilibrium condition mentioned above. If the geometry is spherically symmetric, i.e., V is a sphere of radius R and the local observables depend only on $|\mathbf{x}| = r$, then (4.2.16) can be written as the nonrelativistic **Tolman–Oppenheimer equation**

$$\frac{d}{dr}\tfrac{2}{3}E_k(r) = -\sum_\alpha \frac{\rho_\alpha(r)M_\alpha(r)}{r^2},$$

$$M_\alpha(r) = -\sum_\beta (e_\alpha e_\beta - \kappa M_\alpha M_\beta)\int_0^r dr' 4\pi r'^2 \rho_\beta(r') \tag{4.2.17}$$

(cf. (II: 4.5.11)). The electric and gravitational forces have been expressed in terms of the charges and masses within the sphere.

The Connection between the Local and Global Pressures (4.2.18)

By integrating (4.2.17) by $(4\pi/3)\int_0^R drr^3$ one gets

$$V\tfrac{2}{3}E_k(R) - \tfrac{2}{3}E_k = \frac{4\pi}{3}\int_0^R drr^3 \frac{d}{dr}\tfrac{2}{3}E_k(r)$$

$$= \sum_{\alpha,\beta} \frac{e_\alpha e_\beta - \kappa M_\alpha M_\beta}{3}\int_0^R drr 4\pi\rho_\alpha(r)\int_0^r dr' r'^2 4\pi\rho_\beta(r') = \frac{E_p}{3},$$

so with the virial theorem (4.2.11) the thermodynamic pressure becomes simply the local pressure at the boundary,

$$P = P(R).$$

We see that Thomas–Fermi theory, which begins with the Schrödinger equation, leads eventually to the concepts of classical physics.

A more accurate evaluation of the state functions (4.2.2) requires numerical solutions of Equations (4.2.4) through (4.2.6). In order to lend more physical plausibility to those numbers, let us extend the intuitive arguments of §1.2 to finite temperatures. Since the theory is only valid if gravity is the dominant force, let us simplify by considering only one type of neutral fermion such as neutrons (without nuclear forces). If there were protons and electrons, then the former would provide most of the gravitational force and the latter most of the pressure. This would increase all lengths compared with a system of neutrons by a factor of the ratio of the mass of the neutron to that of the electron, about 2000. Thus, if 10^{57} neutrons are found to have a radius of about 30 km, a similar system made of hydrogen would have a radius of about 6000 km, i.e., that of the earth or of a white dwarf. We begin with the observation that at a nonzero temperature there is a thermal contribution

$$N\tfrac{3}{2}T = N\frac{3}{2}\left(\frac{N}{V}\right)^{2/3}\exp\left(\frac{2S}{3N} - 1\right)$$

in addition to the zero-point energy $N(N/V)^{2/3}$. At high temperatures this is exactly the classical expression. In order to interpolate to intermediate temperatures, we shall simply combine the zero-point energy with the classical expression. This turns out to approximate the energy of free fermions (2.5.32) to within about 20%. It remains to add in the gravitational energy. If the mass of the particles is $\frac{1}{2}$, then up to geometric factors we get

$$\frac{E}{N} = \left(\frac{N}{V}\right)^{2/3}\left(1 + \frac{3}{2e}\exp\left(\frac{2S}{3N}\right)\right) - \frac{\kappa N}{V^{1/3}}$$

in natural units. In checking the properties (2.3.10) of the microcanonical energy density, it becomes readily apparent that, in agreement with (4.2.10; 4),

$$\rho^{-1}\varepsilon = \rho^{2/3}\left(1 + \frac{3}{2e}\exp(2\sigma/3\rho)\right) - \kappa N^{2/3}\rho^{1/3}$$

is independent of N only if $\kappa \sim N^{-2/3}$. Although ε increases as a function of σ, conditions (2.3.10(ii)) and (2.3.10(iii)) are not always satisfied; our ansatz does not do justice to the subadditivity (2.3.5). The reason becomes apparent when it is observed that the pressure

$$P = -\left.\frac{\partial E}{\partial V}\right|_{S,N} = \frac{2}{3}\left(\frac{N}{V}\right)^{5/3}\left(1 + \frac{3}{2e}\exp\left(\frac{2S}{3N}\right)\right) - \frac{\kappa N^2}{3V^{4/3}} = \frac{E - E_p/2}{3V/2},$$

$$E_p = -\frac{\kappa N^2}{V^{1/3}},$$

consists of three parts, from the zero-point, thermal, and gravitational energies. The first two are positive and the last one is negative, and may dominate in the intermediate regime of average densities. However, a negative pressure is physically impossible; the system does not adhere to the walls and pull them inward. What happens is that the system shrinks itself down to such a small radius, $V_0 = (\kappa N^2/-2E)^3$, that it reaches $P = 0$. A better ansatz consists in replacing V with V_0 in E when $P < 0$,

$$\frac{E}{N} = \left(\frac{N}{V}\right)^{2/3}\left(1 + \frac{3}{2e}\exp\left(\frac{2S}{3N}\right) - \frac{\kappa N}{V^{1/3}}\right)\Theta_+ - \frac{\kappa^2 N^{4/3}/2}{2 + 3\exp((2S/3N) - 1)}\Theta_-,$$

$$\Theta_\pm = \Theta\left(\pm\left(2 + 3\exp\left(\frac{2S}{3N} - 1\right) - \kappa(NV)^{1/3}\right)\right).$$

The function $\Theta_\pm$ is also equal to $\Theta(\pm(E + \kappa N^2/2V^{1/3}))$, implying that if the total energy is sufficiently negative, then the system condenses into a volume V_0. As in Example (2.3.32; 1) this brings about a phase transition with negative specific heat: The calculation

$$\begin{aligned}\tfrac{3}{2}T &= \frac{3}{2}\frac{\partial E}{\partial S}\bigg|_{V,N} = \frac{3}{2e}\left(\frac{N}{V}\right)^{2/3}\exp\left(\frac{2S}{3N}\right)\Theta_+ \\ &\quad + \frac{3}{2e}\frac{\kappa^2 N^{4/3}\exp(2S/3N)}{(2+3\exp(2S/3N-1))^2}\Theta_- \\ &= \left(\frac{E}{N}-\left(\frac{N}{V}\right)^{2/3}+\frac{\kappa V}{N^{1/3}}\right)\Theta_+ + \left[-\frac{E}{N}-\left(\frac{2E}{\kappa N^{5/3}}\right)^2\right]\Theta_-\end{aligned}$$

reveals that the classical linear dependence of T on E becomes parabolic in the condensation region (see Figure 32). The temperature begins to rise again as E decreases, and afterwards, when the zero-point energy gets larger than the gravitational energy, it falls to zero. It is in fact observed by astrophysicists that large gaseous masses contract under the influence of gravity, thereby heating up and radiating the gravitational energy that has been set free. This activity, which indicates a range of values for which $S(E)$ is convex and hence microcanonically a negative specific heat, is a direct consequence of the virial theorem and the theorem of equipartition: energy = −kinetic energy = $-(3N/2)$ temperature. Yet this is true only in the intermediate region, since it ignores the external virial (the pressure) and the equipartition theorem is not valid for degenerate gases. This also becomes visible in the computer solution of the Thomas–Fermi equation, as shown in Figures 33 and 34. At the smaller radius $R = 30$ km the zero-point energy predominates and the star acts normally, whereas an intermediate region of negative specific heat shows up at $R = 100$ km.

This phenomenon can not arise in the canonical ensemble, so our next topic will be what the situation is like in that ensemble. In the transition region the Thomas–Fermi equation has many solutions for a given β, and the

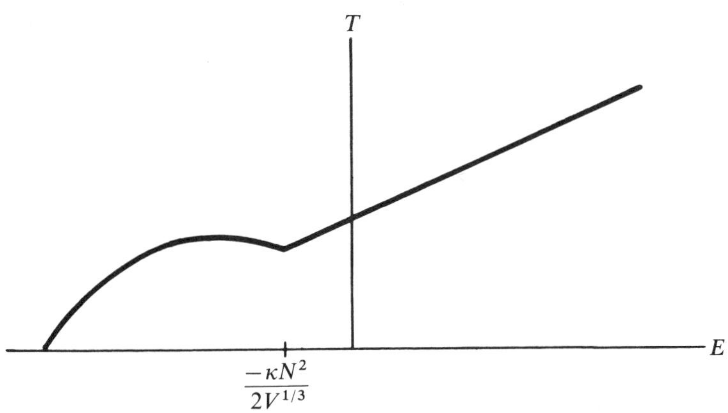

Figure 32 The function $T(E)$ for a conceptual model.

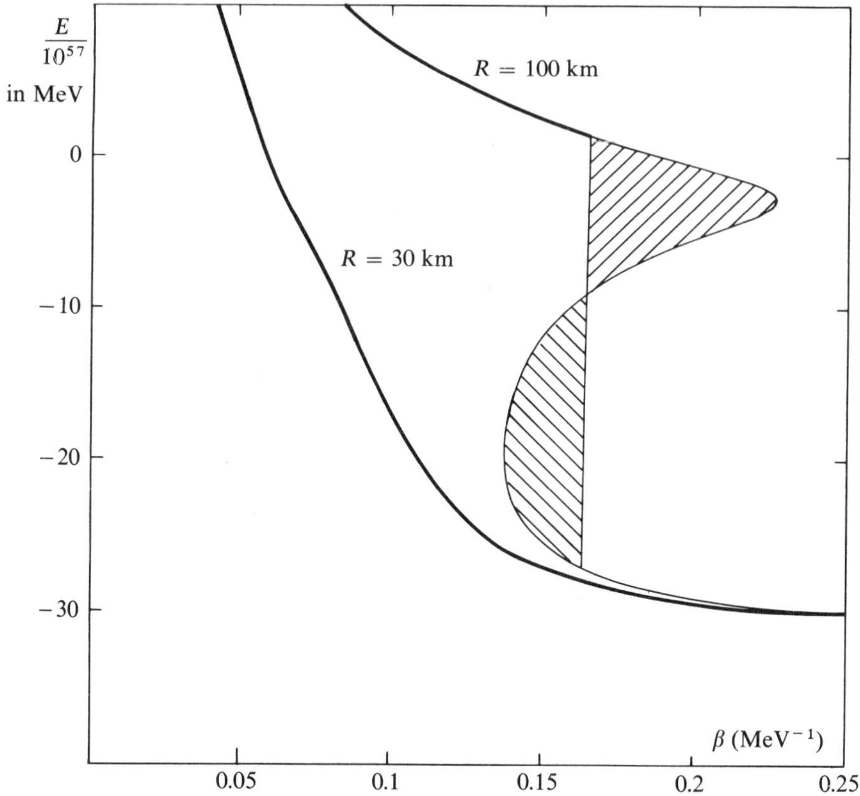

Figure 33 Phase transition in $E(\beta)$.

analysis leading to (4.2.2) shows that the right solution to choose is the one with the smallest value of F. The existence of many different values of F in this situation (for a fixed β) follows from the change in the sign of $P = -\partial F/\partial V$ (see Figure 35). The computed dependence of $-F$ on β is shown in Figure 36. If $R = 100$ km, then F has a sharp bend at some transition temperature; in Figure 33 it shows up as the lines that divide the surface $E(\beta)$ into two equal parts (Problem 1). At this transition temperature the system in the canonical ensemble rises from one branch of the curve to the other. The energy has a nonzero jump ($\sim$30 MeV per particle) at the transition; in the canonical ensemble the region of negative specific heat is bridged over by a phase transition.

Computers have also been used to solve for the local observable $\rho(r)$, which is shown in Figure 37 at various temperatures and with $R = 100$ km. At the transition temperature 1/0.165 MeV an almost homogeneous density becomes strongly concentrated at the center. The picture that emerges is of a star with a rather definite surface and a central density about 10^6

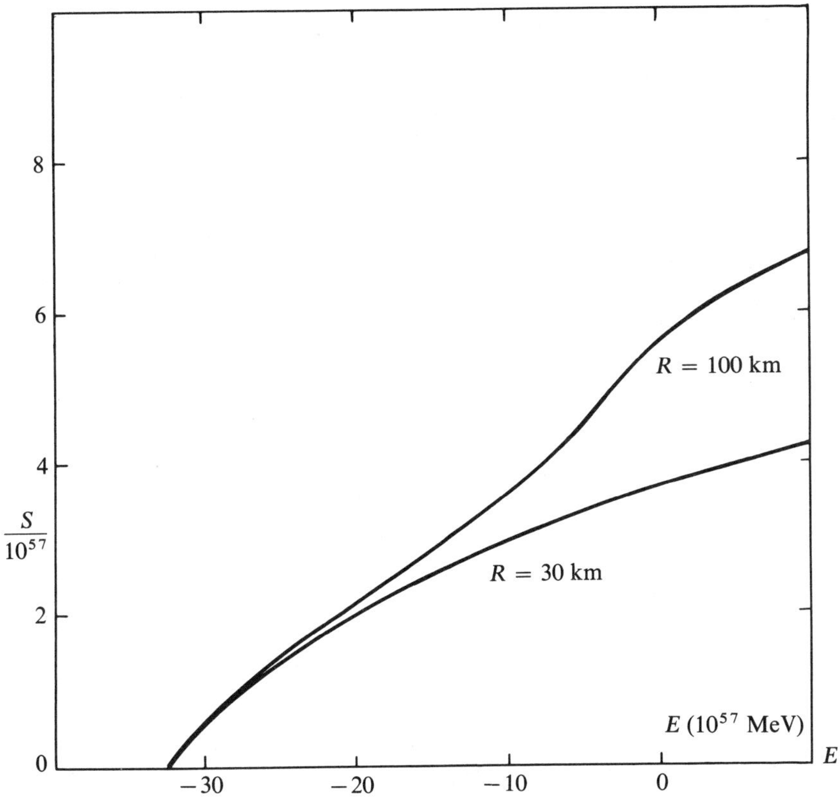

Figure 34 Convex region in $S(E)$.

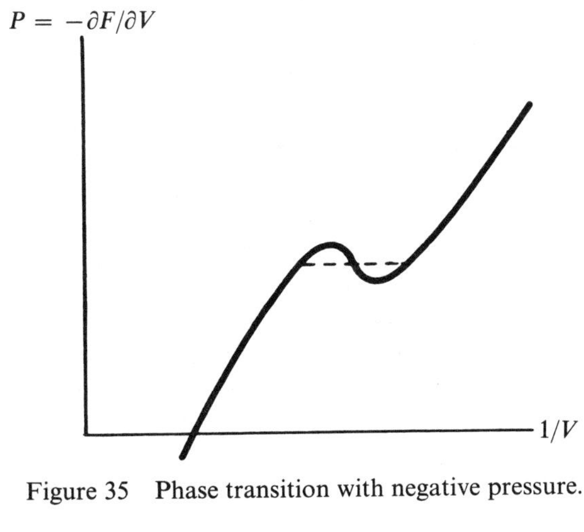

Figure 35 Phase transition with negative pressure.

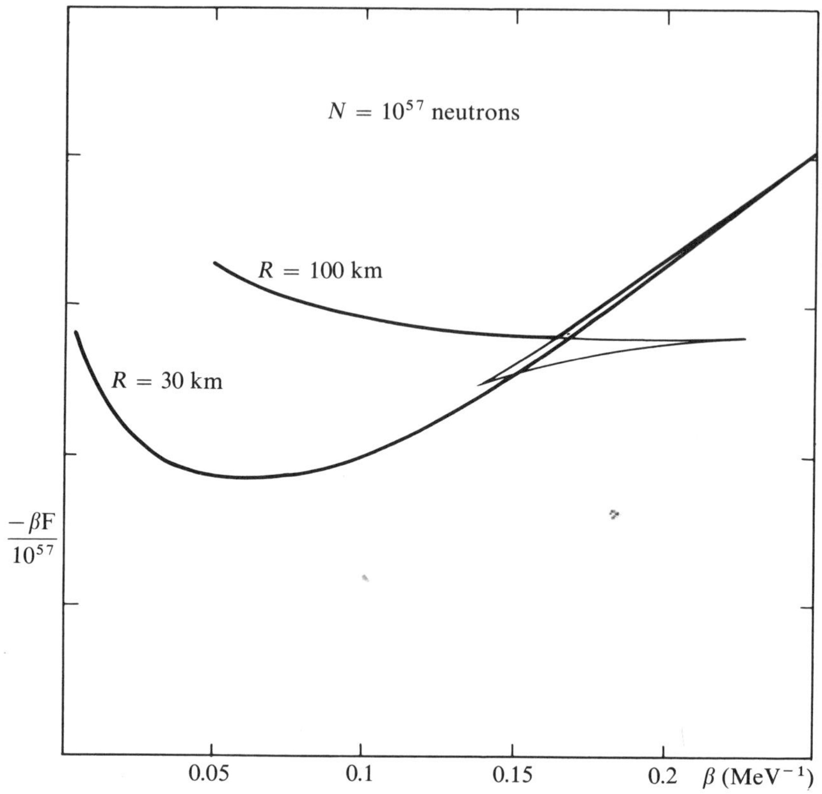

Figure 36 The negative free energy.

times the density of the atmosphere. At still lower temperatures the atmosphere also condenses, but it only increases the density of the star a tiny bit. The radius of a neutron star is only about 10 km at low temperature, which is why at first hardly any difference shows up in S in Figure 34 between the systems at $R = 30$ km and at $R = 100$ km. Only after the transition energy does the star spread out so as to make the entropy rise rapidly enough in a box with $R = 100$ km that $S(E)$ becomes no longer concave.

Another interesting local observable is the degree of degeneracy

$$\xi(r) = \frac{3T}{2} \frac{\rho(r)}{E_k(r)}. \tag{4.2.19}$$

For a classical gas ξ is 1, and for a completely degenerate Fermi gas it is 0. Figure 38 shows $\xi(r)$ for $R = 100$ km and various temperatures. It reveals that the gas becomes degenerate after the phase transition. Only the zero-point energy of the fermions ($\sim \rho^{-5/3}$) can withstand the gravitational pressure ($\sim \rho^{-4/3}$), while the classical pressure is weaker ($\sim \rho^{-1}$). This

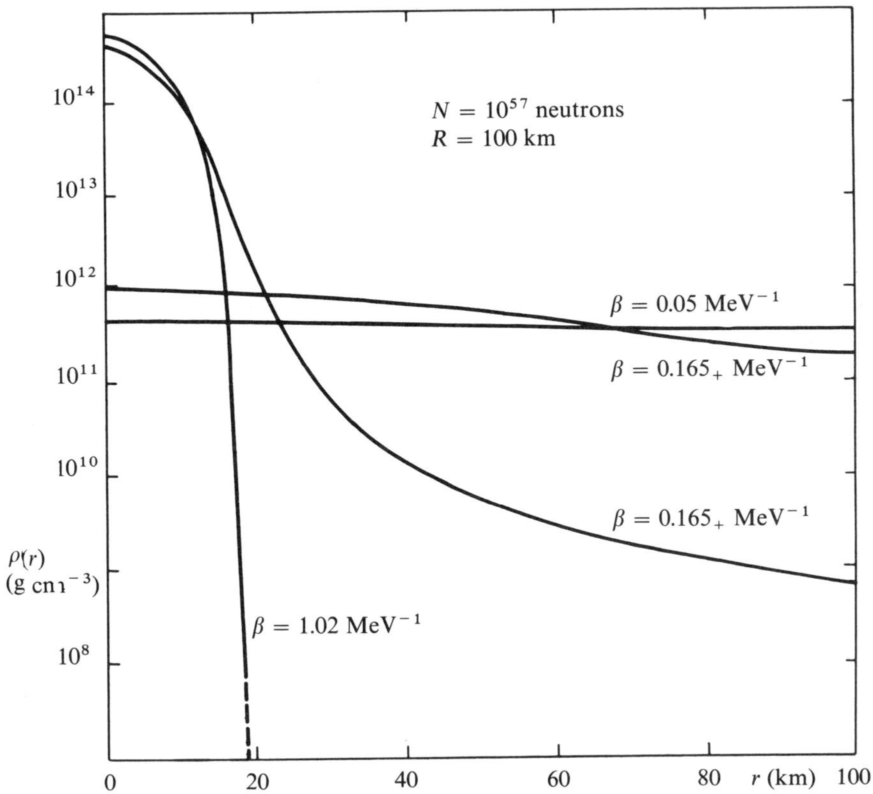

Figure 37 The change in the density at a phase transition.

means that the interior of the star is degenerate, while the atmosphere remains a classical gas.

Problem (4.2.20)

Show that the reciprocal β_c of the transition temperature for the canonical ensemble is determined by

$$0 = \int_{E_2}^{E_1} dE(\beta(E) - \beta_c), \qquad \beta(E_1) = \beta(E_2) = \beta_c.$$

Solution (4.2.21)

Since $\beta = dS/dE$, the condition implies

$$S(E_1) - S(E_2) - \beta_c(E_1 - E_2) = \beta_c(F(E_2) - F(E_1)) = 0.$$

At β_c the two branches of the curves $F(E)$ cross, and the canonical ensemble always selects the lower branch.

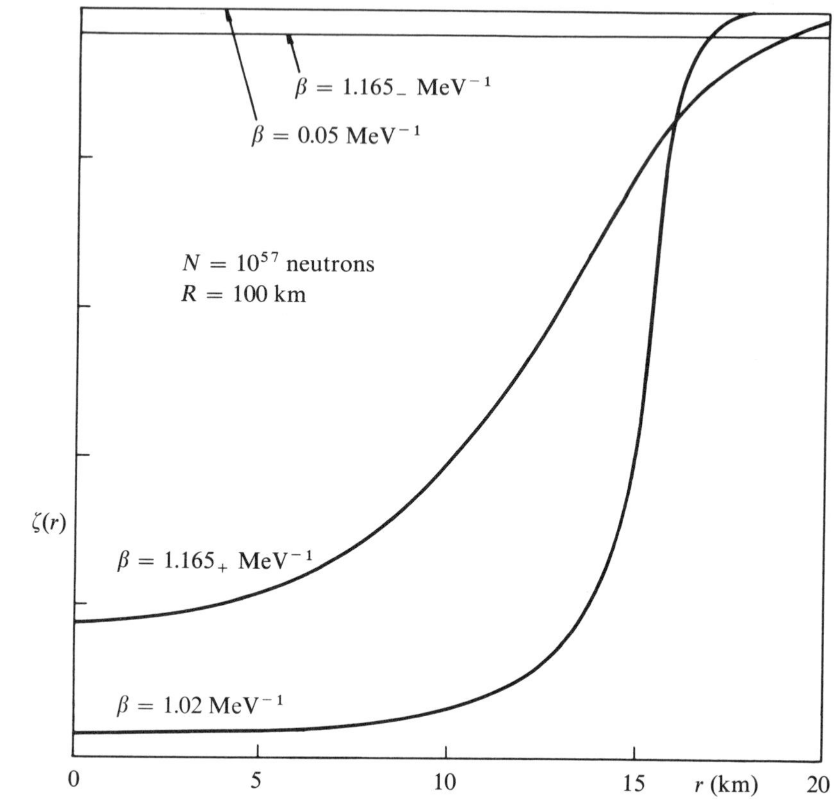

Figure 38 The change in the degree of degeneracy ζ at a phase transition.

4.3 Normal Matter

Although matter consisting of electrons and atomic nuclei exhibits extremely varied and complicated phenomena, some of its essential features can be deduced from the fundamental physical laws.

With the results of §4.1 we are now in a position to cope with a central problem, the stability of matter. As discussed in (1.2.17; 2), it is essential that the electrons follow Fermi statistics, though the statistics of the nuclei should not matter. Moreover, it is the mass of the electron rather than the nucleus that occurs in the basic Rydberg energy $e^4m^2/2$. We shall therefore assume that the nuclei are infinitely massive and use the Hamiltonian H_N of (4.1.2); at any rate it provides a lower bound to (4.1.1) with $\kappa = 1$. The wall W can then also be dispensed with. The question to be confronted is whether a bound $H_N > -AN$ can be found for fixed Z_k but M and $N \to \infty$. With this in mind, write (4.1.6) with $\mu = W = 0$ as

$$\begin{aligned} H_N &= \sum_{i=1}^{N} |\mathbf{p}_i|^2 - \sum_{i=1}^{N}\sum_{k=1}^{M} \frac{Z_k}{|\mathbf{x}_i - \mathbf{X}_k|} + \sum_{i>j} \frac{1}{|\mathbf{x}_i - \mathbf{x}_j|} + \sum_{k>l} \frac{Z_k Z_l}{|\mathbf{X}_k - \mathbf{X}_l|} \\ &\geq \sum_{i=1}^{N} |\mathbf{p}_i|^2 + \sum_{i=1}^{N} \left[-\sum_k \frac{Z_k}{|\mathbf{x}_i - \mathbf{X}_k|} + \int \frac{d^3x' n(\mathbf{x}')}{|\mathbf{x}_i - \mathbf{x}'|} \right] \\ &\quad + \sum_{k>l} \frac{Z_k Z_l}{|\mathbf{X}_k - \mathbf{X}_l|} - 3.68\gamma N \\ &\quad - \frac{1}{2} \int \frac{d^3x\, d^3x'}{|\mathbf{x} - \mathbf{x}'|} n(\mathbf{x}) n(\mathbf{x}') - \frac{3}{5\gamma} \int d^3x n(\mathbf{x}) \equiv H_n. \end{aligned} \tag{4.3.1}$$

The first step is to bound the kinetic energy by $\int \rho^{5/3}$ with the inequality of (4.1.47; 2) and set $n = \rho$. This is a bound for every expectation value with spin $-\frac{1}{2}$ fermions, so, again with the aid of (4.1.46; 2), we obtain

$$\begin{aligned} \langle \psi | H_N | \psi \rangle &\geq \frac{3}{5} \left(\left(\frac{3\pi}{4} \right)^{2/3} - \frac{1}{\gamma} \right) \int d^3x n^{5/3}(\mathbf{x}) - \int d^3x \sum_{k=1}^{M} \frac{Z_k n(\mathbf{x})}{|\mathbf{x} - \mathbf{X}_k|} \\ &\quad + \frac{1}{2} \int \frac{d^3x\, d^3x'}{|\mathbf{x} - \mathbf{x}'|} n(\mathbf{x}) n(\mathbf{x}') + \frac{1}{2} \sum_k \frac{Z_k Z_m}{|\mathbf{X}_k - \mathbf{X}_m|} - 3.68\gamma N \\ &\geq -3.68 \left(\gamma N + \frac{1}{(3\pi/4)^{2/3} - 1/\gamma} \sum_{k=1}^{M} Z_k^{7/3} \right) \text{ for } \|\psi\|^2 = 1. \end{aligned} \tag{4.3.2}$$

If this is optimized in γ, it shows the

Stability of Matter (4.3.3)

$$H_N \geq -2.08N \left[1 + \left(\sum_{k=1}^{M} \frac{Z_k^{7/3}}{N} \right)^{1/2} \right]^2.$$

Remarks (4.3.4)

1. If there were q kinds of electrons instead of the two spin orientations, then the right side would be multiplied by $(q/2)^{2/3}$. Thus there is a bound $\sim N^{5/3}$ independently of the statistics of the electrons.
2. The solution of the Thomas–Fermi equation describes a neutral system, and accordingly the bound is $MZ^{7/3}$ if all Z_k equal $Z = N/M$. The bound is certainly not optimal if $N \ll MZ$, for one would expect $\sim NZ^2$. However, (4.3.3) suffices for our purposes, as we are concerned only with the neutral case.

3. Inequality (4.1.47; 2) is presumably not optimal; on the right the constant should be increased by a factor $(4\pi)^{2/3}$ to $\frac{3}{5}(3\pi^2)^{2/3}$. If this conjecture were proved, then (4.3.3) would be improved by the same factor, reading
$$H \geq -0.385 \sum_{k=1}^{M} Z_k^{7/3}(1 + O(Z_k^{-7/6})).$$
If $Z_k \to \infty$ this approaches the sum of the Thomas–Fermi energies of the atoms. Such an optimal inequality can in fact be proved, although only in the form
$$H \geq -0.385 \sum_{k=1}^{M} Z_k^{7/3}(1 + O(Z_k^{-2/33}))$$
[28].
4. Inequality (4.3.3) holds *a fortiori* for a system in a finite volume.
5. Since the kinetic energy of the nuclei was not used, they may follow either Bose or Fermi statistics.
6. The important property of the Coulomb potential for stability is that $1/r$ is a function of positive type, i.e., $\tilde{v} \geq 0$. The Yukawa potential $v(r) = \exp(-\mu r)/r$ similarly satisfies $\tilde{v} > 0$, and stability can be proved analogously. In contrast the potential $v(r) = (a + br)\exp(-\mu r)$ with $b > a\mu > 0$, $\mu > 0$, which is even finite and of short range, does not lead to stability for the Hamiltonian $\sum_{i=1}^{N} |\mathbf{p}_i|^2 + \sum_{i>j} e_i e_j v(\mathbf{x}_i - \mathbf{x}_j)$, even for fermions: There is an $r_0 > 0$ such that $v(r_0) > v(0)$ (this would be impossible if $\tilde{v} > 0$), so let us confine $N/2$ positive and negative particles to separated balls of radius $r_0 \varepsilon$, $\varepsilon \ll 1$, arrayed at a distance r_0 from one another. Then the interaction between the balls, $-e^2 v(r_0)N^2/4$, wins out over the respulsive energy of the like-charged particles within the balls, $\sim e^2 v(0)N(N-2)/4$, and also wins out over the kinetic energy $\sim N^{5/3}(r_0\varepsilon)^2$ as $N \to \infty$. Thus the total energy goes to $-\infty$ as $-N^2$ when $N \to \infty$. This shows that the problem of the stability of matter has nothing to do with the long range of the Coulomb potential. The proof with the Yukawa potential is not any simpler; in a way it is more difficult, since stability with a Yukawa potential immediately implies stability with a Coulomb potential—as remarked in (1.2.17; 5) the difference produces stability—but not vice versa. However, as we have just seen, the $1/r$ singularity is not the only danger for stability; even regular potentials v with energies $\sum_{i \leq j} e_i e_j v(\mathbf{x}_i - \mathbf{x}_j)$ that take on both signs can lead to instability. This shows the superficiality of the common opinion that stability is not a real physical problem, since actual potentials do not become singular.

The Extensivity of the Volume (4.3.5)

If $H > -cN$ and the expectation value of H in a state is nonpositive, $\langle H \rangle \leq 0$, then no volume $\Omega \leq \varepsilon N$ contains more than $N(\frac{20}{3}c)^{3/5}(4\varepsilon/3\pi)^{2/5}$ particles.

Proof

Let $H = T + V$. Since the energy is proportional to the mass in a Coulomb system, $\frac{1}{2}\langle T\rangle \le -\langle\frac{1}{2}T + V\rangle \le 2cN$. Then it follows from

$$\langle T\rangle \ge \frac{3}{5}\left(\frac{3\pi}{4}\right)^{2/3}\int \rho^{5/3}$$

that

$$\int_\Omega \rho(\mathbf{x})\, d^3x \le \left(\int_\Omega \rho^{5/3}\, d^3x\right)^{3/5}\left(\int_\Omega d^3x\right)^{2/5} \le \left(\frac{5}{3}\left(\frac{4}{3\pi}\right)^{2/3} 4cN\right)^{3/5}(\varepsilon N)^{2/5}$$
$$\le N\left(\frac{20}{3}c\right)^{3/5}\left(\frac{4\varepsilon}{3\pi}\right)^{2/5}. \qquad \square$$

Remarks (4.3.6)

1. If Ω is a ball, then it is possible to derive bounds of the form $\langle r^\nu\rangle \ge cN^{\nu/3}$, in analogy with (III: 4.5.28).
2. The material up to this point does not allow upper bounds of the form $r \sim N^{1/3}$ to be proved. Neutrality does not enter in an important way, and with an excess of electrons the Coulomb potential would cause the system to swell out to infinity. In other words, it has been proved that matter is stable in the sense that it does not implode, but it might still explode.

The Existence of the Thermodynamic Functions (4.3.7)

We are now faced with the question of how to define the energy density when $N \to \infty$ [30]. It clearly follows from (4.3.3) that $(1/V)E(V\sigma, V, \rho V) > -\rho \cdot$ constant for all V, and since it is easy to show that E/V remains bounded above, $\underline{\lim}_{V\to\infty}\,(1/V)E(V\sigma, V, \rho V)$ could be regarded as $\varepsilon(\rho, \sigma)$ (by definition, $\underline{\lim}_{n\to\infty}\, a_n = \sup_{n'}\inf_{n>n'}\, a_n$). This cheap way out is physically unsatisfying, however; one would hope that the limit exists and that the energy density becomes independent of V as the system is made infinitely large. This means that the sequence should be proved monotonic, as was done in (2.3.6). Unfortunately, the inductive procedure followed there, of imagining each cube to consist of smaller cubes, does not work in this case, since it is difficult to estimate the Coulomb interaction between cubes. Balls can be used instead of cubes, however, as their interactions are as if the charges were concentrated at their centers, according to a theorem dating from Newton. In particular, if they are overall neutral, then they do not interact with charges placed outside them. Of course, spheres do not fill space as densely as cubes, but by the use of spheres of different radii the unfilled volume can be made arbitrarily small. The convergence proof consequently proceeds by three steps.

(a) We must first show that the interaction between the spheres is not positive, in order to prove monotony.
(b) It must be shown that the radii of the balls can be chosen so that the fraction of volume outside them goes to zero in the limit.
(c) The distribution of particles in this procedure must lead to a homogeneous density in the limit.

The Interaction between Balls (4.3.8)

We consider

$$H = \sum_{i=1}^{N} \left(|\mathbf{p}_i|^2 - \sum_{k=1}^{M} \frac{Z_k}{|\mathbf{x}_i - \mathbf{X}_k|} \right) + \sum_{i>j} |\mathbf{x}_i - \mathbf{x}_j|^{-1}$$

$$+ \sum_{k>l} \frac{Z_k Z_l}{|\mathbf{X}_k - \mathbf{X}_l|} + \sum_{k=1}^{M} \frac{|\mathbf{p}_k|^2}{2M_k} \tag{4.3.9}$$

in a ball B, such that $\psi|_{\partial B} = 0$, and examine the neutral case with only one kind of nucleus: $N = MZ$, $N_t = N(1 + 1/Z) =$ the total number of particles. The eigenvalues $e_i(V, N_t)$, $i = 1, 2, \ldots$, of H depend on the volume V of B and on N_t, and the microcanonical energy is given by

$$E(S, V, N_t) = \exp(-S) \sum_{i=1}^{\exp(S)} e_i(V, N_t),$$

where E and E_m have been identified in accordance with (2.3.13; 2). Now put k disjoint balls B_α of volumes V_α into B,

$$B \supset \bigcup_{\alpha=1}^{k} B_\alpha,$$

and form a system of trial functions ψ_i by taking tensor products of the eigenfunctions of H_α, defined as H for N_α particles in B_α:

$$\psi_i = \psi_{i_1} \otimes \psi_{i_2} \otimes \cdots \otimes \psi_{i_k}$$

The trial functions then have to be antisymmetrized in the electron variables and either symmetrized or antisymmetrized in the nuclear coordinates, depending on the nuclear statistics. Yet since ψ_{i_α} and ψ_{i_β} have disjoint support, there are no cross terms in their interaction, and the expectation values are the same as those with the unsymmetrized ψ_i. (The subscript i is to be treated as a multi-index $i_1, \ldots, i_k$.) We always choose the first

$\exp(S_\alpha)$ eigenfunctions of the operators H_α (and denote the eigenvalues $e_{\alpha,i}$), so

$$\sum_{i=1}^{\exp(S)} = \sum_{i_1=1}^{\exp(S_1)} \cdots \sum_{i_k=1}^{\exp(S_k)},$$

where $S = \sum_{\alpha=1}^{k} S_\alpha$, $N = \sum_{\alpha=1}^{k} N_\alpha$, and $N_\alpha/Z + 1$ is an integer. Then each B_α can be filled with whole atoms, becoming neutral. As in (2.3.5), with the min–max principle (III: 3.5.21),

$$E(S, V, N) \le \exp(S) \sum_{i=1}^{\exp(S)} \langle \psi_i | H \psi_i \rangle = \sum_{\alpha=1}^{k} \exp(-S_\alpha) \sum_{i_\alpha=1}^{\exp(S_\alpha)} e_{\alpha, i_\alpha}(N_\alpha, N_\alpha) + U$$

$$= \sum_{\alpha=1}^{k} E_\alpha(S_\alpha, V_\alpha, N_\alpha) + U, \tag{4.3.10}$$

but this time there is an energy of the interaction between the balls,

$$U = \sum_{\alpha > \beta} \exp(-S_\alpha - S_\beta) \sum_{i_\alpha=1}^{\exp(S_\alpha)} \sum_{i_\beta=1}^{\exp(S_\beta)} U_{i_\alpha i_\beta},$$

$$U_{i_\alpha i_\beta} \equiv \sum_{j=1}^{N_\alpha} \sum_{m=1}^{N_\beta} e_j e_m \int \frac{d^{3N_\alpha}x \, d^{3N_\beta}y}{|\mathbf{x}_j - \mathbf{y}_m|} |\psi_{i_\alpha}(\mathbf{x}_1, \ldots, \mathbf{x}_{N_\alpha})|^2 |\psi_{i_\beta}(\mathbf{y}_1, \ldots, \mathbf{y}_{N_\beta})|^2.$$

Because of the spherical symmetry of B_α and H_α, the functions ψ_{i_α} can be ordered according to the eigenvalues l_α of the total angular momentum L_α about the center of B_α. The eigenvalues $e_{\alpha,i}$ do not depend on the z-component of the angular momentum (which has quantum numbers m_α, $-l_\alpha \le m_\alpha \le l_\alpha$), and

$$\rho_\alpha(\mathbf{x}) = \sum_{i_\alpha} \int d^3x_2 \cdots d^3x_{N_\alpha} |\psi_i(\mathbf{x}, \mathbf{x}_2, \ldots, \mathbf{x}_{N_\alpha})|^2$$

will be spherically symmetric if the sum runs over a full L-shell. If the limits of summation $\exp(S_\alpha)$ corresponded exactly to full shells, then U would equal zero by Newton's theorem. It will now be shown that the partially filled shells can be chosen to make U negative. Let μ_α, and ν_α be the indices nearest to $\exp(S_\alpha)$ corresponding to filled shells, such that $\mu_\alpha \le \exp(S_\alpha) \le \nu_\alpha$. Thus

$$\sum_{i_\alpha=1}^{\exp(S_\alpha)} \sum_{i_\beta=1}^{\exp(S_\beta)} U_{i_\alpha i_\beta} = \sum_{i_\alpha=\mu_\alpha}^{\exp(S_\alpha)} \sum_{i_\beta=\mu_\beta}^{\exp(S_\beta)} U_{i_\beta i_\alpha},$$

and the interaction energy can be written as

$$U = c \sum_{i_1=\mu_1}^{\exp(S_1)} \sum_{i_2=\mu_2}^{\exp(S_2)} \cdots \sum_{i_k=\mu_k}^{\exp(S_k)} U_{i_1,\ldots,i_k}, \qquad c > 0,$$

$$U_{i_1,\ldots,i_k} = \left\langle \psi_i \middle| \sum_{j>m} \frac{e_j e_m}{|\mathbf{x}_j - \mathbf{x}_m|} \psi_i \right\rangle.$$

We know that

$$\sum_{i_1=\mu_1}^{\nu_1} \sum_{i_2=\mu_2}^{\nu_2} \cdots \sum_{i_k=\mu_k}^{\nu_k} U_{i_1,\ldots,i_k} = 0,$$

and since the eigenvalues e_{i_α} are degenerate if $\mu_\alpha \leq i_\alpha \leq \nu_\alpha$, it is possible to select $\exp(S_1) - \mu_1$ indices i_1 such that

$$\sum_{i_2=\mu_2}^{\nu_2} \cdots \sum_{i_k=\mu_k}^{\nu_k} \sum_{i_1=\mu_1}^{\exp(S_1)} U_{i_1,\ldots,i_k} \leq 0$$

without changing the first sum in (4.3.10). We now proceed inductively and choose $\exp(S_2) - \mu_2$ indices i_2 such that

$$\sum_{i_2=\mu_2}^{\exp(S_2)} \cdots \leq 0$$

and so forth, until finally $U \leq 0$. This proves the

Monotony of the Energy (4.3.11)

If $B \supset \bigcup_{\alpha=1}^{k} B_\alpha$, $N_t = \sum_{\alpha=1}^{k} N_\alpha$, *and* $N_\alpha/Z + 1$ *is integral*, $S = \sum_{\alpha=1}^{k} S_\alpha$, *and E is as defined in* (4.3.8), *then*

$$E(S, V, N_t) \leq \sum_{\alpha=1}^{k} E_\alpha(S_\alpha, V_\alpha, N_\alpha).$$

Remarks (4.3.12)

1. The B_α are required only to be disjoint; how well they fill B does not affect the validity of the equation.
2. All but one of the B_α have to be spherical and electrically neutral, but one of them need not be.
3. The theorem holds regardless of the statistics of the particles, which can affect it only by ensuring the existence of a bound on E/N.

The question of how completely B can be filled by the B_α is a purely geometrical one. It is answered by the

Swiss Cheese Theorem (4.3.13)

Let $R_j = (1 + p)^j R_0$, $p \in \mathbb{Z}^+$, $1 + p \geq 27$, *be the radii of the balls of a given size indexed by j and let* B_m *be a ball of size m. Then for all* $m > 0$, B_m *contains*

the union from $j = 1$ to $m - 1$ of v_j disjoint balls of size j, where

$$v_j = \frac{(1 + p)^{3(m-j)}}{p}\left(\frac{p}{1 + p}\right)^{m-j} \in \mathbb{Z}^+.$$

Remarks (4.3.14)

1. This theorem makes more precise the fact, clear at the intuitive level, that a large ball can be filled extremely well by smaller ones if their radii are chosen suitably. The total volume of the small balls is

$$\sum_{j=0}^{m-1} R_j^3 v_j = ((1 + p)^m R_0)^3\left(1 - \left(\frac{p}{1 + p}\right)^m\right),$$

so that the unfilled fraction is only $(p/(1 + p))^m$, which tends to zero as $m \to \infty$.
2. Of course, the filling of a ball uses more small balls than large ones, but the fraction of volume filled by the balls of size j is $(1/p)(p/(1 + p))^{m-j}$, as the larger balls are much more voluminous.

Proof

See Problem 1. □

The Homogeneity of the Density (4.3.15)

The next step in §2.3 was to consider a sequence of larger and larger cubes, all of which had the same entropy and particle density. Nothing like that is possible in this situation, since to compensate for the gaps some of the balls will have greater densities than the average density overall. Since the unfilled volume gets smaller and smaller, however, it suffices to impose relatively large densities on the balls of size 0 and assign equal densities to all the others. Let us thus choose $N_\alpha/V_\alpha = \rho(p + 1) \equiv \rho_0$ for $\alpha = 1, 2, \ldots, v_0$, so for the balls of size 0, $N_\alpha/V_\alpha = \rho$ for all $\alpha > v_0$. If ρ_j is the density in a ball of size j, and we let $\rho_1, \ldots, \rho_m = \rho$, then the ρ_j satisfy a recursion formula

$$\rho_m = \sum_{j=0}^{m-1} \rho_j v_j\left(\frac{R_j}{R_m}\right)^3 = \frac{\rho_0}{p}\left(\frac{p}{p + 1}\right)^m + \frac{\rho}{p}\sum_{j=1}^{m-1}\left(\frac{p}{p + 1}\right)^{m-j} = \rho$$

for all $m \geq 1$.

In the same way the entropy is distributed so that the entropy density σ_j in the balls of size j satisfies

$$\sigma_0 = \sigma(p + 1), \qquad \sigma_1 = \sigma_2 = \cdots = \sigma_m = \sigma = \frac{1}{p}\sum_{j=0}^{m-1} \sigma_j\left(\frac{p}{p + 1}\right)^{m-j}.$$

If $V_0 = 4\pi R_0^3/3$ and E_j is the energy and ε_j the energy density of the balls of size j, then Proposition (4.3.11) specializes for this particular filling to

$$E_k(S, N) \leq \sum_{j=0}^{k-1} E_j(S_j, N_j)v_j,$$

$$\varepsilon_k(\sigma_k, \rho_k) = [(1+p)^{3k}V_0]^{-1}E_k(S_k, N_k)$$

$$\leq \frac{1}{V_0 p}\sum_{j=0}^{k-1}\left(\frac{p}{1+p}\right)^{k-j}(1+p)^{-3j}E_j(S_j, N_j)$$

$$= \frac{1}{p}\sum_{j=0}^{k-1}\left(\frac{p}{1+p}\right)^{k-j}\varepsilon_j(\sigma_j, \rho_j).$$

This is a modification of (2.3.6) and similarly allows the convergence of $\varepsilon_k \equiv \varepsilon_k(\sigma_k, \rho_k)$ to be demonstrated: There exist numbers $c_k \leq 0$ such that

$$\varepsilon_k = c_k + \frac{1}{p}\sum_{j=0}^{k-1}\left(\frac{p}{1+p}\right)^{k-j}\varepsilon_j.$$

The recursion formula has the solution

$$\varepsilon_k = c_k + \frac{1}{1+p}\left(\varepsilon_0 + \sum_{j=0}^{k-1} c_j\right). \tag{4.3.16}$$

Since the sequence $\{\varepsilon_k\}$ is bounded from below, $\sum_j c_j$ must converge, so $\lim_{k\to\infty} c_k = 0$. Since $\varepsilon_k - c_k$ decreases monotonically as a function of k by (4.3.16), ε_k must tend to a limit. If $k > 0$, then all the densities had the same values (σ, ρ), and we arrive at the

Existence of the Thermodynamic Limit (4.3.17)

For the H of (4.3.19), *the limit* $\varepsilon(\sigma, \rho) \equiv \lim_{V\to\infty}(1/V)E(\sigma V, \rho V)$ *exists.*

Remarks (4.3.18)

1. The theorem has been proved for spherical volumes, but it generalizes to other shapes with a reasonable relationship between volume and surface area.
2. Although the theorem and proof are given here for strictly neutral systems, it is clear that a small excess charge ΔQ can be allowed as long as its electrostatic energy $\sim(\Delta Q)^2/V^{1/3}$ can be neglected in comparison with E.
3. Although we have assumed there was only one kind of nucleus, the case of any number of kinds of nucleus can be covered simply by generalizing the notation.

4. Since $\varepsilon_k - c_k$ is a monotonic sequence, Dini's theorem guarantees that ε_k converges uniformly on compact sets in (σ, ρ); to use this argument it is necessary to extend the definition of the function ε_V, which was initially defined for finite V on a discrete set, to make it continuous. The continuity of ε will follow from the convexity to be proved below.
5. The Hamiltonian (4.3.9) includes the kinetic energy of the nuclei. Strangely, the existence of the thermodynamic limit (4.3.17) has not been proved in the apparently simpler case where $M_k = \infty$.

The existence of the limit means that all systems characterized by N have the same dependence on the averaged quantity ε provided that they are large enough. But does the theory predict a reasonable dependence? The temperature, pressure, specific heat, and compressibility should at least be positive in accordance with our experience. The positivity of the temperature and pressure are ensured by our definition of entropy and by the boundary conditions. With the aid of (2.3.29), the positivity of the other observables is a consequence of the convexity of the function $(\sigma, \rho) \to \varepsilon(\sigma, \rho)$, which, however, does not follow directly from the definitions—recall that the preceding chapter illustrated this with a counter example. Yet it is possible to formulate a theorem on the

Thermodynamic Stability of Coulomb Systems (4.3.19)

The mapping $\mathbb{R} \times \mathbb{R}^+ \to \mathbb{R}: (\sigma, \rho) \to \varepsilon(\sigma, \rho)$ *is*

(i) *convex*;
(ii) *nondecreasing in* σ;
(iii) *bounded below by* $-c\rho$ *for* $c \in \mathbb{R}^+$;
(iv) *such that* $\rho^{-1}\varepsilon(\sigma\rho, \rho)$ *is an increasing function of* ρ.

Proof

(i) Let p be an odd integer, so that $v_j = (1 + p)^{2(k-j)}p^{k-j-1}$ is even for $0 \leq j \leq k - 1$, and fill half of the balls of a given size with densities ρ, σ (or, respectively, $\rho_0 = \rho(1 + p)$, $\sigma_0 = \sigma(1 + p)$) and the other half with ρ', σ' (or, respectively, $\rho'_0 = \rho'(1 + p)$, $\rho'_0 = \sigma'(1 + p)$). Then, since the energy is monotonic as in (4.3.11),

$$\varepsilon_k(\bar{\sigma}_k, \bar{\rho}_k) \leq \frac{1}{2p} \sum_{j=0}^{k-1} \left(\frac{p}{1 + p}\right)^{k-j} [\varepsilon_j(\sigma_j, \rho_j) + \varepsilon_j(\sigma'_j, \rho'_j)],$$

$$\bar{\sigma}_k = \frac{1}{2p} \sum_{j=0}^{k-1} \left(\frac{p}{1 + p}\right)^{k-j} (\sigma_j + \sigma'_j),$$

and

$$\bar{\rho}_k = \frac{1}{2p} \sum_{j=0}^{k-1} \left(\frac{p}{1+p} \right)^{k-j} (\rho_j + \rho'_j),$$

which implies that

$$\varepsilon(\tfrac{1}{2}(\sigma + \sigma'), \tfrac{1}{2}(\rho + \rho')) \le \tfrac{1}{2}(\varepsilon(\sigma, \rho) + \varepsilon(\sigma', \rho'))$$

as $k \to \infty$. Now note that ε is monotonic in σ and $\rho^{-1}\varepsilon(\sigma\rho, \rho)$ is monotonic in ρ, so according to (2.3.11; 1) ε is convex not just with coefficient $\frac{1}{2}$ but with all $\alpha \in [0, 1]$. Hence it is continuous on the interior of $\mathbb{R}^+ \times \mathbb{R}^+$.

(ii) See Remark (2.3.3; 3).

(iii) This follows from the estimate (4.3.3) showing the stability of matter.

(iv) From the monotonic property (2.3.4) of the energy, $\partial E/\partial V|_{S, N=\text{const}} \le 0$. □

Since ε has the right sort of convexity, one of the assumptions needed to prove the existence of the thermodynamic limit of the canonical ensemble is satisfied. More information about the function $\varepsilon(\sigma, \rho)$ is needed to verify the other hypotheses made in Theorem (2.4.14). In particular it needs to be shown that ε increases rapidly enough with σ that the σ_0 introduced in (2.3.11; 4) is finite, and $\lim_{\sigma \to \infty} \varepsilon/\sigma = \infty$. This is shown by the

Lower Bound for the Energy Density (4.3.20)

If $H = H_\alpha \equiv K + \alpha \sum_{i>j} e_i e_j |\mathbf{x}_i - \mathbf{x}_j|^{-1}$ *and* ε_α *are the corresponding energy densities, then*

$$\varepsilon_\alpha(\sigma, \rho) \ge \lambda \varepsilon_0(\sigma, \rho) - \frac{c\rho\alpha^2}{1 - \lambda} \quad \textit{for all } 0 \le \lambda < 1,$$

where

$$c = 2.08(1 + Z^{2/3})^2.$$

Proof

According to (2.3.3; 4), ε_α is concave in α, and $\varepsilon_\alpha \ge \lambda\varepsilon_0 + (1 - \lambda)\varepsilon_{\alpha/(1-\lambda)}$. However, by (4.3.3), $-c\rho\alpha^2$ is a lower bound for all ρ and σ. □

Corollaries (4.3.21)

1. Since it was shown in (2.5.23) that in the case of one kind of particle, $\varepsilon_0(\sigma, \rho) = c'\rho^{5/3} \exp(2\sigma/3\rho)$, $c' > 0$, is the limit as $\sigma \to \infty$, it follows that $\lim_{\sigma \to \infty\, \rho\, \text{fixed}} \varepsilon(\sigma, \rho)/\sigma = \infty$.

2. Even for a finite volume $-c\rho\alpha^2$ is a lower bound, which makes it easy to verify that there exists a function $s(\varepsilon, \rho)$ dominating σ for all volumes, and satisfying $\lim_{\varepsilon\to\infty} s/\varepsilon = 0$.
3. In (4.3.43; 2) we shall find an upper bound on the ground-state energy density, of the form $c_1\rho^{5/3} - \alpha c_2\rho^{4/3}$. When combined with (4.3.20) it yields an upper bound for the σ_0 of (2.3.11; 4) at which $\varepsilon(\sigma)$ starts to move up.

This fact is not yet enough to ensure that thermodynamics works perfectly. Let us write down a

Thermodynamic Wish List (4.3.22)

1. $\sigma_0 = 0$.
2. $\partial\varepsilon/\partial\sigma|_{\sigma=\sigma_0} = 0$.
3. $\lim_{\sigma\to\infty} (\partial\varepsilon/\partial\sigma) = \infty$.
4. The function ε is continuously differentiable.
5. The function ε is strictly convex for large σ and is linear on certain intervals in σ when σ is small.

Open Questions for the Wish List

1. Statement 1 is a strong formulation of the third law of thermodynamics, and is unproved for Coulomb systems. Although there is an upper bound on σ_0 in (4.3.21; 3), it is not sharp enough to show that $\sigma_0 = 0$.
2. The second statement implies that the system does not fall into its ground state if the temperature is higher than absolute zero, and our bounds are likewise too crude to prove it.
3. The third statement means that there is no maximum temperature, and is proved by (4.3.21; 1).
4. Kinks in the graph of ε would correspond to "anti-phase-transitions" at which either the temperature or the pressure shows a discontinuity while the energy remains continuous. The specific heat and the compressibility would be zero at such a point. Such things do not appear to happen in reality, though the arguments we have made do not exclude them.
5. It is known empirically that there are no phase transitions at high temperatures, only at low temperatures. However, this fact has not been proved in the theory.

The equivalence with the canonical ensemble requires only the positivity of the specific heat, which is guaranteed by (4.3.19). The assumptions of

Theorems (2.4.14) are fulfilled because of (4.3.18; 4), (4.3.19(i)), and (4.3.21; 2), so it leads to the

Thermodynamic Limit of the Canonical Ensemble (4.3.23)

The limit

$$\lim_{V\to\infty}\left(-\frac{T}{V}\ln \operatorname{Tr}\exp(-\beta H)\right) = \inf_{\varepsilon}(\varepsilon - T\sigma(\varepsilon, \rho)) = \varphi(T, \rho)$$

exists.

Remarks (4.3.24)

1. The properties of the free-energy density listed in (2.4.16) are also proved.
2. It is possible to prove the existence of the limit as $V \to \infty$ directly, but that is not enough to show the equivalence with the microcanonical ε. In particular it does not show that ε is convex in σ.

Finally, consider the grand canonical ensemble, supposing there are N_e electrons and N_k nuclei with chemical potentials μ_e and μ_k. The function to investigate is

$$P(T, \mu_e, \mu_k) \equiv \lim_{V\to\infty}\frac{T}{V}\ln \operatorname{Tr}\exp[-\beta(H - N_e\mu_e - N_k\mu_k)]. \quad (4.3.25)$$

One difficulty with (4.3.25) is that the trace contains the sum over all possible numbers of particles, and not only the neutral configuration for which $N_e = ZN_k$. Fortunately, it turns out that the non-neutral contributions have such large Coulomb energies that they play no role. Stated without proof [30], here is the resulting proposition on the

Thermodynamic Limit of the Grand Canonical Ensemble (4.3.26)

The limit (4.3.25) *exists, and*

$$P(T, \mu_e, \mu_k) = \sup_{\rho_e = Z\rho_k}(\mu_e\rho_e + \mu_k\rho_k - \varphi(T, \rho)),$$

$$\rho = \frac{N_e + N_k}{V} = \left(1 + \frac{1}{Z}\right)\rho_e.$$

Remarks (4.3.27)

1. Although the supremum is *a priori* over all density configurations, it is attained in the neutral sector.

2. Roughly speaking, to generalize this to cover arbitrarily many components it is only necessary to treat μ and ρ as "isovectors."

Bounds for $\varepsilon(\sigma, \rho)$ (4.3.28)

The question that now arises is to what extent the qualitative propositions that have been derived about $\varepsilon(\sigma, \rho)$ can be sharpened and made quantitative. For instance, it would be desirable to find an upper bound to complement the lower bound (4.3.20); upper bounds are always easy to discover, since with the min–max principle it is only necessary to devise some good trial functions. In the limit $\rho \to 0$ an obvious upper bound for the ground-state energy is the sum of the energies of the individual atoms. If the density is finite, then one would think of using the ground state of the kinetic energy K in the variational principle, and the result is the first-order perturbation-theoretic approximation to $H_\alpha = K + \alpha V$.

Remarks (4.3.29)

1. It is impossible for the expansion in powers of α to converge in the thermodynamic limit; if $\alpha < 0$, then the electrons would attract one another, as would the nuclei, whereas the nuclei would repel the electrons. The ground-state energy of fermions with an attractive $1/r$ potential goes as $-N^{7/3}$, and that of bosons goes as $-N^3$ (see (1.2.22) and (1.2.23; 3)). If a trial function is constructed with all the electrons on one side of the container and all the nuclei on the other, then the expectation value of the energy is greater than $-N^{7/3} + N^2/R \to -N^{7/3}$, so E/N does not remain bounded from below. On the other hand, the convergence of a series in the limit $N \to \infty$ would imply that $\lim_{N\to\infty} E/N$ would be finite on the whole disc of convergence, which would include some negative values of α. In fact the explicit calculation reveals that even the second-order contribution becomes infinite as $N \to \infty$. Even so, the first-order result is useful as an upper bound.
2. According to (III: 3.5.21) the min–max principle applies to finite σ other than the ground state, but it is more difficult to calculate the microcanonical expectation values than the grand canonical ones. Hence, for nonzero temperatures it is better to use (2.1.8; 3) to bound the grand canonical partition function with $-P_\alpha \leq -P_0 + \mathrm{Tr}\, V\rho_{GC}$.

The Ground State (4.3.30)

The simplest case is $T = 0$, so let us see how far we can get with the easiest methods. Take the expectation value of (4.3.1) in the ground state of the electrons; if they are confined in a box Λ with periodic boundary conditions, the ground state is a plane wave, producing a constant electron density ρ_e. If the nuclear charges are all Z and the nuclear masses are all μ, that leaves

$$\begin{aligned}\langle H\rangle = &\sum_{k=1}^{M} \frac{|\mathbf{p}_k|^2}{2\mu} + Z^2 \sum_{k>j} |\mathbf{X}_k - \mathbf{X}_j|^{-1} - \sum_k \int_\Lambda \frac{d^3x \rho_e Z}{|\mathbf{x} - \mathbf{X}_k|} \\ &+ \frac{1}{2}\int_\Lambda \frac{d^3x\, d^3y \rho_e^2}{|\mathbf{x} - \mathbf{y}|} \\ &+ \left\langle \sum_i |\mathbf{p}_i|^2 \right\rangle + \left\langle \sum_{i>k} |\mathbf{x}_i - \mathbf{x}_k|^{-1} - \frac{1}{2}\int_\Lambda \frac{d^3x\, d^3y \rho_e^2}{|\mathbf{x} - \mathbf{y}|} \right\rangle \\ &+ \sum_k \left\langle \int \frac{d^3x \rho_e}{|\mathbf{x} - \mathbf{X}_k|} - \sum_j \frac{1}{|\mathbf{x}_j - \mathbf{X}_k|} \right\rangle. \end{aligned} \tag{4.3.31}$$

The first line of this equation is the Hamiltonian H_J of jellium (1.2.6) in the nuclear variables. If we therefore add the ground-state energy of jellium to the other expectation values, we get an upper bound on the ground-state energy of H, corresponding to a trial function consisting of the tensor product of the ground state of H_J with the electron wave-function. The zero-point energy of the electrons is the next term in (4.3.31), followed by what is referred to as the exchange energy, and the final expectation value is zero. By (2.5.32), if the spin is $\frac{1}{2}$, the zero-point energy goes as

$$\left\langle \sum_{i=1}^{N} |\mathbf{p}_i|^2 \right\rangle = N\tfrac{3}{5}(3\pi^2\rho_e)^{2/3} = N\frac{2.2}{r_s^2}, \qquad r_s = \left(\frac{3}{4\pi\rho_e}\right)^{1/3}, \tag{4.3.32}$$

as $N \to \infty$, and with only a little difficulty the exchange energy can be calculated as

$$\left\langle \sum_{i>j} |\mathbf{x}_i - \mathbf{x}_j|^{-1} - \frac{1}{2}\int \frac{d^3x\, d^3x' \rho_e^2}{|\mathbf{x} - \mathbf{x}'|} \right\rangle = -0.458\frac{N}{r_s} \tag{4.3.33}$$

(Problem 3). It expresses the effect of the correlations among the electrons owing to their having to avoid each other to satisfy the exclusion principle. The result is to lower the Coulomb energy in comparison with that of a homogeneous charge distribution.

The Ground State of Jellium (4.3.34)

As for H_J, an upper bound can be obtained by using plane waves as trial functions, for which $\langle H_J\rangle$ once again consists of zero-point energy and exchange energy. A lower bound comes from the sum of the zero-point energy and the minimum of the potential (1.2.10), and when combined they bound E_J according to

$$\frac{2.2}{2\mu r_s^2} - \frac{0.9}{r_s} \leq \frac{E_J}{N} \leq \frac{2.2}{2\mu r_s^2} - \frac{0.458}{r_s} \tag{4.3.35}$$

if $Z = 1$ and the spin is $\frac{1}{2}$. If the density is large ($r_s \to 0$), then the bounds are close together, but they spread out if the density is small. At small densities it is better to array the nuclei on a lattice; give them wave-functions $\sim \sin(\pi r a)/r$, where r is the distance from the lattice site if it is less than a and otherwise let the wave-function be 0, and take a small enough that the wave-functions will not overlap, and will thus be orthogonal. The most convenient configuration is a body-centered cubic lattice, which consists of two simple cubic lattices, one of which has been displaced along a diagonal so that its corners are at the centers of the other. If the density is 2, i.e., the lattice constant of the simple cubic lattice is 1, then a must be less than $\sqrt{3}/4$ in order that the balls of radius a do not intersect; in terms of r_s, the distance between nuclei,

$$a \leq \left(\frac{8\pi}{3}\right)^{1/3} \frac{\sqrt{3}}{4} r_s. \tag{4.3.36}$$

If the nuclei were concentrated at the points of the lattice, then the Coulomb energy per particle would be $-0.896/r_s$ according to (1.2.11; 2). Provided that they do not overlap, the repulsion between the nuclei will be the same even if they are somewhat spread out. On the other hand, their interaction (per particle) with the background would be affected by

$$\frac{\rho}{2} \int_0^a dr r^2 \sin^2 \frac{r\pi}{a} \bigg/ \int_0^a dr \sin^2 \frac{r\pi}{a} = \frac{\rho}{2} a^2 \left(\frac{1}{3} - \frac{1}{2\pi^2}\right). \tag{4.3.37}$$

If this is added to the kinetic energy $(\pi/a)^2$ (for mass $\frac{1}{2}$), then the minimum

$$\frac{E}{N} = \left(\frac{\pi}{2} - \frac{3}{4\pi}\right)^{1/2} r_s^{-3/2} - \frac{0.896}{r_s} = 1.15 r_s^{-3/2} - \frac{0.896}{r_s}$$

is attained when

$$a = \left[\frac{\rho}{2\pi^2}\left(\frac{1}{3} - \frac{1}{2\pi^2}\right)\right]^{-1/4} = r_s^{3/4}\left[\frac{3}{8\pi^3}\left(\frac{1}{3} - \frac{1}{2\pi^2}\right)\right]^{-1/4}.$$

Condition (4.3.36) means that

$$r_s \geq \frac{8^3\pi^4}{3(2\pi^2 - 3)} \left(\frac{3}{8\pi}\right)^{1/3} \cong 489. \tag{4.3.38}$$

If r_s is smaller, then a must be taken as $(8\pi/3)^{1/3}(\sqrt{3}/4)r_s$, which costs some kinetic energy, $12.75/r_s^2$, and raises the Coulomb interaction above that due to the background by $0.026/r_s$. The figures become more favorable, however, when it is recalled that wave-functions of nuclei with opposite spins do not need to be spatially orthogonal to avoid incurring exchange energy. Suppose that the nuclei have spin $\frac{1}{2}$, as with protons, and put nuclei with spin up on one of the simple cubic lattices and nuclei with spin down on the other. Then the spheres are required only not to overlap with other spheres on the same simple cubic lattice. This weakens the bound (4.3.36) to

$$a \leq \left(\frac{8\pi}{3}\right)^{1/3} \frac{r_s}{2},$$

which weakens the lower bound on r_s (4.3.38) by a factor $\frac{9}{16}$, so

$$r_s \geq 275, \tag{4.3.39}$$

and also diminishes the zero-point energy by $\frac{3}{4}$ to $9.54/r_s^2$ and increases the interaction with the background by the same factor. The Coulomb repulsion between neighboring nuclei decreases, but only by an insignificant amount $10^{-3}/r_s$ The net effect is to produce

Bounds on the Ground-State Energy of Spin $-\frac{1}{2}$ Jellium (4.3.40)

$$\leq \frac{2.2}{r_s^2} - \frac{0.458}{r_s} \qquad \text{(i)}$$

$$\frac{2.2}{r_s^2} - \frac{0.9}{r_s} \leq \frac{E}{N} \leq \frac{9.58}{r_s^2} - \frac{0.85}{r_s} \qquad \text{(ii)}$$

$$\leq \frac{1.15}{r_s^{3/2}} - \frac{0.89}{r_s} \quad \text{if } r_s > 275, \qquad \text{(iii)}$$

where $e = 2\mu = 1$. (See Figure 39).

Remarks (4.3.41)

1. The distance between particles as measured in Bohr radii with the appropriate mass is r_s. If H_J is the Hamiltonian of the nuclei, and the pressure is not too huge, then r_s is on the order of the ratio of the mass of the nucleus to that of the elctron, which is at least 2000. This means that (4.3.40(i)) will be the best of the bounds. If jellium is taken as a model of electrons in a metal, then $r_s \sim 1$, and (4.3.40(i)) is best.
2. There are conjectures that the transition from homogeneity to a lattice structure as r_s increases is accompanied by a phase transition. It is even believed that the exchange energy, which favors parallel spins, causes ferromagnetism. Despite the simple form of H_J it has not been possible to prove these speculations.

If we focus attention again on real matter, we must add the contribution from the electrons to that of the protons. Observe first that for nuclei the parameter $r_s \sim \rho^{-1/3}$/Bohr radius is increased by a factor μZ^2, but at the same time the energies in (4.3.40) are multiplied by μZ^2. Since the zero-point energy obtains an extra factor $1/\mu$, it can be neglected. For the densities of interest, $r_s > 275/\mu Z^2$, so (4.3.40(iii)) applies to nuclei. Of course, the trial function with a homogeneous electron distribution is poor when Z is large, and does not contribute the right dependence on Z. If $Z = 1$, our earlier results on the energy per electron are only

Crude Bounds (4.3.42)

$$-8.32 \leq \frac{E}{N} \leq \frac{2.2}{r_s^2} - \frac{1.34}{r_s}.$$

Refinements (4.3.43)

1. The lower bound. The Birman–Schwinger bound (III: 3.5.36) can be improved with the methods of functional integration [31], sharpening Inequality (4.1.47; 2) by a factor of 1.5. Then with (4.3.20), if the density is finite, λ is chosen optimally, and $\varepsilon_0 = 5.74\rho^{5/3}$, or equivalently $E_0/N = 2.2/r_s^2$, there results

$$\frac{2.2}{r_s^2} - \frac{5.5}{r_s} \leq E.$$

2. The upper bound. The ground-state energy in a box of volume V is of the form

$$E = V^{-2/3} f(V^{1/3}\alpha).$$

The facts that $\partial E/\partial V \leq 0$ and $\partial^2 \varepsilon/\partial \rho^2 \geq 0$ and the convexity in α are expressed by the inequalities

$$f(x) \geq \frac{x}{2} f'(x) \quad \text{and} \quad 6xf'(x) - 10f(x) \leq x^2 f''(x) \leq 0.$$

Since $\partial E/\partial V \leq 0$, a linear bound $f(x)/f(0) \leq 1 - \gamma x$ for $x > 2/\gamma$ can be improved by a parabolic bound $f(x)/f(0) \leq -x^2(\gamma/2)^2$. By (4.3.43; 1) $\gamma^{-1} = 2.2/1.34$, so if $r_s > 2\gamma^{-1} = 3.28$, then f is less than $-f(0)x^2 \cdot 1.34/4(2.2)^2$. It follows that

$$\frac{E}{N} \leq \begin{cases} 2.2/r_s^2 - 1.34/r_s, & \text{if } r_s < 3.28 \\ -0.204, & \text{if } r_s \geq 3.28. \end{cases}$$

These bounds are far from satisfactory. Not only do they fail to allow finer details to be discerned, but indeed they do not even prove that hydrogen holds together at $T = 0$ rather than breaking up into separated atoms. In these units the energy of a separated hydrogen atom is $-\frac{1}{4}$, i.e., less than the upper bound, which only shows how large a territory still remains open to exploration with exact methods in physics.

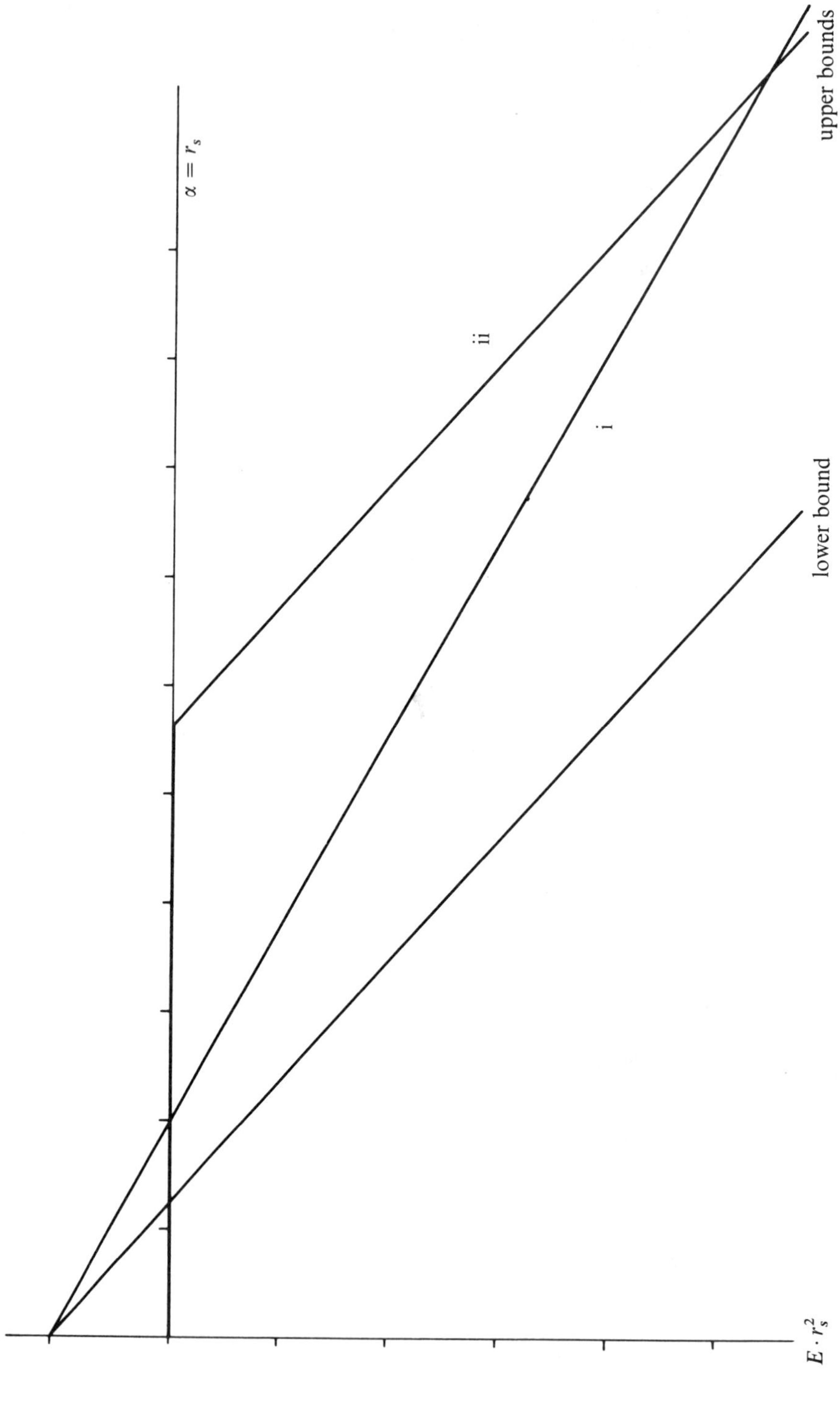

Figure 39 Bounds (4.3.40) for the energy of jellium.

Problems (4.3.44)

1. Prove the Swiss cheese theorem (4.3.13): For any region $\Lambda \subset \mathbb{R}^3$ and any real number h let $\Lambda_h = \{\mathbf{x} \in \Lambda: d(\mathbf{x}, \Lambda^c) < h\}$, if $h > 0$ and $\Lambda_h = \{\mathbf{x} \in \Lambda^c: d(\mathbf{x}, \Lambda) \leq -h\}$, if $h \leq 0$, and denote the volume of Λ_h by $V(h, \Lambda)$.
   Then prove the following two lemmas: (i) Suppose Λ is covered by closed cubes of side l, the interiors of which do not intersect, and let v be the number of cubes entirely contained in Λ. Then the volume of Λ not covered by these v cubes is at most $V(l\sqrt{3}, \Lambda)$. (ii) Let $B \subset \mathbb{R}^3$ be an open ball of radius R and y a number satisfying the inequality $R \geq 2\sqrt{3}\,y \geq 0$. Then $V(2\sqrt{3}\,y, B) \leq V(-2\sqrt{3}\,y, B) \leq 56\pi R^2 y/\sqrt{3}$. Finish the proof of the theorem by covering B_0 with a cubic lattice of spacing $2R_1$, and in each cube of the lattice place a ball of radius R_1, then cover the balls with a cubic lattice of spacing R_2, etc. Use the lemmas to estimate v_j and the fraction of volume taken up by the balls of size j.

2. Use Inequalities (III: 4.5.24) and (4.1.5) to find a lower bound for the potential energy of jellium,

$$U = \sum_{j>k} |\mathbf{x}_j - \mathbf{x}_k|^{-1} - \sum_i \int d^3x \rho(\mathbf{x}) |\mathbf{x} - \mathbf{x}_i|^{-1} + \frac{1}{2} \int \frac{d^3x\, d^3y}{|\mathbf{x} - \mathbf{y}|} \rho(\mathbf{x})\rho(\mathbf{y})$$

   and compare with (1.2.10). (Let ρ be constant in any ball.)

3. Calculate

$$\lim_{V \to \infty} V^{-1} \langle \psi | \left(\sum_{i>k} |\mathbf{x}_i - \mathbf{x}_k|^{-1} - \frac{1}{2} \int_\Lambda \frac{d^3x\, d^3y}{|\mathbf{x} - \mathbf{y}|} \rho_e^2 \right) | \psi \rangle$$

   if ψ is the ground state of a system of free electrons in a box of volume V. (The momentum states in both spin orientations are occupied up to a maximum momentum p such that $p^3/3\pi^2 = N/V = 3/4\pi r_s^3$.)

4. Verify that the concavity of E as a function of $(1/m, \alpha)$ is no more severe a restriction than the concavity of f in (4.3.43; 2).

Solutions (4.3.45)

1. (i) If Λ is covered by cubes of length l, but all cubes intersecting Λ^c are removed, then the uncovered portion of Λ is contained in $\Lambda_{l\sqrt{3}}$. (Hence the number v_{2l} of cubes of length $2l$ that can be packed entirely into Λ is at least $(2l)^{-3}[V(\Lambda) - V(2\sqrt{3}l, \Lambda)]$.)
   (ii) If $0 \leq h \leq R$, then

$$V(h, B) = \frac{4\pi}{3}[R^3 - (R - h)^3] \leq V(-h, B) = \frac{4\pi}{3}[(R + h)^3 - R^3].$$

   The lemma is then a consequence of the convexity of the function $f(\varepsilon) \equiv (1 + \varepsilon)^3 - 1$, which implies that $f(\varepsilon) \leq f(0) + \varepsilon[f(1) - f(0)] = \varepsilon[2^3 - 1] = 7\varepsilon$.

Proof of the packing estimates. For simplicity assume that $R_0 = 1$, and let $\nu_j = p^{j-1}(1+p)^{2j}$. If a unit ball is covered by cubes of length $2R_1 = 2(1+p)^{-1}$, then it contains ν_1 cubes, as we shall show. If we then cover the unit ball with a lattice of length $2R_2$, then there are ν_2 cubes contained in the unit ball and not intersecting the first ν_1 balls of size 1. The general fact will follow by induction. Therefore it needs to be shown that when the ball has been filled with smaller balls up to size j, it is still possible to pack ν_{j+1} balls of radius R_{j+1} into the remaining space $B - \bigcup_{k=1}^{j}$ (balls of size k) $\equiv \Omega_j$:

$$V(\Omega_j) = \frac{4\pi}{3}\left(1 - \sum \nu_k R_k^3\right) = \frac{4\pi}{3}\left(\frac{p}{p+1}\right)^j.$$

$V(2\sqrt{3}R_{j+1}, \Omega_j) \leq M_j$, defined as the sum of $V(-2\sqrt{3}R_{j+1}, B)$ for all balls of size $\leq j$ and $V(2\sqrt{3}R_{j+1}, B)$, where B is the unit ball. Because of (ii) and the inequality $2\sqrt{3}R_{j+1} < R_j$,

$$\begin{aligned} V(2\sqrt{3}R_{j+1}, \Omega_j) \leq M_j &\leq \frac{56\pi}{\sqrt{3}} R_{j+1}\left[1 + \sum \nu_k R_k^2\right] \\ &= (p^j + p - 2)(p-1)^{-1}(1+p)^{-(j+1)}\frac{56\pi}{\sqrt{3}} \equiv \tilde{M}_j. \end{aligned}$$

Therefore it suffices to show that

$$(2R_{j+1})^3 \nu_{j+1} \leq [V_j - \tilde{M}_j] \leq [V(\Omega_j) - V(2\sqrt{3}R_{j+1}, \Omega_j)],$$

i.e.,

$$1 \leq \frac{\pi}{6}\left[p + 1 - 14\sqrt{3}\,\frac{1 + p^{-j}(p-2)}{p-1}\right].$$

Since $p^{-j}(p-2) \leq (p-2)$, this reduces to

$$1 \leq \frac{\pi}{6}\left[p + 1 - 14\sqrt{3}\right],$$

which is true when $p + 1 \geq 27$. The fraction of the volume taken up by the balls of radius R_j is

$$\frac{p^{j-1}}{(1+p)^j},$$

which shows that the packing fills the original ball exponentially fast.

2. From (III; 4.5.24),

$$U \geq -\tfrac{3}{4}\left[8\pi N^2 \int d^3x \rho^2(\mathbf{x})\right]^{1/3} = -1.35N/r_s,$$

and from (4.1.5),

$$U \geq -2\left[3.68N\,\frac{3}{5}\int \rho^{5/3}\right]^{1/2} = -1.84N/r_s.$$

3. As N and $V \to \infty$, make the replacements

$$\frac{1}{V} \sum_{|k| \leq p} \to \int_{|k| \leq p} \frac{d^3k}{(2\pi)^3},$$

$$v(\mathbf{k}) \equiv \frac{1}{V} \int \frac{d^3x\, d^3x'}{|\mathbf{x} - \mathbf{x}'|} \exp[i\mathbf{k} \cdot (\mathbf{x} - \mathbf{x}')] \to 4\pi/|\mathbf{k}|^2,$$

to find that

$$\sum_{\substack{|\mathbf{k}| \leq p \\ |\mathbf{k}'| \leq p}} v(\mathbf{k} - \mathbf{k}') = \int \frac{d^3k}{(2\pi)^3} \frac{4\pi}{|\mathbf{k}|^2} \int \frac{d^3q}{(2\pi)^3} \Theta(p - |\mathbf{q}|)\Theta(p - |\mathbf{k} - \mathbf{q}|)$$

$$= \frac{2}{\pi} \int_0^{2p} dk \frac{p^3}{6\pi^2} \left[1 - \frac{3k}{4p} + \frac{k^3}{16p^3}\right] = \frac{p^3}{3\pi^2} \frac{p}{\pi} \frac{3}{4}$$

$$= \frac{N}{V} \left(\frac{9\pi}{4}\right)^{1/3} \frac{1}{r_s} \frac{3}{4\pi} = \frac{N}{V} \frac{0.458}{r_s}.$$

In order to justify this formal calculation, make a convolution so that

$$v(\mathbf{k}) = \frac{4\pi}{|\mathbf{k}|^2} * F(\mathbf{k}),$$

where

$$F(\mathbf{k}) = \frac{1}{V} \int_{\mathbf{x} \in V, \mathbf{x}' \in V} d^3x\, d^3x' \exp[i\mathbf{k} \cdot (\mathbf{x} - \mathbf{x}')]$$

$$= L^{-3} \left(\frac{\sin k_1 L/2}{L/2}\right)^2 \left(\frac{\sin k_2 L/2}{L/2}\right)^2 \left(\frac{\sin k_3 L/2}{L/2}\right)^2$$

is the Fourier transform of the characteristic function of the box, and use Lebesgue's dominated convergence theorem to show that the integrals have the limits given above.

4. With $1/m = v$: $E = vf(\alpha/v)$,

$$E_{,\alpha\alpha} = \frac{1}{v} f'', \qquad E_{,vv} = \frac{\alpha^2}{v^3} f'',$$

$$E_{,v\alpha} = -\frac{\alpha}{v^2} f'', \qquad E_{,\alpha\alpha} E_{,vv} - (E_{,v\alpha})^2 = 0.$$

Bibliography

Works Cited in the Text

[1] F. J. Dyson and A. Lenard. Stability of Matter, I. *J. Math. Phys.* **8**, 423–433, 1967; Stability of Matter, II. *Ibid.* **9**, 698–711, 1968.

[2] E. H. Lieb. The $N^{5/3}$ Law for Bosons. *Phys. Lett.* **70A**, 71–73, 1979.

[3] R. A. Goldwell–Horstall and A. A. Maradudin. Zero-Point Energy of an Electron Lattice. *J. Math. Phys.* **1**, 395–404, 1960.

[4] J. Dixmier. Les Algèbres d'Opérateurs dans l'Espace Hilbertien. Paris, Gauthier-Villars, 1969.

[5] A. Wehrl. General Properties of Entropy. *Rev. Mod. Phys.* **50**, 221–260, 1978. E. H. Lieb. Convex Trace Functions and the Wigner–Yanase–Dyson Conjecture. *Adv. Math.* **11**, 267–288, 1973. B. Simon. Trace Ideals and their Applications. London and New York, Cambridge Univ. Press, 1979. A. Uhlmann. Relative Entropy and the Wigner–Yanase–Dyson–Lieb Concavity in an Interpolation Theory. *Commun. Math. Phys.* **40**, 147–151, 1975; Sätze über Dichtematrizen. *Wiss. Z. Karl-Marx-Univ. Leipzig* **20**, 633, 1971; The Order Structure of States. In: Proc. Intl. Symp. on Selected Topics in Statistical Mechanics. JINR-Publ. D17-11490. Dubna USSR, 1978.

[6] M. B. Ruskai. Inequalities for Traces on von Neumann Algebras. *Commun. Math. Physics* **26**, 280–289, 1972. M. Breitenecker, H.-R. Grümm. Note on Trace Inequalities. *Commun. Math. Phys.* **26**, 276–279, 1972. K. Symanzik. Proof and Refinements of an Inequality of Feynman. *J. Math. Phys.* **6**, 1155–1156, 1965.

[7] J. Aczel, B. Forte, and C. T. Ng. Why the Shannon and Hartley Entropies are "Natural." *Adv. Appl. Prob.* **6**, 131–146, 1974.

[8] E. H. Lieb and M. B. Ruskai. Proof of the Strong Subadditivity of Quantum-Mechanical Entropy. *J. Math. Phys.* **14**, 1938–1941, 1973. H. Araki and E. H. Lieb. Entropy Inequalities. *Commun. Math. Phys.* **18**, 160–170, 1970.

[9] P. C. Martin and J. Schwinger. Theory of Many-Particle Systems, I. *Phys. Rev.* **115**, 1342–1373, 1959.

[10] R. Peierls. Surprises in Theoretical Physics. Princeton, Princeton Univ. Press. 1976.
[11] E. H. Lieb and W. Thirring. Inequalities for the Moments of the Eigenvalues of the Schrödinger Hamiltonian and their Relation to Sobolev Inequalities. In: Studies in Mathematical Physics, Essays in Honor of Valentine Bargmann, A. S. Wightman, E. H. Lieb, and B. Simon, eds. Princeton, Princeton Univ. Press, 1976.
[12] E. T. Whittaker and G. N. Watson. A Course of Modern Analysis. Cambridge, at the University Press, 1969.
[13] J. T. Cannon. Infinite Volume Limits of the Canonical Free Bose Gas States on the Weyl Algebra. *Commun. Math. Phys.* **29**, 89–104, 1973.
[14] G. Lindblad. On the Generators of Quantum Dynamical Semigroups. *Commun. Math. Phys.* **48**, 119–130, 1976.
[15] A. Kossakowski and E. C. G. Sudarshan. Completely Positive Dynamical Semigroups of N-Level Systems. *J. Math. Phys.* **17**, 821–825, 1976.
[16] T. L. Saaty and J. Bram. Nonlinear Mathematics. New York, McGraw-Hill, 1964.
[17] D. Ruelle. Statistical Mechanics, Rigorous Results. New York, Benjamin, 1969.
[18] A. Guichardet. Systèmes Dynamiques non Commutatifs. *Astérisque* **13-14**, 1974.
[19] I. M. Gel'fand, R. A. Minlos, and Z. Ya. Shapiro. Representations of the Rotation and Lorenz Group and their Applications. Oxford, Pergamon Press, 1963.
[20] R. B. Israel, ed. Convexity in the Theory of Lattice Gases. Princeton, Princeton Univ. Press, 1979.
[21] O. Bratteli and D. W. Robinson. Operator Algebras and Quantum Statistical Mechanics, in two volumes. New York, Springer, 1979, 1980.
[22] H. Narnhofer. Kommutative Automorphismen und Gleichgewichtszustände. *Acta Phys. Austriaca* **47**, 1–29, 1977.
[23] H. Narnhofer. Scattering Theory for Quasi-Free Time Automorphisms of C^* Algebras and von Neumann Algebras. *Rep. Math. Phys.* **16**, 1–8, 1979.
[24] H. Araki and G. L. Sewell. KMS Conditions and Local Thermodynamic Stability of Quantum Lattice Systems. *Commun. Math. Phys.* **52**, 103–109, 1977.
[25] G. L. Sewell. Relaxation, Amplification, and the KMS Conditions. *Ann. Phys.* (N.Y.) **85**, 336–377, 1974.
[26] H. Narnhofer and G. L. Sewell. Vlasov Hydrodynamics of a Quantum Mechanical Model. *Commun. Math. Phys.* **79**, 9–24, 1981.
[27] J. Messer. The Pressure of Fermions with Gravitational Interaction. *Z. Phys.* **B33**, 313–316, 1979.
[28] W. Thirring. A Lower Bound with the Best Possible Constant for Coulomb Hamiltonians. *Commun. Math. Phys.* **79**, 1–7, 1981.
[29] E. H. Lieb. The Stability of Matter. *Rev. Mod. Phys.* **48**, 553–569, 1976.
[30] J. Lebowitz and E. H. Lieb. The Constitution of Matter: Existence of Thermodynamics for Systems Composed of Electrons and Nuclei. *Adv. Math.* **9**, 316–398, 1972.
[31] E. H. Lieb. The Number of Bound States of One-Body Schroedinger Operators and the Weyl Problem. *Proc. Symposia in Pure Math.* **36**, 241–252, 1980.
[32] E. H. Lieb. Proof of an Entropy Conjecture of Wehrl. *Commun. Math. Phys.* **62**, 35–41, 1978.
[33] N. Dunford and J. T. Schwartz. Linear Operators, part I. New York, Wiley-Interscience, 1967.
[34] E. H. Lieb. Thomas–Fermi and Related Theories of Atoms and Molecules. *Rev. Mod. Phys.* **53**, 603–641, 1981.

[35] S. Chandrasekhar. An Introduction to the Study of Stellar Structure. New York, Dover, 1967.

Further Reading

Section 1.1

H. Wergeland. Irreversibility in Many-Body Systems. In: Irreversibility in the Many-Body Problem, J. Biel and J. Rae, eds. New York and London, Plenum, 1972.

(1.1.1)

G. Emch. Non-Markovian Model for the Approach to Equilibrium. *J. Math. Phys.* **7**, 1198–1206, 1966.

(1.1.13)

E. Schrödinger. Zur Dynamik elastisch gekoppelter Punktsysteme. *Ann. Phys.* (Leipzig) **44**, 916–934, 1914.

I. Prigogine and G. Klein. Sur la Mécanique Statistique des Phénomènes Irréversibles, III. *Physica* **19**, 1053–1071, 1953.

(1.2.10)

E. H. Lieb and H. Narnhofer. The Thermodynamic Limit for Jellium. *J. Stat. Phys.* **12**, 291–310, 1975.

Section 1.3

F. A. Berezin. Method of Second Quantization. New York, Academic Press, 1966.

M. Reed and B. Simon. Methods of Modern Mathematical Physics, vol. II: Fourier Analysis, Self-Adjointness. New York, Academic Press, 1975.

(1.4.2)

J. von Neumann. On Infinite Direct Products. *Compositio Math.* **6**, 1–77, 1939.

(1.4.9)

O. Bratteli and D. W. Robinson. Operator Algebras and Quantum Statistical Mechanics, vol. I. New York, Springer, 1979.

Section 2.1

A. Wehrl. How Chaotic is a State of a Quantum System? *Rep. Math. Phys.* **6**, 15–28, 1974.

A. Uhlmann. Sätze über Dichtematrizen. *Wiss. Z. Karl-Marx-Univ. Leipzig* **20**, 633, 1971; Endlichdimensionale Dichtematrizen, I. *Ibid.* **21**, 421; 1972; Endlichdimensionale Dichtematrizen, II. *Ibid.* **22**, 139, 1973.

(2.1.7)

E. H. Lieb. Some Convexity and Subadditivity Properties of Entropy. *Bull. Amer. Math. Soc.* **81**, 1–13, 1975.

B. Simon. Trace Ideals and their Applications. London and New York, Cambridge Univ. Press, 1979.

(2.2.9)

B. Baumgartner. Classical Bounds on Quantum Partition Functions. *Commun. Math. Phys.* **75**, 25–41, 1980.

(2.2.11)

F. A. Berezin. Wick and Anti-Wick Operator Symbols. *Math. USSR Sbornik* **15**, 577–606, 1971. (Translation of Vikovskie i antivikovskie simboly operatorov. *Mat. Sbornik*. **86(128)**, 578–610, 1971.)

(2.2.22)

G. Lindblad. Entropy, Information, and Quantum Measurements. *Commun. Math. Phys.* **33**, 305–322, 1973.

H. Umegaki. Conditional Expectation in an Operator Algebra. *Kodai Math. Seminar. Rep.* **14**, 59–85, 1962.

H. Araki. RIMS preprint 190, Kyoto, 1975.

Section 2.3

R. Griffiths. Microcanonical Ensemble in Quantum Statistical Mechanics. *J. Math. Phys.* **6**, 1447–1461, 1965.

(2.3.39)

A. S. Wightman. Convexity and the Notion of Equilibrium State in Thermodynamics and Statistical Mechanics. In: Convexity in the Theory of Lattice Gases, R. Israel, ed. Princeton, Princeton Univ. Press, 1979.

(2.4.7)

H. D. Maison. Analyticity of the Partition Function for Finite Quantum Systems. *Commun. Math. Phys.* **22**, 166–172, 1971.

(2.4.9)

E. H. Lieb. The Classical Limit of Quantum Spin Systems. *Commun. Math. Phys.* **31**, 326–341, 1973.

(2.4.15; 2)

A. S. Wightman, *op. cit.* in (2.3.39)

(2.5.15)

W. Thirring. Bounds on the Entropy in Terms of One Particle Distributions. *Lett. Math. Phys.* **4**, 67–70, 1980.

(2.5.26)

K. Huang. Statistical Mechanics. New York; Wiley, 1963.
F. Reif. Fundamentals of Statistical and Thermal Physics. New York, McGraw-Hill, 1965.

Section 3.1

D. Ruelle. Statistical Mechanics, Rigorous Results. New York, Benjamin, 1969.
G. Emch. Algebraic Methods in Statistical Mechanics and Quantum Field Theory. New York, Wiley, 1971.

(3.1.2)

E. B. Davies. Quantum Theory of Open Systems. New York, Academic Press, 1976.

(3.1.4)

V. Gorini, A. Frigerio, M. Verri, A. Kossakowski, and E. C. G. Sudarshan. Properties of Quantum Markovian Master Equations. *Rep. Math. Phys.* **13**, 149–173, 1978.
P. Martin. Modèles en Mécanique Statistique des Processus Irréversibles. New York and Berlin, Springer, 1979.

(3.1.12)

G. Lindblad. Completely Positive Maps and Entropy Inequalities. *Commun. Math. Phys.* **40**, 147–151, 1975.
A. Uhlmann. Relative Entropy and The Wigner–Yanase–Dyson–Lieb Concavity in an Interpolation Theory. *Commun. Math. Phys.* **54**, 21–32, 1970.

(3.1.14)

F. Greenleaf. Invariant Means on Topological Groups. New York, Van Nostrand, 1966.
A. Guichardet. Systèmes Dynamiques non Commutatifs. *Astérisque* **13–14**, 1974.

(3.1.25)

V. I. Arnol'd and A. Avez. Ergodic Problems of Classical Mechanics. New York, Benjamin, 1968.

(3.2.1)

N. M. Hugenholtz. Article in: Mathematics of Contemporary Physics, R. Streater, ed. New York and London, Academic Press, 1972.

(3.2.6)

R. Haag, N. M. Hugenholtz, and M. Winnink. On the Equilibrium States in Quantum Statistical Mechanics. *Commun. Math. Phys.* **5**, 215–236, 1967.

(3.2.13)

M. Takesaki. Tomita's Theory of Modular Hilbert Algebras and Its Applications, Lecture Notes in Mathematics, vol. 128. New York and Berlin, Springer, 1970.

(3.3.7)

H. Narnhofer and D. W. Robinson. Dynamical Stability and Pure Thermodynamic Phases. *Commun. Math. Phys.* **41**, 89–97, 1975.

(3.3.10)

R. Haag, D. Kastler, and E. B. Trych-Pohlmeyer. Stability and Equilibrium States. *Commun. Math. Phys.* **38**, 173–193, 1974.

(3.3.22)

W. Pusz and S. L. Woronovicz. Passive States and KMS States for General Quantum Systems. *Commun. Math. Phys.* **58**, 273–290, 1978.

A. Lenard. Thermodynamical Proof of the Gibbs Formula for Elementary Quantum Systems. *J. Stat. Phys.* **19**, 575–586, 1978.

Section 4.1

E. H. Lieb and B. Simon. The Thomas–Fermi Theory of Atoms, Molecules, and Solids. *Adv. Math.* **23**, 22–116, 1977.

H. Narnhofer and W. Thirring. Asymptotic Exactness of Finite Temperature Thomas–Fermi Theory. *Ann. Phys.* (N.Y.) **134**, 128–140, 1981.

B. Baumgartner. The Thomas–Fermi Theory as Result of a Strong-Coupling Limit. *Commun. Math. Phys.* **47**, 215–219, 1976.

Section 4.2

P. Hertel, H. Narnhofer, and W. Thirring. Thermodynamic Functions for Fermions with Gravostatic and Electrostatic Interactions. *Commun. Math. Phys.* **28**, 159–176, 1972.

P. Hertel and W. Thirring. Article in: Quanten und Felder, H. Dürr, ed. Brunswick, Vieweg, 1971.

J. Messer. Temperature Dependent Thomas–Fermi Theory, Lecture Notes in Physics, vol. 147. New York and Berlin, Springer, 1979.

B. Baumgartner. Thermodynamic Limit of Correlation Functions in a System of Gravitating Fermions. *Commun. Math. Phys.* **48**, 207–213, 1976.

Section 4.3

E. H. Lieb. The Stability of Matter. *Rev. Mod. Phys.* **48**, 553–569, 1976.
W. Thirring. Stability of Matter. In: Current Problems in Elementary Particle and Mathematical Physics, P. Urban, ed. *Acta Phys. Austriaca Suppl.* **XV**, 337–354, 1976.

(4.3.22)

R. Griffiths. Microanonical Ensemble in Quantum Statistical Mechanics. *J. Math. Phys.* **6**, 1447–1461, 1965.

(4.3.40)

H. Narnhofer and W. Thirring. Convexity Properties for Coulomb Systems. *Acta Phys. Austriaca.* **41**, 281–297, 1975.

Index